T0205876

Graduate Texts in Physics

Series Editors
Gene Stanley
Boston, USA

William Rhodes
Florida, USA

Richard Needs
Cambridge, UK

Susan Scott
Canberra, Australia

Martin Stutzmann
München, Germany

Kurt H. Becker
New York, USA

Bill Munro
Kanagawa, Japan

Sadri D. Hassani
Normal, USA

Andreas Wipf
Fürstengraben, Germany

Graduate Texts in Physics publishes core learning/teaching material for graduate- and advanced-level undergraduate courses on topics of current and emerging fields within physics, both pure and applied. These textbooks serve students at the MS- or PhD-level and their instructors as comprehensive sources of principles, definitions, derivations, experiments and applications (as relevant) for their mastery and teaching, respectively. International in scope and relevance, the textbooks correspond to course syllabi sufficiently to serve as required reading. Their didactic style, comprehensiveness and coverage of fundamental material also make them suitable as introductions or references for scientists entering, or requiring timely knowledge of, a research field.

More information about this series at http://www.springer.com/series/8431

Takafumi Kita

Statistical Mechanics of Superconductivity

Springer

Takafumi Kita
Department of Physics
Hokkaido University
Sapporo, Japan

ISSN 1868-4513 ISSN 1868-4521 (electronic)
Graduate Texts in Physics
ISBN 978-4-431-56414-0 ISBN 978-4-431-55405-9 (eBook)
DOI 10.1007/978-4-431-55405-9

Springer Tokyo Heidelberg New York Dordrecht London
TOUKEIRIKIGAKU KARA RIKAI SURU TYODENDOURIRON by Takafumi Kita.
Original Japanese language edition published by Saiensu-sha Co., Ltd.
1-3-25, Sendagaya, Shibuya-ku, Tokyo 151-0051, Japan
Copyright © 2013, Saiensu-sha Co., Ltd. All Rights reserved.
© Springer Japan 2015
Softcover reprint of the hardcover 1st edition 2015
This work is subject to copyright. All rights are reserved by the Publisher, whether the whole or part of the material is concerned, specifically the rights of translation, reprinting, reuse of illustrations, recitation, broadcasting, reproduction on microfilms or in any other physical way, and transmission or information storage and retrieval, electronic adaptation, computer software, or by similar or dissimilar methodology now known or hereafter developed.
The use of general descriptive names, registered names, trademarks, service marks, etc. in this publication does not imply, even in the absence of a specific statement, that such names are exempt from the relevant protective laws and regulations and therefore free for general use.
The publisher, the authors and the editors are safe to assume that the advice and information in this book are believed to be true and accurate at the date of publication. Neither the publisher nor the authors or the editors give a warranty, express or implied, with respect to the material contained herein or for any errors or omissions that may have been made.

Printed on acid-free paper

Springer Japan KK is part of Springer Science+Business Media (www.springer.com)

Preface

The purpose of this book is to present the fundamentals of the theory of superconductivity in a self-contained manner by developing and illustrating every required technique of advanced equilibrium statistical mechanics. It is addressed to graduate and undergraduate students who have finished elementary courses of thermodynamics and quantum mechanics. No further background knowledge is required in reading through all the chapters.

Superconductivity is one of the most spectacular phenomena in nature and typical of broken symmetries. The Bardeen–Cooper–Schrieffer (BCS) theory that has clarified it has had a tremendous impact on the whole field of physics, ranging from condensed-matter physics itself to nuclear and particle physics. Hence, one may expect that learning superconductivity enables one to reach and acquire key concepts and techniques of modern theoretical physics. This book treats this fascinating topic from the viewpoint of statistical mechanics to clarify both mathematical and logical structures of the theory as transparently as possible.

Standard textbooks on the topic usually begin by describing basic experimental results such as the Meissner effect, show subsequently that electron–phonon interactions may establish virtual attractive forces between electrons, and proceed to present the BCS theory for homogeneous systems. Descriptions of the phenomenological London and Ginzburg–Landau theories are often inserted prior to the microscopic BCS theory. In this way, one may see that these theories can describe experiments exceedingly well and also acquire basic skills to use them for one's own purposes. However, it may not be entirely clear in this standard approach where superfluidity (flow without dissipation) originates, what causes the Meissner effect to expel the magnetic field from the bulk, or how phase coherence responsible for superfluidity is established. There is also a high threshold in learning about superconductivity for those who are not well acquainted with electromagnetism or especially well versed in topics in solid-state physics.

With these observations, this book adopts an alternative approach based on statistical mechanics. Specifically, it starts from statistical mechanics of quantum ideal gases, adding one by one every new element that is required in understanding superconductivity together with relevant techniques of modern statistical mechanics. The theory of superconductivity is developed on this basis by taking full advantage of the second-quantization method so that macroscopic condensation into a two-particle bound state is manifest. The starting point is the BCS wave function in real space, which is closely connected with the coherent state for lasers and Bose–Einstein condensates. A definite advantage of this approach is that phase coherence is quite apparent. The basic formulation is thereby performed in real space to derive the Bogoliubov–de Gennes equations so that inhomogeneous cases and arbitrary pairing symmetry can be studied on an equal footing. The BCS theory is presented subsequently as an application of it to homogeneous s-wave pairing.

It would bring great pleasure to me, the author, if the book is helpful for students full of curiosity and pioneering spirit. Finally, I would like to express my gratitude to Professor Koh Wada for a critical and careful reading of the manuscript and consequent useful comments.

Sapporo, Japan Takafumi Kita
February 2015

Contents

1 Review of Thermodynamics 1
1.1 Thermodynamics and Hiking 1
1.2 Equation of State 3
1.3 Laws of Thermodynamics 4
1.4 Equilibrium Thermodynamics 5
1.4.1 Basic Equation 6
1.4.2 Equilibrium Conditions 6
1.4.3 Legendre Transformation and Free Energy 7
1.4.4 Particle Number as a Variable 8
1.5 Thermodynamic Construction of Entropy and Internal Energy 10
Problems 11

2 Basics of Equilibrium Statistical Mechanics 13
2.1 Entropy in Statistical Mechanics 13
2.2 Deriving Equilibrium Distributions 16
2.2.1 Microcanonical Distribution 17
2.2.2 Canonical Distribution 18
2.2.3 Grand Canonical Distribution 21
Problems 23
References 23

3 Quantum Mechanics of Identical Particles 25
3.1 Permutation 25
3.2 Permutation Symmetry of Identical Particles 26
3.3 Eigenspace of Permutation 29
3.4 Bra-Kets for Many-Body Wave Functions 31
3.5 Orthonormality and Completeness of Bra-Kets 32
3.6 Matrix Elements of Operators 33
3.7 Summary of Two Equivalent Descriptions 33
3.8 Second Quantization for Ideal Gases 34
3.9 Coherent State 39

Problems 41
References 41

4 Statistical Mechanics of Ideal Gases 43
4.1 Bose and Fermi Distributions 43
4.2 Single-Particle Density of States 45
4.3 Monoatomic Gases in Three Dimensions 46
4.3.1 Single-Particle Density of States 46
4.3.2 Connection Between Internal Energy and Pressure 47
4.3.3 Introducing Dimensionless Variables 48
4.3.4 Temperature Dependences of Thermodynamic Quantities 50
4.4 High-Temperature Expansions 51
4.5 Fermions at Low Temperatures 52
4.5.1 Fermi Energy and Fermi Wave Number 52
4.5.2 Sommerfeld Expansion 53
4.5.3 Chemical Potential and Heat Capacity 55
4.6 Bosons at Low Temperatures 56
4.6.1 Critical Temperature of Condensation 56
4.6.2 Thermodynamic Quantities of $T < T_0$ 57
4.6.3 Chemical Potential and Heat Capacity for $T \gtrsim T_0$ 57
4.7 Bose-Einstein Condensation and Density of States 58
Problems 59
References 60

5 Density Matrices and Two-Particle Correlations 61
5.1 Density Matrices 61
5.2 Bloch–De Dominicis Theorem 62
5.3 Two-Particle Correlations of Monoatomic Ideal Gases 67
Problems 71
References 71

6 Hartree–Fock Equations and Landau's Fermi-Liquid Theory 73
6.1 Variational Principle in Statistical Mechanics 73
6.2 Hartree–Fock Equations 74
6.2.1 Derivation Based on the Variational Principle 74
6.2.2 Derivation Based on Wick Decomposition 77
6.2.3 Homogeneous Cases 78
6.3 Application to Low-Temperature Fermions 80
6.3.1 Fermi Wave Number and Fermi Energy 80
6.3.2 Effective Mass, Density of States, and Heat Capacity 80
6.3.3 Effective Mass and Landau Parameter 82
6.3.4 Spin Susceptibility 84
6.3.5 Compressibility 86
6.3.6 Landau Parameters 87

Problems ... 89
References ... 89

7 Attractive Interaction and Bound States ... 91
7.1 Attractive Potential in Two and Three Dimensions ... 91
7.1.1 Bound State in Three Dimensions ... 92
7.1.2 Bound State in Two Dimensions ... 93
7.2 Consideration in Wave Vector Domain ... 94
7.3 Cooper's Problem ... 97
Problems ... 98
References ... 99

8 Mean-Field Equations of Superconductivity ... 101
8.1 BCS Wave Function for Cooper-Pair Condensation ... 101
8.2 Quasiparticle Field for Excitations ... 103
8.3 Bogoliubov–de Gennes Equations ... 105
8.3.1 Derivation Based on Variational Principle ... 106
8.3.2 Derivation Based on Wick Decomposition ... 113
8.3.3 Matrix Representation of Spin Variables ... 115
8.3.4 BdG Equations for Homogeneous Cases ... 117
8.4 Expansion of Pairing Interaction ... 119
8.4.1 Isotropic Cases ... 119
8.4.2 Anisotropic Cases ... 120
Problems ... 122
References ... 122

9 BCS Theory ... 125
9.1 Self-Consistency Equations ... 125
9.2 Effective Pairing Interaction ... 128
9.3 Gap Equation and Its Solution ... 132
9.4 Thermodynamic Properties ... 135
9.4.1 Heat Capacity ... 135
9.4.2 Chemical Potential ... 138
9.4.3 Free Energy ... 138
9.5 Landau Theory of Second-Order Phase Transition ... 139
Problems ... 141
References ... 141

10 Superfluidity, Meissner Effect, and Flux Quantization ... 143
10.1 Superfluid Density and Spin Susceptibility ... 143
10.1.1 Spin Susceptibility ... 146
10.1.2 Superfluid Density ... 147
10.1.3 Leggett's Theory of Superfluid Fermi Liquids ... 149
10.2 Meissner Effect and Flux Quantization ... 151
10.2.1 Ampère's Law ... 152
10.2.2 London Equation ... 153

10.2.3 Meissner Effect 154
10.2.4 Flux Quantization 155
Problems 156
References 157

11 Responses to External Perturbations 159
11.1 Linear-Response Theory 159
11.1.1 Response in Time Domain 159
11.1.2 Response in Frequency Domain 161
11.1.3 Energy Dissipation 162
11.2 Ultrasonic Attenuation 163
11.3 Nuclear-Spin Relaxation 168
Problems 174
References 174

12 Tunneling, Density of States, and Josephson Effect 175
12.1 Formula for Tunneling Current 175
12.2 NN Junction 182
12.3 SN Junction and Density of States 182
12.4 SS Junction and Josephson Effect 183
Problems 187
References 187

13 *P*-Wave Superfluidity 189
13.1 Effective Pairing Interaction 189
13.2 Gap Matrix 190
13.3 Two Bulk Phases 191
13.3.1 B Phase 191
13.3.2 A Phase 194
13.4 Gap Anisotropy and Quasiparticle Density of States 197
Problems 199
References 199

14 Gor'kov, Eilenberger, and Ginzburg–Landau Equations 201
14.1 Matsubara Green's Function 201
14.2 Gor'kov Equations 204
14.2.1 Equation of Motion for Field Operators 204
14.2.2 Derivation of the Gor'kov Equations 206
14.2.3 Matrix Representation of Spin Variables 209
14.2.4 Gauge Invariance 210
14.2.5 Gauge-Covariant Wigner Transform 211
14.3 Eilenberger Equations 213
14.3.1 Quasiclassical Green's Function 213
14.3.2 Pair Potential 217
14.3.3 Current Density 219
14.3.4 Summary of the Eilenberger Equations 219
14.4 Ginzburg–Landau Equations 221

Problem 225
References 227

15 Abrikosov's Flux-Line Lattice 229
15.1 Ginzburg-Landau Equations 229
15.2 Microscopic Flux Density and Magnetization 230
15.3 Dimensionless Equations 231
15.4 Upper Critical Field and Distinction Between Type-I and II 233
15.5 Flux-Line Lattice Near H_{c2} 234
15.5.1 Constructing Basis Functions 234
15.5.2 Minimization of the Free Energy Functional 238
15.6 Lower Critical Field H_{c1} 241
Problems 245
References 245

16 Surfaces and Vortex Cores 247
16.1 Andreev Reflection 247
16.2 Vortex-Core States 251
16.3 Quasiclassical Study of an Isolated Vortex 254
16.3.1 Eilenberger Equations in Magnetic Fields 254
16.3.2 Transformation to a Riccati-Type Equation 256
16.3.3 Equations for an Isolated Vortex 257
16.3.4 Numerical Procedures 259
16.3.5 Results 261
Problems 263
References 263

17 Solutions to Problems 265
References 286

Index 287

Chapter 1
Review of Thermodynamics

Abstract We summarize the fundamentals of thermodynamics that will be indispensable for describing superconductivity. Advanced readers familiar with the topic may wish to skim through to the next chapter.

1.1 Thermodynamics and Hiking

It may be useful for those who feel *thermodynamics is too abstract* to imagine hiking with a geographical map. A position on the map is specified by a two-dimensional coordinate vector $\mathbf{r} \equiv (x, y)$ with which an altitude $z = z(x, y)$ is associated. *A contour line* defines a continuous curve that connects positions of a given altitude. As illustrated in Fig. 1.1, consider walking up to the peak P from the trailhead A by following either trail C_1 or C_2. The distances one has to walk along C_1 and C_2 are generally different. However, the acquired elevation does not depend on the path; it is expressible solely in terms of the initial and final altitudes as $\Delta z = z(x_P, y_P) - z(x_A, y_A)$.

This altitude $z = z(x, y)$ is given uniquely as a function of (x, y). In thermodynamics, we call such a quantity a *potential* or *state quantity*, for which the infinitesimal increment will be denoted with symbol d as dz. In contrast, the distance $\Delta \ell$ from trailhead A to peak P depends on the path. We call such a quantity a *non-potential* and distinguish its infinitesimal increment with symbol d$'$ as d$'\ell$. Note that we also have to specify the direction when determining d$'\ell$.

When we know that there is a potential, how is that potential constructed? For altitude $z = z(x, y)$, one may (i) combine angle θ determining the elevation with distance L to the target to obtain $\Delta z = L \sin \theta$; distance L can be found fairly easily using an instrument such as a laser rangefinder. Alternatively, one may (ii) measure the gradient:

$$\nabla z \equiv \left(\frac{\partial z}{\partial x}, \frac{\partial z}{\partial y} \right), \tag{1.1}$$

© Springer Japan 2015
T. Kita, *Statistical Mechanics of Superconductivity*, Graduate Texts in Physics, DOI 10.1007/978-4-431-55405-9_1

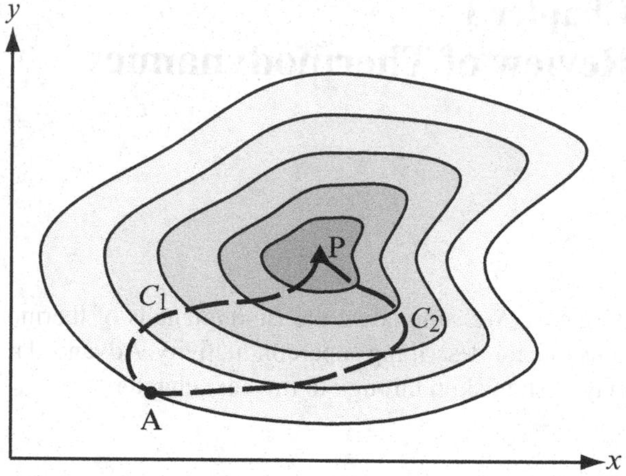

Fig. 1.1 A pair of trails C_1 and C_2 that start from trailhead A up to peak P

at every point and subsequently integrate the total derivative:

$$dz \equiv \frac{\partial z}{\partial x} dx + \frac{\partial z}{\partial y} dy, \tag{1.2}$$

along a convenient path; here "≡" signifies definition.

In thermodynamics, we encounter proper potentials such as *internal energy* and *entropy*. However, we seldom have a direct method to find them as for method (i) above. Hence, we are almost always obliged to rely on method (ii) to construct them. The required integration of (1.2) is carried out as follows; see **Problem** 1.1 for a specific example. First, we integrate it along the x axis as

$$z(x, y) = \int_{x_0}^{x} \frac{\partial z(x_1, y)}{\partial x_1} dx_1 + g(y), \tag{1.3}$$

where x_0 is an arbitrary lower limit and $g(y)$ denotes an unknown function of y. To determine $g(y)$, we differentiate (1.3) in terms of y to obtain an ordinary first-order differential equation for $g(y)$ as

$$\frac{dg(y)}{dy} = \frac{\partial z(x, y)}{\partial y} - \int_{x_0}^{x} \frac{\partial^2 z(x_1, y)}{\partial y \partial x_1} dx_1, \tag{1.4}$$

which can be integrated easily. Substituting the resulting $g(y)$ into (1.3), we obtain $z = z(x, y)$. The integration constant, which may be chosen arbitrarily, becomes irrelevant when we consider differences Δz.

The right-hand side of (1.4) has a formal x dependence that is absent on the left-hand side. Hence, we differentiate (1.4) in terms of x to obtain an integrability condition for z,

$$\frac{\partial^2 z(x, y)}{\partial x \partial y} = \frac{\partial^2 z(x, y)}{\partial y \partial x}, \tag{1.5}$$

which is known as *Maxwell's relation* in thermodynamics. It is quite useful for confirming the consistency of experimental data as well as for predicting new results without relying on experiments.

1.2 Equation of State

Thermodynamics describes macroscopic behaviors of systems with many particles based on state quantities. Those quantities familiar to us include volume V, pressure P, and absolute temperature T. It has been established experimentally that the three variables are not independent but generally obey a constraint:

$$P = P(T, V), \tag{1.6}$$

which is called the *equation of state*. Quantities (T, V, P) here correspond to (x, y, z) on the map, and one may draw a contour line based on (1.6). The independent variables (T, P) in (1.6) are sometimes called *state variables* to distinguish them from the dependent variable P called *state function*.

Among the well-known equations of state are the ideal gas law:

$$P = \frac{nRT}{V}, \tag{1.7}$$

and van der Waals' equation of state:

$$P = \frac{nRT}{V - nb} - a\frac{n^2}{V^2}, \tag{1.8}$$

where n is the number of moles of a substance, $R = 8.31$ J/(mol·K) denotes the *gas constant*, and a and b are some positive constants. Equation (1.8) has a definite advantage over (1.7) in that it can describe both gas and liquid phases on an equal footing; parameters a and b represent measures of attraction between molecules and the finite extension of the constituent molecules, respectively. Derived in 1873, van der Waals' equation of state provided a theoretical guideline around the turn of the twentieth century in realizing the condensation of various gases by cooling, thereby stimulating the development of low-temperature physics toward the discovery of superconductivity in mercury at 4.2 K by Kamerlingh Onnes in 1911.

Equation (1.6) may be converted to $T = T(P, V)$ or $V = V(T, P)$. For example, (1.7) can be expressed alternatively as $T = PV/nR$ and $V = nRT/P$. However, this change of independent variables may not always be possible analytically. In general, the equation of state forms a constraint $f(T, V, P) = 0$ among (T, V, P).

State quantities are also divided into *extensive* and *intensive variables*. To see this, let us express (1.7) and (1.8) as $P = RT/(V/n)$ and $P = RT/(V/n - b) - a(n/V)^2$. We thereby realize that the volume V for a fixed (T, P) increases in proportion to the number of moles n. We call thermodynamic variables that are proportional to (independent of) n *extensive (intensive) variables*. Therefore, V is extensive whereas T and P are intensive.

A couple of comments are in order before closing the section. First, volume V can be defined mechanically, whereas (T, P) are not. Indeed, we shall see below in Sect. 1.4.2 that intensive variables (T, P) are true thermodynamic variables that define the state of equilibrium. Second, all state quantities are defined within the limited regime of *thermodynamic equilibrium* being connected by reversible processes, which may be realized experimentally as *quasistatic processes*. Such processes change state variables so slowly that one can define (T, P) unambiguously at every moment.

1.3 Laws of Thermodynamics

Thermodynamics has established the existence of a couple of novel state quantities called *internal energy* and *entropy*. We present the basic laws of thermodynamics concerning the state quantities in their mathematical forms.

Consider a closed system that does not exchange matter with its surroundings, as depicted in Fig. 1.2. The first law of thermodynamics states that infinitesimal heat $d'Q$ flowing into the system and work $d'W$ performed on it, which are both non-potentials, add up to an increment dU of a potential U of the system called *internal energy* as

First Law of Thermodynamics

$$dU = d'Q + d'W. \tag{1.9}$$

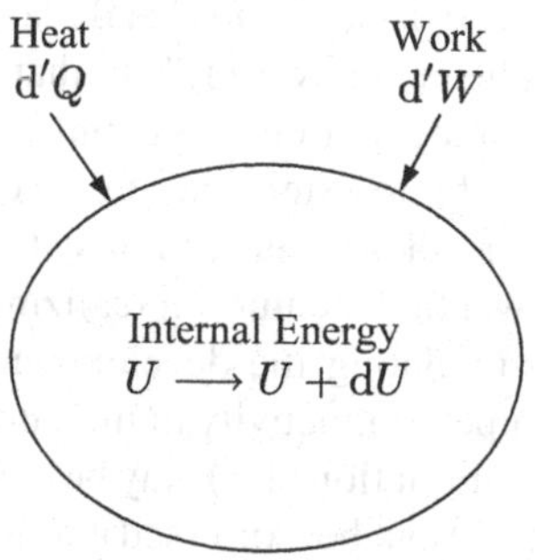

Fig. 1.2 Change in internal energy dU of a closed system due to infinitesimal heat $d'Q$ entering the system and work $d'W$ performed on the system

Historically, the first law thereby identified heat as a form of energy and established the law of energy conservation (1.9).

The second law of thermodynamics characterizes heat in (1.9) as an inferior form of energy in comparison with work in that it is bounded by the *Clausius inequality*:

Second Law of Thermodynamics

$$d'Q \leq T dS, \tag{1.10}$$

where T is the temperature of the surroundings from which heat flows, and S is a new state quantity called *entropy*. Equality holds in reversible processes along which we can trace changes in entropy. The second law plays a central role in thermodynamics. It cannot be over-emphasized that entropy is defined here with regard to thermodynamic equilibrium.

The third law of thermodynamics, which is also referred to as *Nernst's theorem*, reads

Third Law of Thermodynamics

$$\lim_{T \to 0} S = 0. \tag{1.11}$$

It also implies that entropy in the limit $T \to 0$ does not depend on other state variables such as P and V; that is,

$$\lim_{T \to 0} \frac{\partial S}{\partial P} = 0, \qquad \lim_{T \to 0} \frac{\partial S}{\partial V} = 0.$$

Thermodynamics can only state that entropy at $T = 0$ is a constant that does not depend on other state variables. Statistical mechanics reveals that the zero on the right-hand side of (1.11) corresponds to the fact that every system at $T = 0$ is in a single *pure state* with the lowest energy, the details of which appear around (2.8) below.

1.4 Equilibrium Thermodynamics

We now focus on reversible processes for which $d'Q = T dS$ in (1.10) and discuss the fundamentals of equilibrium thermodynamics. To be specific, we shall consider gases as a typical example, where the work performed on the system is given by

$$d'W = -P dV. \tag{1.12}$$

1.4.1 Basic Equation

Substitution of (1.12) and $d'Q = TdS$ into (1.9) yields

$$dU = TdS - PdV. \tag{1.13}$$

This is the basic equation of equilibrium thermodynamics for gases. It may be expressed alternatively in terms of entropy as

$$dS = \frac{1}{T}dU + \frac{P}{T}dV, \tag{1.14}$$

from which we can identify

$$\frac{1}{T} = \left(\frac{\partial S}{\partial U}\right)_V, \qquad \frac{P}{T} = \left(\frac{\partial S}{\partial V}\right)_U, \tag{1.15}$$

where we have used a standard notation $(\partial S/\partial U)_V \equiv \partial S(U, V)/\partial U$ in thermodynamics.

1.4.2 Equilibrium Conditions

A system with no exchange of heat and work with its surroundings is termed an *isolated system.* Substitution of its defining condition $d'Q = d'W = 0$ into (1.10) yields inequality $0 \le dS(U, V)$ for arbitrary processes, including irreversible ones such as the free expansion into an insulated evacuated chamber. This inequality tells us that entropy of an isolated system should increase up to its maximum value, whereupon the variation stops. To put it another way, the equilibrium of every isolated system is identified as the state with maximum entropy. We shall use this statement to explain the meaning of intensive variables (T, P).

Consider the isolated system depicted in Fig. 1.3, which is partitioned into subsystems 1 and 2 by a wall that moves smoothly and transmits heat. The total internal energy U and volume V are extensive variables expressible in terms of those of the subsystems:

$$U = U_1 + U_2, \qquad V = V_1 + V_2, \tag{1.16}$$

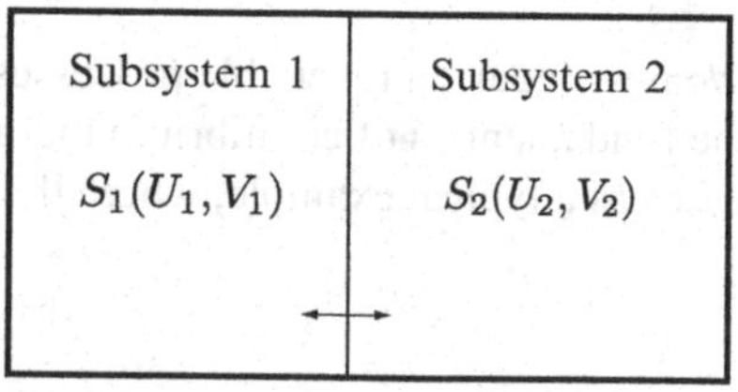

Fig. 1.3 An isolated system $1 + 2$ partitioned by a wall that moves smoothly and transmits heat

each of which remains constant for an isolated system. The total entropy $S = S(U, V)$ is also extensive so that we can write it as

$$S = S_1(U_1, V_1) + S_2(U_2, V_2). \tag{1.17}$$

Now, suppose that the system has reached equilibrium at (U_j, V_j) $(j = 1, 2)$, the state having maximum entropy. This implies that entropy should decrease upon a virtual variation $(U_j, V_j) \to (U_j + \delta U_j, V_j + \delta V_j)$ as

$$\Delta S \equiv \sum_{j=1}^{2} [S_j(U_j + \delta U_j, V_j + \delta V_j) - S_j(U_j, V_j)] < 0. \tag{1.18}$$

A necessary condition for this inequality is that ΔS remains invariant up to first order in $(\delta U_j, \delta V_j)$; that is,

$$0 = \sum_{j=1}^{2} \left[\left(\frac{\partial S_j}{\partial U_j} \right)_{V_j} \delta U_j + \left(\frac{\partial S_j}{\partial V_j} \right)_{U_j} \delta V_j \right] = \left(\frac{1}{T_1} - \frac{1}{T_2} \right) \delta U_1 + \left(\frac{P_1}{T_1} - \frac{P_2}{T_2} \right) \delta V_1. \tag{1.19}$$

In the second equality, we have substituted (1.15) and subsequently used

$$\delta U_1 + \delta U_2 = 0, \qquad \delta V_1 + \delta V_2 = 0$$

that result from (1.16). Noting that (1.19) should hold for an arbitrary pair of $(\delta U_1, \delta V_1)$, we conclude

$$T_1 = T_2, \qquad P_1 = P_2. \tag{1.20}$$

Thus, both temperature and pressure are the same between subsystems 1 and 2 in equilibrium. Dividing the system into more subsystems and repeating the argument, one may conclude that the temperature and pressure remain constant throughout the isolated system in equilibrium. To put it another way, the temperature and pressure are true thermodynamic variables that specify the state of equilibrium.

1.4.3 Legendre Transformation and Free Energy

In experiments, the temperature is easier to control than the internal energy or entropy, because we only need to supply a contact with a heat bath of a known temperature and wait until the system comes to an equilibrium. Correspondingly, we now introduce a state quantity, called the *Helmholtz free energy*, which has temperature as a natural independent variable.

Equation (1.13) implies that temperature is a gradient of the internal energy given by $T = (\partial U/\partial S)_V$. Using it, we define a state function F by

Helmholtz Free Energy

$$F \equiv U - TS, \qquad T \equiv \left(\frac{\partial U}{\partial S}\right)_V. \tag{1.21}$$

Its infinitesimal increment $\mathrm{d}F = \mathrm{d}U - \mathrm{d}(TS)$ is rewritten using (1.13) and $\mathrm{d}(TS) = S\mathrm{d}T + T\mathrm{d}S$ into

$$\mathrm{d}F = -S\mathrm{d}T - P\mathrm{d}V. \tag{1.22}$$

Given $S = S(T, V)$ and $P = P(T, V)$, we can integrate (1.22) to obtain $F = F(T, V)$. Thus, we have successfully introduced a relevant state function. Because U and S are both extensive whereas T is intensive, F is classified as an extensive quantity.

The procedure of (1.21) is mathematically called a *Legendre transformation* from $U(S, V)$ to $F(T, V)$ with $T = (\partial U/\partial S)_V$. We frequently encounter this transformation in other fields of physics. A typical example is that encountered in classical mechanics in mapping Lagrangian $L(\mathbf{r}, \mathbf{v}, t)$ to Hamiltonian $H(\mathbf{r}, \mathbf{p}, t) = \mathbf{p}\cdot\mathbf{v} - L$ with $\mathbf{p} \equiv \partial L/\partial \mathbf{v}$, where $\mathbf{r} = \mathbf{r}(t)$ and $\mathbf{v}(t) \equiv \mathrm{d}\mathbf{r}(t)/\mathrm{d}t$ denote the position and velocity, respectively, of a particle at time t.

Finally, we trace the origin of the term *free energy* for F. Substituting (1.10) into (1.9), imposing the isothermal condition $\mathrm{d}T = 0$, and using (1.21), we obtain

$$-\mathrm{d}'W = -\mathrm{d}U + \mathrm{d}'Q \le -\mathrm{d}U + T\mathrm{d}S = -\mathrm{d}(U - TS) = -\mathrm{d}F. \tag{1.23}$$

That is, work performed by a system in contact with a heat bath has an upper bound given by the decrease in F. Expressed another way, F represents the maximum energy that can be extracted *freely* from a system in contact with a heat bath.

1.4.4 Particle Number as a Variable

We extend the above analysis to cases where the number of particles varies. Such a situation is realized when vapor is in equilibrium with water in a closed vessel, for example. The associated formalism is also indispensable as a purely mathematical tool for describing many-particle systems that obey quantum mechanics.

We extend the basic equation (1.13) to include a contribution to work due to a change in the particle number N,

$$\mathrm{d}U = T\mathrm{d}S - P\mathrm{d}V + \mu\mathrm{d}N. \tag{1.24}$$

The third term on the right-hand side denotes the new contribution that embodies an increment of internal energy due to a change $\mathrm{d}N$ in the particle number. The coefficient μ is called the *chemical potential*, which is classified as an intensive quantity as N is extensive. This μ is kept constant throughout for a system in equilibrium, as may be realized by repeating the argument in Sect. 1.4.2 for a permeable wall.

Correspondingly, (1.22) for an infinitesimal increment of F is modified into

$$\mathrm{d}F = -S\mathrm{d}T - P\mathrm{d}V + \mu\mathrm{d}N, \tag{1.25}$$

where the chemical potential is given by $\mu \equiv (\partial F/\partial N)_{T,V}$. It is convenient for later purposes to define another state quantity $\Omega = \Omega(T, V, \mu)$ by

Grand Potential

$$\Omega \equiv F - \mu N, \qquad \mu \equiv \left(\frac{\partial F}{\partial N}\right)_{T,V}. \tag{1.26}$$

Alternatively called the *thermodynamic potential*, its infinitesimal increment can be expressed, using (1.25) and $\mathrm{d}(\mu N) = \mu\mathrm{d}N + N\mathrm{d}\mu$, as

$$\mathrm{d}\Omega = -S\mathrm{d}T - P\mathrm{d}V - N\mathrm{d}\mu. \tag{1.27}$$

Given (S, P, N) as a function of (T, V, μ), we can integrate (1.27) to obtain $\Omega(T, V, \mu)$.

Let us express Ω in terms of quantities that are more familiar. To this end, we notice that $\Omega(T, V, \mu)$ is an extensive quantity with V as the only extensive variable. Hence, if the volume is multiplied by λ with (T, μ) fixed, the grand potential should also be increased by factor λ, i.e.,

$$\Omega(T, \lambda V, \mu) = \lambda\,\Omega(T, V, \mu).$$

Let us express $\lambda V \equiv V_\lambda$ and differentiate the above equation in terms of λ. We then obtain

$$\frac{\partial\Omega(T, V_\lambda, \mu)}{\partial V_\lambda}\frac{\partial V_\lambda}{\partial\lambda} = \Omega(T, V, \mu).$$

Setting $\lambda = 1$ subsequently yields

$$\left(\frac{\partial\Omega}{\partial V}\right)_{T,\mu} V = \Omega.$$

As $(\partial\Omega/\partial V)_{T,\mu} = -P$ according to (1.27), we obtain the desired relation:

$$\Omega = -PV. \tag{1.28}$$

1.5 Thermodynamic Construction of Entropy and Internal Energy

We finally clarify the minimal information required for constructing entropy S and internal energy U thermodynamically as a function of (T, V) for a system with a fixed number of particles. This consideration also reveals how thermodynamics is useful.

Choosing (T, V) as state variables, one can write the infinitesimal change in S formally as

$$dS = \left(\frac{\partial S}{\partial T}\right)_V dT + \left(\frac{\partial S}{\partial V}\right)_T dV. \tag{1.29}$$

Substitution of this equation into (1.13) yields

$$dU = T\left(\frac{\partial S}{\partial T}\right)_V dT + \left[T\left(\frac{\partial S}{\partial V}\right)_T - P\right] dV. \tag{1.30}$$

It also follows from (1.10) for reversible processes that the first term in (1.30) represents infinitesimal heat entering the system under a fixed volume. Dividing the term by dT yields the heat required to raise the temperature by 1 K at a fixed volume,

$$C_V \equiv T\left(\frac{\partial S}{\partial T}\right)_V = \left(\frac{\partial U}{\partial T}\right)_V, \tag{1.31}$$

which is called the *heat capacity at constant volume*. Meanwhile, Maxwell's relation (1.5) for (1.30) reads

$$\frac{\partial}{\partial V}\left(T\frac{\partial S}{\partial T}\right) = \frac{\partial}{\partial T}\left(T\frac{\partial S}{\partial V} - P\right),$$

i.e.,

$$\left(\frac{\partial S}{\partial V}\right)_T = \left(\frac{\partial P}{\partial T}\right)_V. \tag{1.32}$$

Substitution of (1.31) and (1.32) into (1.29) and (1.30) yields

$$dS = \frac{C_V}{T}dT + \left(\frac{\partial P}{\partial T}\right)_V dV, \tag{1.33}$$

$$dU = C_V dT + \left[T\left(\frac{\partial P}{\partial T}\right)_V - P\right] dV. \tag{1.34}$$

Heat capacity (1.31) in these expressions is a basic quantity that is measurable in experiments. Further, the coefficients for dV can be calculated once the equation of state $P = P(T, V)$ is known accurately. Hence, apart from $P = P(T, V)$, all that is required experimentally to construct entropy and internal energy is $C_V = C_V(T, V)$. Moreover, Maxwell's relations for (1.33) and (1.34) both read

$$\left(\frac{\partial C_V}{\partial V}\right)_T = T\left(\frac{\partial^2 P}{\partial T^2}\right)_V. \tag{1.35}$$

The inference is that we need only know the temperature dependence of C_V once $P = P(T, V)$ is measured accurately. Equation (1.35) may also be used as a consistency check between independent experiments used to obtain $C_V = C_V(T, V)$ and $P = P(T, V)$.

Thus, thermodynamics has clarified that we need not perform direct measurements of $(\partial S/\partial V)_T$ and $(\partial U/\partial V)_T$. Indeed, they can be calculated alternatively based on the equation of state by (1.32) and

$$\left(\frac{\partial U}{\partial V}\right)_T = T\left(\frac{\partial P}{\partial T}\right)_V - P. \tag{1.36}$$

As an example, consider the internal energy for n moles of an ideal gas. It follows from the ideal gas law $P = nRT/V$ that $(\partial P/\partial T)_V = nR/V = P/T$. Substitution into (1.36) yields $(\partial U/\partial V)_T = 0$. Thus, the internal energy of an ideal gas should be independent of volume; that is, expressible as $U = U(T)$. This fact was found by Joule experimentally in 1844 when thermodynamics had not yet been established. In contrast, we can reach this statement theoretically with the help of thermodynamics using only the ideal gas law.

Problems

1.1. Consider the case where the gradients in (1.2) are given by

$$\frac{\partial z}{\partial x} = 2xy + 1, \qquad \frac{\partial z}{\partial y} = x^2 + 2y.$$

(a) Check that the integrability condition (1.5) is satisfied.
(b) Integrate the corresponding (1.2) based on the procedure described around (1.3) and (1.4).

1.2. Use (1.9) and (1.10) to show that *it is impossible to do work outside using a heat engine with only a single heat bath.*

1.3. Suppose that there are n moles of a gas that obey van der Waals' equation (1.8), for which the heat capacity C_V is known experimentally to be temperature independent.

(a) Write expressions for the infinitesimal increments in entropy $\mathrm{d}S$ and internal energy $\mathrm{d}U$ based on (1.33) and (1.34).
(b) Show that C_V does not depend on volume V.
(c) Obtain expressions for entropy S and internal energy U.
(d) Show that quantity $T(V - nb)^{nR/C_V}$ does not change in reversible adiabatic processes.
(e) Calculate the temperature change during an adiabatic free expansion from volume V_1 to volume V_2.

Chapter 2
Basics of Equilibrium Statistical Mechanics

Abstract The basics of equilibrium statistical mechanics are developed. We first derive a statistical-mechanical expression for entropy, (2.10) called the Gibbs or von Neumann entropy, that is compatible with the laws of thermodynamics. It is used subsequently to find the equilibrium statistical distributions, namely, microcanonical, canonical, and grand canonical distributions as (2.12), (2.18), and (2.26), respectively, based on the principle of maximum entropy by Jaynes.

2.1 Entropy in Statistical Mechanics

We first derive a statistical-mechanical expression for entropy in terms of probability w_ν for the state ν to occur as (2.10) below based on three plausible assumptions. This forms our foundation in formulating equilibrium statistical mechanics and will be used subsequently to obtain the equilibrium probability distributions.

Statistical mechanics aims at providing a theoretical framework that is consistent with thermodynamics and also enables us to perform theoretical calculations of macroscopic quantities such as the heat capacity and equation of state, which in thermodynamics can only be obtained by experiments. Such a system contains a large number of particles or degrees of freedom comparable with the *Avogadro constant* $N_\mathrm{A} = 6.02 \times 10^{23}\,\mathrm{mol}^{-1}$, for which it is practically impossible to solve the Newtonian equations of motion or the Schrödinger equation. If it were possible, the resulting data for the points of phase space or the wave function of N_A particles in configuration space, which accumulate at every moment, could only exhaust computer memories without ever being accessed usefully. Instead, what we need here is information on their average motion and position, which are closely related to the observable quantities such as pressure and temperature. Hence, statistical mechanics is designed to calculate average quantities concisely with the help of statistics and probability theory.

Therefore, we begin by summarizing the basics of statistics and probability theory. Quantum mechanics [3, 5] tells us that every microscopic state may be specified by a set of discrete quantum numbers, which we denote by the Greek

© Springer Japan 2015
T. Kita, *Statistical Mechanics of Superconductivity*, Graduate Texts in Physics,
DOI 10.1007/978-4-431-55405-9_2

symbol ν. Let us assume that (i) there are $\nu = 1, 2, \cdots, W$ states and (ii) each state occurs with probability $w_\nu \geq 0$, which is normalized as

$$\sum_{\nu=1}^{W} w_\nu = 1. \tag{2.1}$$

Consider a quantity g_ν of state ν; its average over the whole states, called the *expectation*, is defined as

$$\langle g \rangle \equiv \sum_\nu w_\nu g_\nu. \tag{2.2}$$

Sometimes $\langle g \rangle$ is denoted alternatively as $\bar{g}$. We also introduce its *standard deviation* by

$$\sigma_g \equiv \sqrt{\sum_\nu w_\nu (g_\nu - \langle g \rangle)^2} = \sqrt{\langle g^2 \rangle - \langle g \rangle^2}, \tag{2.3}$$

where the second expression is obtained using (2.1) and (2.2). The standard deviation is also called the *fluctuation* in statistical mechanics.

With these preliminaries, we construct the fundamentals of statistical mechanics. The second law of thermodynamics establishes the existence of *entropy*. Unlike the internal energy or volume, it is a true thermodynamic quantity that cannot be expressed mechanically. Thus, we start by deriving a statistical-mechanical expression for entropy. To this end, we adopt three plausible assumptions:

(a) *Entropy is an extensive quantity.*
In other words, the total entropy of a composite of subsystems 1 and 2, as given in Fig. 1.3, can be written as the sum of the entropy for each subsystem,

$$S^{(1+2)} = S^{(1)} + S^{(2)}. \tag{2.4}$$

(b) *Entropy is given as a functional of probability w_ν alone.*
The statement implies that we can express entropy as

$$S = \sum_\nu w_\nu f(w_\nu), \tag{2.5}$$

where $f(w)$ is an unknown function. Here we ignore all correlations like $w_\nu w_{\nu'}$ between different states ν and ν' by assuming that each state is realized independently.

(c) *Two subsystems of a composite system are statistically independent.*
This may be expressed mathematically as

$$w^{(1+2)}_{\nu_1,\nu_2} = w^{(1)}_{\nu_1} w^{(2)}_{\nu_2}, \tag{2.6}$$

where $w_{\nu_j}^{(j)}$ and $w_{\nu_1,\nu_2}^{(1+2)}$ denote the probabilities that subsystem j and total system $1+2$ are in state ν_j and (ν_1, ν_2), respectively.

Conditions (a)–(c) determine the expression for entropy uniquely as follows. First, let us substitute (2.5) into (2.4) and use (2.1) and (2.6). We can thereby rewrite (2.4) as

$$
\begin{aligned}
0 &= S^{(1+2)} - S^{(1)} - S^{(2)} \\
&= \sum_{\nu_1}\sum_{\nu_2} w_{(\nu_1,\nu_2)}^{(1+2)} f\left(w_{(\nu_1,\nu_2)}^{(1+2)}\right) - \sum_{\nu_1} w_{\nu_1}^{(1)} f\left(w_{\nu_1}^{(1)}\right) - \sum_{\nu_2} w_{\nu_2}^{(2)} f\left(w_{\nu_2}^{(2)}\right) \\
&= \sum_{\nu_1}\sum_{\nu_2} w_{\nu_1}^{(1)} w_{\nu_2}^{(2)} f\left(w_{\nu_1}^{(1)} w_{\nu_2}^{(2)}\right) - \sum_{\nu_1} w_{\nu_1}^{(1)} f\left(w_{\nu_1}^{(1)}\right) \sum_{\nu_2} w_{\nu_2}^{(2)} - \sum_{\nu_1} w_{\nu_1}^{(1)} \sum_{\nu_2} w_{\nu_2}^{(2)} f\left(w_{\nu_2}^{(2)}\right) \\
&= \sum_{\nu_1}\sum_{\nu_2} w_{\nu_1}^{(1)} w_{\nu_2}^{(2)} \left[f\left(w_{\nu_1}^{(1)} w_{\nu_2}^{(2)}\right) - f\left(w_{\nu_1}^{(1)}\right) - f\left(w_{\nu_2}^{(2)}\right) \right].
\end{aligned}
$$

Requiring that this equality holds for an arbitrary pair of $(w_{\nu_1}^{(1)}, w_{\nu_2}^{(2)})$, we obtain $f(uw) = f(u) + f(w)$. Its differentiation with respect to u yields $wf'(uw) = f'(u)$. Setting $u = 1$, we then obtain $f'(w) = f'(1)/w$. The equation can be integrated easily,

$$
f(w) = -k_B \ln w + C, \tag{2.7}
$$

where $k_B \equiv -f'(1)$ and C are constants and $\ln w$ denotes the natural logarithm of w. Thus, $f(w)$ has been obtained explicitly.

Let us substitute (2.7) into (2.5) and use (2.1). We thereby obtain the expression sought for the entropy

$$
S = -k_B \sum_{\nu} w_\nu \ln w_\nu + C. \tag{2.8}
$$

We also note that $-w_\nu \ln w_\nu \geq 0$ for $0 \leq w_\nu \leq 1$, where equality $-w_\nu \ln w_\nu = 0$ holds for $w_\nu = 0, 1$. Hence, the first term of (2.8) satisfies $-k_B \sum_\nu w_\nu \ln w_\nu \geq 0$, for which the lowest value 0 corresponds to a *pure state* where some single state ν_0 is realized with probability 1. This pure state can be expressed as $w_\nu = \delta_{\nu\nu_0}$, where $\delta_{\nu\nu_0}$ denotes the *Kronecker delta* defined by

$$
\delta_{\nu\nu_0} \equiv \begin{cases} 1 & : \nu = \nu_0 \\ 0 & : \nu \neq \nu_0 \end{cases}. \tag{2.9}
$$

Combining the above with the third law of thermodynamics (1.11), we conclude $C = 0$ in (2.8). Thus, we obtain the statistical-mechanical expression for entropy:

Entropy in Statistical Mechanics

$$S = -k_{\mathrm{B}} \sum_{\nu} w_\nu \ln w_\nu . \tag{2.10}$$

Further, the requirement that (2.10) be compatible with the thermodynamic entropy enables us to identify k_{B} with the gas constant $R = 8.31$ J/(mol·K) divided by the Avogadro constant $N_{\mathrm{A}} = 6.02 \times 10^{23}\,\mathrm{mol}^{-1}$, i.e.,

$$k_{\mathrm{B}} = R/N_{\mathrm{A}} = 1.38 \times 10^{-23}\,\mathrm{J/K}, \tag{2.11}$$

which is known as the *Boltzmann constant.*

Gibbs pointed out in his seminal book [1] that entropy for the canonical distribution is given as the expectation of the logarithm of the probability. Later, von Neumann derived (2.10) based on a quantum-mechanical consideration [7]. Thus, (2.10) may be called the *Gibbs entropy* [4] or *von Neumann entropy*. It is essentially equivalent to the *Shannon entropy* of information theory [6].

Expression (2.10) still contains unknown parameters $\{w_\nu\}$, which will be determined appropriately below given three distinct external conditions. In this sense, (2.10) is a form of *nonequilibrium entropy*. However, it cannot be used to describe nonequilibrium time developments. Because we have no knowledge of how w_ν changes in time, we usually regard w_ν as constant. Correspondingly, (2.10) is invariant in time even for an isolated system [4], in contradiction with the second law of thermodynamics. Indeed, we do not have a widely accepted general expression of nonequilibrium entropy that develops in time. It should be remembered in this context that entropy in thermodynamics is defined, as in (1.10), only for systems in equilibrium. Despite this obvious defect, the great advantage of (2.10) is that it enables a concise and transparent derivation of three representative equilibrium probability distributions in statistical mechanics, to be described shortly below.

2.2 Deriving Equilibrium Distributions

We have seen in Sect. 1.4.2 that the thermodynamic equilibrium of an isolated system corresponds to the state of maximum entropy. In addition, (1.23) states that the inequality $\mathrm{d}(-U + TS) \geq 0$ holds for a system of fixed volume (i.e., $\mathrm{d}'W = 0$) in contact with a heat bath of temperature T, which is equivalent to $\mathrm{d}(S - T^{-1}U) \geq 0$. Thus, the state of equilibrium for this system corresponds to a maximum of $S - T^{-1}U$, where T^{-1} may be regarded mathematically as a Lagrange multiplier used in maximizing S subject to a fixed U.

On the basis of these observations, we here adopt the *principle of maximum entropy*, which was proposed by Jaynes in 1957 [2], to derive equilibrium probabilities for various external conditions.

2.2.1 Microcanonical Distribution

First, we consider an isolated system of fixed volume V with no exchange of energy or matter with its surroundings and derive its equilibrium probability distribution $\{w_\nu\}$ as (2.12) below.

Following the principle of maximum entropy and noting that there are no external conditions for an isolated system, we maximize (2.10) subject to (2.1). In accordance with the *method of Lagrange multipliers*, this is equivalent to the optimization problem of

$$\tilde{S}(\{w_\nu\}, \lambda) \equiv S - \lambda\left(\sum_\nu w_\nu - 1\right) = -k_\mathrm{B} \sum_\nu w_\nu \ln w_\nu - \lambda\left(\sum_\nu w_\nu - 1\right),$$

where λ is the Lagrange multiplier. Indeed, (2.1) is equivalent to $\partial\tilde{S}/\partial\lambda = 0$.

A necessary condition for $\tilde{S}$ to take its maximum at $w_\nu = w_\nu^\mathrm{eq}$ is given by

$$0 = \left.\frac{\partial\tilde{S}}{\partial w_\nu}\right|_{w_\nu = w_\nu^\mathrm{eq}} = -k_\mathrm{B}(\ln w_\nu^\mathrm{eq} + 1) - \lambda.$$

Its solution immediately follows, i.e., $w_\nu^\mathrm{eq} = \mathrm{e}^{-\lambda/k_\mathrm{B}-1}$. Thus, w_ν^eq does not depend on ν but has a common value, which is determined by (2.1) as

$$w_\nu^\mathrm{eq} = W^{-1}.$$

This probability distribution, called the *microcanonical distribution*, underscores precisely the *postulate or principle of equal a priori probabilities*. Note, however, that it has been *derived* here based on the principle of maximum entropy in terms of (2.10). Substitution of $w_\nu^\mathrm{eq} = W^{-1}$ into (2.10) yields the entropy for a system in equilibrium, $S^\mathrm{eq} \equiv S[w_\nu^\mathrm{eq}]$, as

$$S^\mathrm{eq} = k_\mathrm{B} \ln W,$$

which is known as *Boltzmann's principle*.

Let us call a system that obeys the microcanonical distribution a *microcanonical ensemble*. Using this terminology and removing the superscript $^\mathrm{eq}$ for simplicity, we can summarize the above results as follows:

Microcanonical Ensemble

$$w_\nu = \frac{1}{W} \quad \text{: microcanonical distribution} \tag{2.12}$$

$$W(U, V, N) \quad \text{: number of states} \tag{2.13}$$

$$S = k_B \ln W \quad \text{: entropy (Boltzmann's principle)} \tag{2.14}$$

Thus, a fundamental quantity for a microcanonical ensemble is the number of states W, with which we obtain its entropy using (2.14). Subsequently, we can use the thermodynamic relation (1.24) to calculate the temperature, pressure, and chemical potential:

$$\left(\frac{\partial S}{\partial U}\right)_{V,N} = \frac{1}{T}, \quad \left(\frac{\partial S}{\partial V}\right)_{U,N} = \frac{P}{T}, \quad \left(\frac{\partial S}{\partial N}\right)_{U,V} = -\frac{\mu}{T}. \tag{2.15}$$

In general, $S(U, V, N)$ is a monotonically increasing function of U so that $T > 0$ holds. However, it should be kept in mind that this condition $T > 0$ may not hold for some theoretical models such as a spin system where there is an upper bound in the available energy.

2.2.2 *Canonical Distribution*

Next, we consider a system of fixed volume V that is in contact with a heat bath without any exchange of matter and obtain its equilibrium probability distribution $\{w_\nu\}$ as (2.18) below.

The total energy of such a system in equilibrium is expected to remain constant in time on average, but fluctuations may occur. The condition can be written as

$$\langle \mathscr{E} \rangle \equiv \sum_\nu w_\nu \mathscr{E}_\nu = U, \tag{2.16}$$

where $\mathscr{E}_\nu$ is the energy of the state ν and U is a constant called *internal energy*.

Given the principle of maximum entropy with (2.10), the equilibrium of this system can be found by minimizing

$$\tilde{S}(\{w_\nu\}, \lambda, \lambda_U) \equiv S - \lambda\left(\sum_\nu w_\nu - 1\right) - \lambda_U \sum_\nu w_\nu \mathscr{E}_\nu$$

in terms of $\{w_\nu\}$ under conditions $\partial S/\partial\lambda = 0$, $\partial S/\partial\lambda_U = -U$. However, it is more convenient to consider $F \equiv -\tilde{S}/\lambda_U$ instead of $\tilde{S}$ to make the correspondence with thermodynamics transparent. Indeed, F can be written in terms of new Lagrange multipliers $T \equiv \lambda_U^{-1}$ and $\lambda_S \equiv \lambda/\lambda_U$,

$$F \equiv \langle\mathscr{E}\rangle - TS + \lambda_S\left(\sum_\nu w_\nu - 1\right) = \sum_\nu w_\nu\left(\mathscr{E}_\nu + k_{\mathrm{B}}T\ln w_\nu + \lambda_S\right) - \lambda_S, \tag{2.17}$$

which is exactly the Helmholtz free energy for a system in *nonequilibrium*, as may be realized from a comparison with (1.21). The unknown Lagrange multipliers (T, λ_S) can be determined using (2.1) and (2.16), which are also expressible in terms of F as $\partial F/\partial\lambda_S = 0$ and $F\big|_{T=\lambda_S=0} = U$, respectively. Thus, we have transformed the optimization problem of $\tilde{S}$ into one for F.

A necessary condition that F be extremal at $w_\nu = w_\nu^{\mathrm{eq}}$ is given by

$$0 = \left.\frac{\partial F}{\partial w_\nu}\right|_{w_\nu = w_\nu^{\mathrm{eq}}} = \mathscr{E}_\nu + k_{\mathrm{B}}T(\ln w_\nu^{\mathrm{eq}} + 1) + \lambda_S,$$

which yields $w_\nu^{\mathrm{eq}} = \mathrm{e}^{-(\mathscr{E}_\nu+\lambda_S)/k_{\mathrm{B}}T-1}$. Introducing a new constant $Z \equiv \mathrm{e}^{\lambda_S/k_{\mathrm{B}}T+1}$, we can write w_ν^{eq} alternatively as

$$w_\nu^{\mathrm{eq}} = \mathrm{e}^{-\mathscr{E}_\nu/k_{\mathrm{B}}T}/Z,$$

which is known as the *canonical distribution*. We call a system obeying this distribution a *canonical ensemble*. The physical condition that $w_\nu \to 0$ as $\mathscr{E}_\nu \to \infty$ yields $T > 0$. In addition, (2.1) enables us to express Z as

$$Z = \sum_\nu \mathrm{e}^{-\mathscr{E}_\nu/k_{\mathrm{B}}T},$$

which is known as the *partition function*. Let us substitute the expression of w_ν^{eq} above into (2.17). We thereby obtain $F^{\mathrm{eq}} \equiv F[w_\nu^{\mathrm{eq}}]$ for a system in equilibrium as

$$F^{\mathrm{eq}} = \sum_\nu w_\nu^{\mathrm{eq}}\left[\mathscr{E}_\nu + k_{\mathrm{B}}T\left(-\frac{\mathscr{E}_\nu}{k_{\mathrm{B}}T} - \ln Z\right)\right] = -k_{\mathrm{B}}T\ln Z.$$

It follows from the condition $T > 0$ that F^{eq} with maximum entropy corresponds to the minimum of (2.17).

Removing superscripts ^{eq} for simplicity, the above results are summarized as follows:

Canonical Ensemble ($\beta \equiv 1/k_B T$)

$$w_\nu = \frac{e^{-\beta \mathscr{E}_\nu}}{Z} \quad \text{: canonical distribution} \tag{2.18}$$

$$Z(T, V, N) = \sum_\nu e^{-\beta \mathscr{E}_\nu} \quad \text{: partition function} \tag{2.19}$$

$$F = -\beta^{-1} \ln Z \quad \text{: Helmholtz free energy} \tag{2.20}$$

Thus, the fundamental quantity in a canonical ensemble is the partition function Z, with which we obtain the Helmholtz free energy using (2.20). Subsequently, we can use the thermodynamic relation (1.25) to calculate the entropy, pressure, and chemical potential from

$$\left(\frac{\partial F}{\partial T}\right)_{V,N} = -S, \quad \left(\frac{\partial F}{\partial V}\right)_{T,N} = -P, \quad \left(\frac{\partial F}{\partial N}\right)_{T,V} = \mu. \tag{2.21}$$

Moreover, substitution of (2.18) into (2.16) yields

$$U = \frac{1}{Z} \sum_\nu e^{-\beta \mathscr{E}_\nu} \mathscr{E}_\nu = -\frac{\partial}{\partial \beta} \ln Z, \tag{2.22}$$

where we have used (2.19) in the second equality. Thus, one obtains the internal energy U directly from the partition function. Regarding heat capacity (1.31), we use $\partial/\partial T = -k_B \beta^2 (\partial/\partial \beta)$ and the definition of the expectation in (2.2) to express $C_V = (\partial U/\partial T)_{V,N}$ in two different ways:

$$C_V = \frac{\langle \mathscr{E}^2 \rangle - \langle \mathscr{E} \rangle^2}{k_B T^2} = \frac{1}{k_B T^2} \frac{\partial^2}{\partial \beta^2} \ln Z. \tag{2.23}$$

Hence, we obtain the internal energy and its fluctuation by successive differentiations of the logarithm of the partition function with respect to β, the latter of which is directly related to the heat capacity.

The internal energy U is an extensive quantity, as is the heat capacity C_V obtained by differentiating U in terms of the intensive quantity T. Equation (2.23) enables us now to estimate the fluctuation in energy as $\Delta U \equiv \sqrt{\langle \mathscr{E}^2 \rangle - \langle \mathscr{E} \rangle^2} \propto N^{1/2}$. Hence, we conclude that $\Delta U/U \propto N^{-1/2}$, i.e., the relative magnitude of the energy fluctuation decreases as a function of the particle number in the system as $N^{-1/2}$. This statement holds true generally for fluctuations in thermodynamics and statistical mechanics. It also informs us that the microcanonical and canonical distributions should yield identical expectations.

2.2.3 Grand Canonical Distribution

Finally, we consider a system of fixed volume V exchanging particles as well as heat with a reservoir and derive its equilibrium probability distribution $\{w_\nu\}$ as (2.26) below.

Besides the internal energy, the total particle number of such a system is expected to remain constant on average in equilibrium, although it may fluctuate. The condition can be expressed as

$$\langle \mathscr{N} \rangle \equiv \sum_\nu w_\nu \mathscr{N}_\nu = N, \tag{2.24}$$

where $\mathscr{N}_\nu$ is the number of particles in state ν, and N is some constant, i.e., the average particle number.

The principle of maximum entropy tells us that equilibrium occurs when (2.10) takes its maximum in terms of $\{w_\nu\}$ subject to the conditions given by (2.1), (2.16), and (2.24). This optimization problem can be solved by introducing a functional:

$$\begin{aligned}\Omega &\equiv \langle \mathscr{E} \rangle - \mu \langle \mathscr{N} \rangle - TS + \lambda_S \left(\sum_\nu w_\nu - 1 \right) \\ &= \sum_\nu w_\nu \left(\mathscr{E}_\nu - \mu \mathscr{N}_\nu + k_\mathrm{B} T \ln w_\nu + \lambda_S \right) - \lambda_S,\end{aligned} \tag{2.25}$$

where T, μ, and λ_S are the associated Lagrange multipliers determined by (2.1), (2.16), and (2.24). Recalling (1.26), we realize that (2.25) is the *nonequilibrium* grand potential with T and μ denoting temperature and chemical potential, respectively.

A necessary condition that Ω be extremal at $w_\nu = w_\nu^\mathrm{eq}$ is given by

$$0 = \left. \frac{\partial \Omega}{\partial w_\nu} \right|_{w_\nu = w_\nu^\mathrm{eq}} = \mathscr{E}_\nu - \mu \mathscr{N}_\nu + k_\mathrm{B} T (\ln w_\nu^\mathrm{eq} + 1) + \lambda_S,$$

which is immediately solved to yield $w_\nu^\mathrm{eq} = \mathrm{e}^{-(\mathscr{E}_\nu - \mu \mathscr{N}_\nu + \lambda_S)/k_\mathrm{B} T - 1}$. Introducing a new constant by $Z_\mathrm{G} \equiv \mathrm{e}^{\lambda_S / k_\mathrm{B} T + 1}$, we write w_ν^eq in the form

$$w_\nu^\mathrm{eq} = \mathrm{e}^{-(\mathscr{E}_\nu - \mu \mathscr{N}_\nu)/k_\mathrm{B} T} / Z_\mathrm{G},$$

which is known as the *grand canonical distribution*. We call a system that obeys this distribution a *grand canonical ensemble*. The physical condition that $w_\nu \to 0$ as $\mathscr{E}_\nu \to \infty$ yields $T > 0$. In addition, the normalization condition (2.1) enables us to express $Z_{\rm G}$ as

$$Z_{\rm G} = \sum_\nu {\rm e}^{-(\mathscr{E}_\nu - \mu \mathscr{N}_\nu)/k_{\rm B}T},$$

which is known as the *grand partition function*. Let us substitute the expression of $w_\nu^{\rm eq}$ above into (2.25). We thereby obtain $\Omega^{\rm eq} \equiv \Omega[w_\nu^{\rm eq}]$ in equilibrium as

$$\Omega^{\rm eq} = -k_{\rm B}T \ln Z_{\rm G}.$$

It follows from the condition $T > 0$ that $\Omega^{\rm eq}$ with the maximum entropy corresponds to the minimum of (2.25).

Removing superscript $^{\rm eq}$ for simplicity, the above results are summarized as follows.

Grand Canonical Ensemble $(\beta \equiv 1/k_{\rm B}T)$

$$w_\nu = \frac{{\rm e}^{-\beta(\mathscr{E}_\nu - \mu\mathscr{N}_\nu)}}{Z_{\rm G}} \quad \text{: grand canonical distribution} \tag{2.26}$$

$$Z_{\rm G}(T, V, \mu) = \sum_\nu {\rm e}^{-\beta(\mathscr{E}_\nu - \mu\mathscr{N}_\nu)} \quad \text{: grand partition function} \tag{2.27}$$

$$\Omega = -\beta^{-1} \ln Z_{\rm G} \quad \text{: grand potential} \tag{2.28}$$

Thus, a fundamental quantity associated with grand canonical ensembles is the grand partition function $Z_{\rm G}$, with which we obtain the grand potential using (2.28). Subsequently, we can use the thermodynamic relation (1.27) to calculate entropy, pressure, and particle number:

$$\left(\frac{\partial \Omega}{\partial T}\right)_{V,\mu} = -S, \quad \left(\frac{\partial \Omega}{\partial V}\right)_{T,\mu} = -P, \quad \left(\frac{\partial \Omega}{\partial \mu}\right)_{T,V} = -N. \tag{2.29}$$

Moreover, substitution of (2.26) into (2.16) yields

$$U = \frac{1}{Z_{\rm G}} \sum_\nu {\rm e}^{-\beta(\mathscr{E}_\nu - \mu\mathscr{N}_\nu)}(\mathscr{E}_\nu - \mu\mathscr{N}_\nu + \mu\mathscr{N}_\nu) = -\frac{\partial}{\partial\beta} \ln Z_{\rm G} + \mu N, \tag{2.30}$$

where we have used (2.24) and (2.27) in the second equality. In regard to the fluctuation of the particle number inherent in the grand canonical distribution, a calculation of the variance gives

$$\sigma_{\mathcal{N}}^2 \equiv \langle \mathcal{N}^2 \rangle - \langle \mathcal{N} \rangle^2 = \frac{1}{\beta^2} \frac{\partial^2}{\partial \mu^2} \ln Z_{\mathrm{G}}, \tag{2.31}$$

in exactly the same way as the energy fluctuation, (2.23).

Problems

2.1. Let p and $1-p$ be the probabilities for success and failure, respectively, in a trial. Assuming that the probability of k $(\leq n)$ successes for n independent trials obeys the binomial distribution

$$P_k^n = \frac{n!}{k!(n-k)!} p^k (1-p)^{n-k},$$

answer the following questions.

(a) Show that $kP_k^n = npP_{k-1}^{n-1}$.
(b) Obtain the expectation and standard deviation for the number of successes.

2.2. Use the method of Lagrange multipliers to find the point on a unit circle of $x^2 + y^2 = 1$ that gives the maximum for the function $f(x, y) \equiv x + y$.

References

1. J.W. Gibbs, *Elementary Principles in Statistical Mechanics* (Yale University Press, New Haven, 1902), pp. 44–45 and 168
2. E.T. Jaynes, Phys. Rev. **106**, 620 (1957)
3. L.D. Landau, E.M. Lifshitz, *Quantum Mechanics: Non-relativistic Theory*, 3rd edn. (Butterworth-Heinemann, Oxford, 1991)
4. O. Penrose, *Foundations of Statistical Mechanics* (Pergamon, Oxford, 1970). The footnote of p. 213
5. J.J. Sakurai, *Modern Quantum Mechanics*, rev. edn. (Addison-Wesley, New York, 1994)
6. C.E. Shannon, Bell Syst. Tech. J. **27**, 379 (1948)
7. J. von Neumann, *Mathematical Foundations of Quantum Mechanics* (Princeton University Press, Princeton, 1955)

Chapter 3
Quantum Mechanics of Identical Particles

Abstract In general, superconductivity occurs in a system of identical particles; specifically, the conduction electrons in metals. Being indistinguishable from each other, swapping any two electrons leaves the system unchanged. This feature, which is associated with invariance under permutations, has a profound implication for every system composed of identical particles. We study a crucial connection between the spin of a particle and permutation symmetry of many-particle wave functions. We also develop a special technique called *second quantization* that enables us to describe such a system concisely and conveniently. The results are summarized generally in Sect. 3.7 and specifically for ideal gases in (3.61)–(3.65).

3.1 Permutation

Imagine N children sitting on N chairs that are arranged in a circle and numbered in a clockwise manner from 1 to N. A *permutation* is an action by which the children mutually exchange their seats [5]. Denoted $\hat{P}$, the associated operator is written in the form

$$\hat{P} = \begin{pmatrix} 1 & 2 & 3 & \cdots & N \\ p_1 & p_2 & p_3 & \cdots & p_N \end{pmatrix}, \tag{3.1}$$

where $1 \le p_i \le N$ with no duplication among the p_i's. Hence, the child previously on chair i moves to chair p_i. The number of distinct permutations is easily identified to be $N!$.

Among these permutations, there is a special subset called the *cyclic permutations*, in which the children move in a cyclic fashion, e.g.,

$$\begin{pmatrix} 1\,2\,3\,4 \\ 4\,3\,1\,2 \end{pmatrix} \equiv (1\,4\,2\,3) \tag{3.2}$$

for $N = 4$, where the last identity represents a shorthand notation and is to be read 'the child on chair 1 goes to chair 4', 'the child on chair 4 goes to chair 2', 'the child on chair 2 goes to chair 3', and 'the child on chair 3 goes to chair 1'. A cyclic permutation of length two is called a *transposition*, which is given by

© Springer Japan 2015
T. Kita, *Statistical Mechanics of Superconductivity*, Graduate Texts in Physics,
DOI 10.1007/978-4-431-55405-9_3

$$\hat{P}_{12} \equiv \begin{pmatrix} 1\,2 \\ 2\,1 \end{pmatrix} \equiv (1\,2). \tag{3.3}$$

We briefly give statements of some elementary theorems (without proof) related to permutations. See, for example, [5] for relevant proofs. First, every permutation can be expressed as a product of cyclic permutations that are disjoint to each other with equal or fewer elements. For example,

$$\hat{P}_a \equiv \begin{pmatrix} 1\,2\,3\,4\,5\,6 \\ 2\,5\,6\,4\,1\,3 \end{pmatrix} = (3\,6)\,(1\,2\,5) = (1\,2\,5)\,(3\,6), \tag{3.4}$$

where we have omitted the identity permutation (4); the order of application is to proceed from right to left by definition, but we can change it arbitrarily for disjoint permutations. Second, every cyclic permutation can be composed of a product of transpositions; for example

$$\begin{pmatrix} 1\,2\cdots k-2\;k-1\;k \\ 2\,3\cdots k-1\quad k\quad 1 \end{pmatrix} = (1\,k)\,(1\,k-1)\,\cdots\,(1\,3)(1\,2),$$

where again operations are applied from right to left. Combining the two statements, we realize that every permutation can be expressed as a product of transpositions. For example, (3.4) may be decomposed into

$$\hat{P}_a = (3\,6)\,(1\,5)\,(1\,2).$$

Another example is

$$\hat{P}_b \equiv \begin{pmatrix} 1\,2\,3\,4\,5\,6 \\ 2\,5\,6\,3\,1\,4 \end{pmatrix} = (3\,6\,4)\,(1\,2\,5) = (3\,4)\,(3\,6)\,(1\,5)\,(1\,2).$$

An *odd (even) permutation* describes a permutation that can be expressed as a product of an odd (even) number of transpositions, as for $\hat{P}_a$ ($\hat{P}_b$) above. Indeed, every permutation is either even or odd.

3.2 Permutation Symmetry of Identical Particles

We turn our attention to a physical system of N identical particles. Quantum mechanics tells us that every species of particle has an internal degree of freedom called *spin*, whose magnitude s takes a proper value from the series $0, \frac{1}{2}, 1, \frac{3}{2}, 2, \cdots$ [7, 10]. Accordingly, we need an additional index to specify the spin state of the particle besides its coordinate $\mathbf{r}$. A complete set of indices follows from solving the eigenvalue problem $\hat{s}_z|\alpha\rangle = \alpha|\alpha\rangle$, where $\hat{s}_z$ denotes the z component of the spin operator; it yields the eigenvalue $\alpha = s, s-1, \cdots, -s$. Because every spin state is describable as a linear combination of $\{|\alpha\rangle\}$, index α may be used as the 'coordinate' for the spin degrees of freedom. On the basis of these observations, we combine $\mathbf{r}$ and α to form a set of coordinates $\xi \equiv (\mathbf{r}, \alpha)$ to describe the complete set of states for a single particle.

An N-particle wave function can be expressed generally in terms of this index variable as

$$\Phi_\nu(\xi_1, \xi_2, \cdots, \xi_N), \tag{3.5}$$

where ν denotes a set of quantum numbers to specify the N-particle state. Operating with permutation (3.1) on this wave function yields

$$\hat{P}\Phi_\nu(\xi_1, \xi_2, \cdots, \xi_N) \equiv \Phi_\nu(\xi_{p_1}, \xi_{p_2}, \cdots, \xi_{p_N}), \tag{3.6}$$

by definition.

A many-particle system composed of N identical particles has a special symmetry related to permuting the particles. To see this, consider a system described by Hamiltonian

$$\hat{\mathscr{H}} = \sum_{j=1}^{N} \frac{\hat{\mathbf{p}}_j^2}{2m} + \sum_{i=1}^{N-1} \sum_{j=i+1}^{N} \mathscr{V}(|\mathbf{r}_i - \mathbf{r}_j|) \equiv \sum_{j=1}^{N} \hat{h}_j^{(1)} + \sum_{i<j} \hat{h}_{ij}^{(2)}, \tag{3.7}$$

where $\hat{\mathbf{p}}_j$ is the momentum operator [7, 10] and $\mathscr{V}$ denotes a two-body potential. The corresponding time-independent Schrödinger equation is given by

$$\hat{\mathscr{H}}\Phi_\nu(\xi_1, \xi_2, \cdots, \xi_N) = \mathscr{E}_\nu \Phi_\nu(\xi_1, \xi_2, \cdots, \xi_N), \tag{3.8}$$

with $\mathscr{E}_\nu$ the eigenenergy. Operating with permutation (3.1) from the left of (3.8) and inserting the identity operator $\hat{P}^{-1}\hat{P}$ between $\hat{\mathscr{H}}$ and Φ_ν, we obtain

$$(\hat{P}\hat{\mathscr{H}}\hat{P}^{-1})\hat{P}\Phi_\nu(\xi_1, \xi_2, \cdots, \xi_N) = \mathscr{E}_\nu \hat{P}\Phi_\nu(\xi_1, \xi_2, \cdots, \xi_N). \tag{3.9}$$

Now, operator $\hat{P}\hat{\mathscr{H}}\hat{P}^{-1}$ generally satisfies $\hat{P}\hat{\mathscr{H}}\hat{P}^{-1} = \hat{\mathscr{H}}$. This is exemplified using $\hat{\mathscr{H}}$ for $N = 2$:

$$\begin{aligned}\hat{P}_{12}\hat{\mathscr{H}}\hat{P}_{12}^{-1} &= \hat{P}_{12}\left[\frac{\hat{\mathbf{p}}_1^2}{2m} + \frac{\hat{\mathbf{p}}_2^2}{2m} + \mathscr{V}(|\mathbf{r}_1 - \mathbf{r}_2|)\right]\hat{P}_{12}^{-1} \\ &= \left[\frac{\hat{\mathbf{p}}_2^2}{2m} + \frac{\hat{\mathbf{p}}_1^2}{2m} + \mathscr{V}(|\mathbf{r}_2 - \mathbf{r}_1|)\right]\hat{P}_{12}\hat{P}_{12}^{-1} = \hat{\mathscr{H}}\hat{P}_{12}\hat{P}_{12}^{-1} = \hat{\mathscr{H}}.\end{aligned}$$

Thus, any permutation on $\hat{\mathscr{H}}$ only changes the order of the summation, leaving $\hat{\mathscr{H}}$ itself invariant. Let us rewrite $\hat{P}\hat{\mathscr{H}}\hat{P}^{-1} = \hat{\mathscr{H}}$ as

$$\hat{P}\hat{\mathscr{H}} = \hat{\mathscr{H}}\hat{P}. \tag{3.10}$$

Hence, we conclude by virtue of the Heisenberg equation of motion

$$\mathrm{i}\hbar \frac{\mathrm{d}\hat{P}_{\mathrm{H}}(t)}{\mathrm{d}t} = P_{\mathrm{H}}(t)\hat{\mathscr{H}} - \hat{\mathscr{H}} P_{\mathrm{H}}(t) \tag{3.11}$$

for the operator $\hat{P}_{\mathrm{H}}(t) \equiv \mathrm{e}^{\mathrm{i}\hat{\mathscr{H}}t/\hbar}\,\hat{P}\,\mathrm{e}^{-\mathrm{i}\hat{\mathscr{H}}t/\hbar}$ that the expectation value of $\hat{P}$ does not change in time [10]. In addition, $\hat{P}$ and $\hat{\mathscr{H}}$ can be diagonalized simultaneously [10] assuming $\hat{P}$ is Hermitian; this last point will be proved in the final paragraph of this section.

Let us find the eigenvalues of the permutation operators. We start with the simplest case of transposition (3.3). Because $\hat{P}_{12}^2$ is the identity permutation, an eigenvalue σ of $\hat{P}_{12}$ should satisfy $\sigma^2 = 1$. Hence, σ is either 1 or -1. Next, we consider the general permutation of (3.1) and denote its eigenvalue by σ^P, i.e.,

$$\hat{P}\Phi_\nu(\xi_1, \xi_2, \cdots, \xi_N) = \sigma^P \Phi_\nu(\xi_1, \xi_2, \cdots, \xi_N). \tag{3.12}$$

Now, recalling earlier statements in Sect. 3.1 that "every permutation can be expressed as a product of transpositions" and "every permutation is either even or odd," we find the eigenvalue of $\hat{P}$ to be

$$\sigma^P = \begin{cases} 1 & \text{if } \hat{P} \text{ is even} \\ \sigma & \text{if } \hat{P} \text{ is odd} \end{cases}. \tag{3.13}$$

It has been established that every stationary wave function for a system of identical particles belongs to an eigenstate of the permutation operators. This implies that, upon application of an odd permutation, a wave function either remains invariant ($\sigma = 1$) or changes sign ($\sigma = -1$). The two categories here are connected with the spin magnitude s of the constituent individual particle, specifically

Spin-Statistics Theorem

$$\begin{aligned} s &= 0, 1, 2, \cdots && \longleftrightarrow \sigma = +1 \\ s &= \frac{1}{2}, \frac{3}{2}, \frac{5}{2}, \cdots && \longleftrightarrow \sigma = -1 \end{aligned}. \tag{3.14}$$

A particle with an integer (a half-integer) spin is called *boson* (*fermion*) after Bose (Fermi), who introduced the rule upon studying statistical mechanics of photons with $s = 1$ (electrons with $s = \frac{1}{2}$) without referring to the connection with spin. The remarkable relationship was formulated by Fiertz [3] and proved by Pauli [9] in the context of relativistic quantum field theory. Although the statement itself has been confirmed experimentally without doubt, a direct proof in the non-relativistic framework seems yet to be performed. Incidentally, it is worth pointing out that the connection, which has been known as the "spin-statistics theorem" generally, has nothing to do with "statistics" or "probability" in the present context. Instead, it is relevant here to the permutation symmetry of many-particle wave functions, as we have seen.

The electron, proton, and neutron are all fermions with $s = \frac{1}{2}$, whereas the photon is a boson with $s = 1$. The rule applies also to every atom composed of protons, neutrons, and electrons that moves as a whole in our condensed-matter world. For example, the hydrogen atom consists of one electron and one proton, so that its total spin can be either 0 or 1 according to the addition rule of angular momenta [2, 7, 10]; thus, the hydrogen is a boson. In general, a neutral atom with an equal number of protons and electrons is classified as a boson or fermion according to whether the number of neutrons is even or odd. Thus, the ^{4}He atom with two neutrons is a boson with total spin $s = 0$ in the ground state, whereas its isotope ^{3}He with a single neutron is a fermion with total spin $s = 1/2$ in the ground state.

Finally, we confirm that $\hat{P}$ is Hermitian for a set of wave functions $\{\Phi_\nu\}$ satisfying (3.12). Let us introduce the inner product by

$$\langle \Phi_{\nu'} | \Phi_\nu \rangle \equiv \int d\xi_1 \cdots \int d\xi_N \, \Phi^*_{\nu'}(\xi_1, \xi_2, \cdots, \xi_N) \Phi_\nu(\xi_1, \xi_2, \cdots, \xi_N), \tag{3.15}$$

where the 'integration' over ξ_j signifies an integration and a summation

$$\int d\xi_j \equiv \sum_{\alpha_j = -s}^{s} \int d^3 r_j . \tag{3.16}$$

With this definition, $\langle \Phi_{\nu'} | \hat{P} \Phi_\nu \rangle$ is transformed as

$$\langle \Phi_{\nu'} | \hat{P} \Phi_\nu \rangle = \langle \hat{P}^{-1} \Phi_{\nu'} | \hat{P}^{-1} \hat{P} \Phi_\nu \rangle = \langle \hat{P}^{-1} \Phi_{\nu'} | \Phi_\nu \rangle = \sigma^P \langle \Phi_{\nu'} | \Phi_\nu \rangle = \langle \hat{P} \Phi_{\nu'} | \Phi_\nu \rangle,$$

where we have performed a change of variables corresponding to an operation of $\hat{P}^{-1}$ in the first equality, and subsequently used the fact that the eigenvalue of $\hat{P}^{-1}$ is identical to that of $\hat{P}$. Thus, $\hat{P}$ is indeed Hermitian.

3.3 Eigenspace of Permutation

Equation (3.12) implies that a wave function of identical particles should be a permutational eigenstate that is either symmetric (bosons) or antisymmetric (fermions). Here, we construct the eigenspace of permutations to which the many-particle wave functions belong.

In analogy to the treatment of the harmonic oscillator in quantum mechanics using creation and annihilation operators [10], we first introduce a pair of operators $\hat{\psi}$ and $\hat{\psi}^\dagger$ that satisfy the commutation relations:

$$[\hat{\psi}(\xi), \hat{\psi}^\dagger(\xi')]_\sigma \equiv \hat{\psi}(\xi)\hat{\psi}^\dagger(\xi') - \sigma\hat{\psi}^\dagger(\xi')\hat{\psi}(\xi) = \delta(\xi, \xi'),$$
$$[\hat{\psi}(\xi), \hat{\psi}(\xi')]_\sigma = [\hat{\psi}^\dagger(\xi), \hat{\psi}^\dagger(\xi')]_\sigma = 0, \tag{3.17}$$

with $\sigma = \pm 1$ and $\delta(\xi, \xi') \equiv \delta(\mathbf{r} - \mathbf{r}')\delta_{\alpha\alpha'}$. In addition, we define the ket $|0\rangle$ and bra $\langle 0|$ via the right (left) action of the annihilation (creation) operator,

$$\hat{\psi}(\xi)|0\rangle = 0, \qquad 0 = \langle 0|\hat{\psi}^\dagger(\xi) = [\hat{\psi}(\xi)|0\rangle]^\dagger, \qquad \langle 0|0\rangle = 1. \tag{3.18}$$

These are the basic ingredients needed to construct the eigenspace of permutations.

Next, we introduce the ket $|\xi_1, \xi_2, \cdots, \xi_N\rangle$ and its Hermitian conjugate by

$$|\xi_1, \xi_2, \cdots, \xi_N\rangle \equiv \frac{1}{\sqrt{N!}}\hat{\psi}^\dagger(\xi_1)\hat{\psi}^\dagger(\xi_2)\cdots\hat{\psi}^\dagger(\xi_N)|0\rangle, \tag{3.19}$$

$$\langle \xi_1, \xi_2, \cdots, \xi_N| \equiv \frac{1}{\sqrt{N!}}\langle 0|\hat{\psi}(\xi_N)\cdots\hat{\psi}(\xi_2)\hat{\psi}(\xi_1) = |\xi_1, \xi_2, \cdots, \xi_N\rangle^\dagger. \tag{3.20}$$

The space spanned by (3.19) naturally forms the eigenspace of $\hat{P}$. To see this, we start with transpositions. Let us operate with $\hat{P}_{ij}$ $(i < j)$ on (3.19) and transform the ket by repeatedly using the operator commutation relation (3.17),

$$\begin{aligned}
\hat{P}_{ij}|\xi_1, \xi_2, \cdots, \xi_N\rangle \equiv& |\xi_1, \cdots, \xi_{i-1}, \xi_j, \xi_{i+1}, \cdots \xi_{j-1}, \xi_i, \xi_{j+1}, \cdots, \xi_N\rangle \\
=& \sigma^{j-i}|\xi_1, \cdots, \xi_{i-1}, \xi_i, \xi_j, \xi_{i+1}, \cdots, \xi_{j-1}, \xi_{j+1}, \cdots, \xi_N\rangle \\
=& \sigma^{(j-i)+(j-i-1)}|\xi_1, \cdots \xi_{i-1}, \xi_i, \xi_{i+1}, \cdots, \xi_{j-1}, \xi_j, \xi_{j+1}, \cdots, \xi_N\rangle \\
=& \sigma|\xi_1, \cdots, \xi_{i-1}, \xi_i, \xi_{i+1}, \cdots, \xi_{j-1}, \xi_j, \xi_{j+1}, \cdots, \xi_N\rangle.
\end{aligned}$$

Thus, the states defined by (3.19) are eigenstates of transpositions. Second, as any $\hat{P}$ is either even or odd, we easily conclude that

$$\hat{P}|\xi_1, \xi_2, \cdots, \xi_N\rangle = \sigma^P|\xi_1, \xi_2, \cdots, \xi_N\rangle \tag{3.21}$$

holds generally.

Ket $|\xi_1, \xi_2, \cdots, \xi_N\rangle$ satisfies the normalization condition:

$$\langle \xi_1', \xi_2', \cdots, \xi_{N'}'|\xi_1, \xi_2, \cdots, \xi_N\rangle = \frac{\delta_{N'N}}{N!}\sum_{\hat{P}}\sigma^P\delta(\xi_1', \xi_{p_1})\delta(\xi_2', \xi_{p_2})\cdots\delta(\xi_N', \xi_{p_N}). \tag{3.22}$$

To prove this, we make use of the identity:

$$\begin{aligned}
\hat{\psi}(\xi_1')\hat{\psi}^\dagger(\xi_1)\hat{\psi}^\dagger(\xi_2)\cdots\hat{\psi}^\dagger(\xi_N) =& \delta(\xi_1', \xi_1)\hat{\psi}^\dagger(\xi_2)\cdots\hat{\psi}^\dagger(\xi_N) \\
&+ \sigma^1\delta(\xi_1', \xi_2)\hat{\psi}^\dagger(\xi_1)\hat{\psi}^\dagger(\xi_3)\cdots\hat{\psi}^\dagger(\xi_N)
\end{aligned}$$

$$
\begin{aligned}
&+\cdots \\
&+\sigma^{N-1}\delta(\xi_1', \xi_N)\hat{\psi}^\dagger(\xi_1)\cdots\hat{\psi}^\dagger(\xi_{N-1}) \\
&+\sigma^{N}\hat{\psi}^\dagger(\xi_1)\hat{\psi}^\dagger(\xi_2)\cdots\hat{\psi}^\dagger(\xi_N)\hat{\psi}(\xi_1'),
\end{aligned}
\tag{3.23}
$$

to move the annihilation operator $\hat{\psi}(\xi_j')$ in (3.22) to the left of $|0\rangle$ and use $\hat{\psi}(\xi_j')|0\rangle = 0$ repeatedly for $j = 1, \cdots, N'$. For example, with $N = N' = 2$, (3.22) is shown to hold as

$$
\begin{aligned}
\langle \xi_1', \xi_2' | \xi_1, \xi_2 \rangle &= \frac{1}{2!}\langle 0|\hat{\psi}(\xi_2')\hat{\psi}(\xi_1')\hat{\psi}^\dagger(\xi_1)\hat{\psi}^\dagger(\xi_2)|0\rangle \\
&= \frac{1}{2!}\langle 0|\hat{\psi}(\xi_2')\big[\delta(\xi_1', \xi_1)\hat{\psi}^\dagger(\xi_2) + \sigma\delta(\xi_1', \xi_2)\hat{\psi}^\dagger(\xi_1) \\
&\quad + \sigma^2\hat{\psi}^\dagger(\xi_1)\hat{\psi}^\dagger(\xi_2)\hat{\psi}(\xi_1')\big]|0\rangle \\
&= \frac{1}{2!}\big[\delta(\xi_1', \xi_1)\langle 0|\hat{\psi}(\xi_2')\hat{\psi}^\dagger(\xi_2)|0\rangle + \sigma\delta(\xi_1', \xi_2)\langle 0|\hat{\psi}(\xi_2')\hat{\psi}^\dagger(\xi_1)|0\rangle\big] \\
&= \frac{1}{2!}\big[\delta(\xi_1', \xi_1)\delta(\xi_2', \xi_2) + \sigma\delta(\xi_1', \xi_2)\delta(\xi_2', \xi_1)\big].
\end{aligned}
$$

Thus, we have constructed the eigenspace of $\hat{P}$.

3.4 Bra-Kets for Many-Body Wave Functions

Let us define ket $|\Phi_\nu\rangle$ for the wave function (3.5) by

$$
|\Phi_\nu\rangle \equiv \int d\xi_1 \int d\xi_2 \cdots \int d\xi_N \, |\xi_1, \xi_2, \cdots, \xi_N\rangle \Phi_\nu(\xi_1, \xi_2, \cdots, \xi_N). \tag{3.24}
$$

The corresponding bra is given by

$$
\langle \Phi_\nu| \equiv \int d\xi_1 \int d\xi_2 \cdots \int d\xi_N \, \langle \xi_1, \xi_2, \cdots, \xi_N| \Phi_\nu^*(\xi_1, \xi_2, \cdots, \xi_N) = |\Phi_\nu\rangle^\dagger. \tag{3.25}
$$

The ket has the following properties:

$$
\hat{\psi}(\xi_1)|\Phi_\nu\rangle = \sqrt{N}\int d\xi_2' \cdots \int d\xi_N' \, |\xi_2', \cdots, \xi_N'\rangle \Phi_\nu(\xi_1, \xi_2', \cdots, \xi_N'), \tag{3.26}
$$

$$
\hat{\psi}(\xi_2)\hat{\psi}(\xi_1)|\Phi_\nu\rangle = \sqrt{N(N-1)}\int d\xi_3' \cdots \int d\xi_N' \, |\xi_3', \cdots, \xi_N'\rangle \Phi_\nu(\xi_1, \xi_2, \xi_3', \cdots, \xi_N'), \tag{3.27}
$$

$$
\hat{\psi}(\xi_N)\cdots\hat{\psi}(\xi_2)\hat{\psi}(\xi_1)|\Phi_\nu\rangle = \sqrt{N!}\,|0\rangle \Phi_\nu(\xi_1, \xi_2, \cdots, \xi_N). \tag{3.28}
$$

Thus, the action of $\hat{\psi}(\xi)$ on $|\Phi_\nu\rangle$ extracts argument ξ from $|\Phi_\nu\rangle$. Equations (3.26)–(3.28) may be proved straightforwardly by substituting (3.24) on the left-hand side and using (3.19), (3.23), (3.18), and (3.12) successively (**Problem** 3.1). It follows from (3.18), (3.20), and (3.28) that

$$\langle \xi_1, \xi_2, \cdots, \xi_N | \Phi_\nu \rangle = \Phi_\nu(\xi_1, \xi_2, \cdots, \xi_N) \,. \tag{3.29}$$

Substitution of (3.29) into (3.24) yields

$$|\Phi_\nu\rangle \equiv \int d\xi_1 \int d\xi_2 \cdots \int d\xi_N \, |\xi_1, \xi_2, \cdots, \xi_N\rangle\langle \xi_1, \xi_2, \cdots, \xi_N | \Phi_\nu\rangle,$$

which is equivalent to

$$\int d\xi_1 \int d\xi_2 \cdots \int d\xi_N \, |\xi_1, \xi_2, \cdots, \xi_N\rangle\langle \xi_1, \xi_2, \cdots, \xi_N | = 1. \tag{3.30}$$

Thus, kets $\{|\xi_1, \xi_2, \cdots, \xi_N\rangle\}$ form a complete set for the eigenspace of $\hat{P}$.

Equation (3.24) can also be used to symmetrize or antisymmetrize any wave function $\tilde{\Phi}(\xi_1, \xi_2, \cdots, \xi_N)$ that is not an eigenstate of $\hat{P}$. Indeed, we only need to construct

$$|\Phi\rangle = A_N \int d\xi_1' \int d\xi_2' \cdots \int d\xi_N' |\xi_1', \xi_2', \cdots, \xi_N'\rangle \tilde{\Phi}(\xi_1', \xi_2', \cdots, \xi_N'), \tag{3.31}$$

with A_N the normalization constant. Ket $|\xi_1', \xi_2', \cdots, \xi_N'\rangle$ in the integrand naturally extracts the symmetric or antisymmetric contribution from $\tilde{\Phi}(\xi_1', \xi_2', \cdots, \xi_N')$. Equation (3.31) may be regarded as projecting $\tilde{\Phi}$ onto the eigenspace of $\hat{P}$. The wave function for (3.31) is obtained using (3.22) and (3.29), giving

$$\Phi(\xi_1, \xi_2, \cdots, \xi_N) \equiv \langle \xi_1, \xi_2, \cdots, \xi_N | \Phi\rangle = \frac{A_N}{N!} \sum_{\hat{P}} \sigma^P \tilde{\Phi}(\xi_{p_1}, \xi_{p_2}, \cdots, \xi_{p_N}). \tag{3.32}$$

3.5 Orthonormality and Completeness of Bra-Kets

Let us assume that wave functions $\{\Phi_\nu(\xi_1, \cdots, \xi_N)\}$ satisfy orthonormality and completeness given by

$$\int d\xi_1 \cdots \int d\xi_N \, \Phi_{\nu'}^*(\xi_1, \cdots, \xi_N)\Phi_\nu(\xi_1, \cdots, \xi_N) = \delta_{\nu'\nu}, \tag{3.33}$$

$$\sum_\nu \Phi_\nu(\xi_1, \cdots, \xi_N)\Phi_\nu^*(\xi_1', \cdots, \xi_N') = \frac{1}{N!} \sum_{\hat{P}} \sigma^P \delta(\xi_1', \xi_{p_1}) \cdots \delta(\xi_N', \xi_{p_N}), \tag{3.34}$$

respectively. These relations can be expressed alternatively in terms of the ket and bra of (3.24) and (3.25) as

$$\langle \Phi_{\nu'} | \Phi_\nu \rangle = \delta_{\nu'\nu}, \qquad \sum_\nu |\Phi_\nu\rangle\langle\Phi_\nu| = 1. \tag{3.35}$$

The proof proceeds straightforwardly by substituting (3.24) and (3.25) into (3.35) and using (3.12), (3.21), (3.22), (3.30), (3.33) and (3.34) (**Problem** 3.2).

3.6 Matrix Elements of Operators

Equation (3.7) tells us that every one-particle operator $\hat{\mathscr{H}}^{(1)}$ and two-particle operator $\hat{\mathscr{H}}^{(2)}$ for a system of identical particles may be written generally as

$$\hat{\mathscr{H}}^{(1)} \equiv \sum_{j=1}^{N} \hat{h}_j^{(1)}, \qquad \hat{\mathscr{H}}^{(2)} \equiv \sum_{i<j} \hat{h}_{ij}^{(2)}. \tag{3.36}$$

Matrix elements of these operators between $\Phi_{\nu'}^*$ and Φ_ν are alternatively expressible in terms of $\langle \Phi_{\nu'}|$ and $|\Phi_\nu\rangle$ as

$$\int d\xi_1 \cdots \int d\xi_N \, \Phi_{\nu'}^*(\xi_1, \cdots, \xi_N) \hat{\mathscr{H}}^{(1)} \Phi_\nu(\xi_1, \cdots, \xi_N)$$
$$= \int d\xi_1 \langle \Phi_{\nu'} | \hat{\psi}^\dagger(\xi_1) \hat{h}_1^{(1)} \hat{\psi}(\xi_1) | \Phi_\nu \rangle, \tag{3.37}$$

$$\int d\xi_1 \cdots \int d\xi_N \, \Phi_{\nu'}^*(\xi_1, \cdots, \xi_N) \hat{\mathscr{H}}^{(2)} \Phi_\nu(\xi_1, \cdots, \xi_N)$$
$$= \frac{1}{2} \int d\xi_1 \int d\xi_2 \langle \Phi_{\nu'} | \hat{\psi}^\dagger(\xi_1) \hat{\psi}^\dagger(\xi_2) \hat{h}_{12}^{(2)} \hat{\psi}(\xi_2) \hat{\psi}(\xi_1) | \Phi_\nu \rangle. \tag{3.38}$$

The equalities can be proved by substituting (3.26) and (3.27) into the right-hand sides of (3.37) and (3.38), respectively, and successively using (3.22) and (3.12) (**Problem** 3.3). Note that the expressions on the right-hand sides are free from the sums over the particle indices i and (i, j); each of them is given using only a single one-particle operator $\hat{h}_1^{(1)}$ and two-particle operator $\hat{h}_{12}^{(2)}$ in (3.36), thereby enabling us to simplify the notation considerably.

3.7 Summary of Two Equivalent Descriptions

The preceding considerations have shown that there are at least two equivalent descriptions for many-particle systems in that they yield the same matrix elements and hence the same probabilities for observables quantum mechanically.

In the first description, the Hamiltonian is given for example by

$$\hat{\mathscr{H}} = \sum_{j=1}^{N} \frac{\hat{\mathbf{p}}_j^2}{2m} + \sum_{i=1}^{N-1} \sum_{j=i+1}^{N} \mathscr{V}(|\mathbf{r}_i - \mathbf{r}_j|). \tag{3.39}$$

Its eigenfunctions are obtained by simultaneously solving the Schrödinger equation:

$$\hat{\mathscr{H}} \Phi_\nu(\xi_1, \xi_2, \cdots, \xi_N) = \mathscr{E}_\nu \Phi_\nu(\xi_1, \xi_2, \cdots, \xi_N), \tag{3.40}$$

and the eigenvalue problem:

$$\hat{P} \Phi_\nu(\xi_1, \xi_2, \cdots, \xi_N) = \sigma^P \Phi_\nu(\xi_1, \xi_2, \cdots, \xi_N). \tag{3.41}$$

Here, we have introduced $\xi \equiv (\mathbf{r}, \alpha)$ to express the space coordinate $\mathbf{r}$ and spin variable $\alpha = s, s-1, \cdots, -s$ in a unified way, $\hat{P} \Phi_\nu(\xi_1, \xi_2, \cdots, \xi_N)$ is defined by (3.6), and eigenvalue σ^P is given by (3.13) and (3.14). The orthonormality and completeness of eigenstates $\{\Phi_\nu\}$ are expressible as (3.33) and (3.34).

The alternative description is called *second quantization*, where the Hamiltonian can be written as

$$\hat{H} = \int d\xi_1 \hat{\psi}^\dagger(\xi_1) \frac{\hat{\mathbf{p}}_1^2}{2m} \hat{\psi}(\xi_1) + \frac{1}{2} \int d\xi_1 \int d\xi_2 \hat{\psi}^\dagger(\xi_1) \hat{\psi}^\dagger(\xi_2) \mathscr{V}(|\mathbf{r}_1 - \mathbf{r}_2|) \hat{\psi}(\xi_2) \hat{\psi}(\xi_1), \tag{3.42}$$

and the Schrödinger equation is expressible as

$$\hat{H} |\Phi_\nu\rangle = \mathscr{E}_\nu |\Phi_\nu\rangle. \tag{3.43}$$

The orthonormality and completeness of $\{|\Phi_\nu\rangle\}$ are given by (3.35).

It follows from (3.37) and (3.38) that (3.39)–(3.41) and (3.42)–(3.43) yield the same eigenvalues $\{\mathscr{E}_\nu\}$ and the same matrix elements for every observable of identical many-particle systems. Ket $|\Phi_\nu\rangle$ and wave function $\Phi_\nu(\xi_1, \xi_2, \cdots, \xi_N)$ are connected by

$$|\Phi_\nu\rangle = \int d\xi_1 \cdots \int d\xi_N |\xi_1, \cdots, \xi_N\rangle \Phi_\nu(\xi_1, \cdots, \xi_N), \tag{3.44}$$

where the basis kets $\{|\xi_1, \cdots, \xi_N\rangle\}$ are defined by (3.19) and satisfy (3.21) and (3.22).

3.8 Second Quantization for Ideal Gases

We now focus on non-interacting many-particle systems and express the Hamiltonian and eigenkets in terms of one-particle eigenstates. The eigenkets thereby obtained also form a convenient starting point for a perturbation expansion with respect to the interaction.

Let us consider Hamiltonian:

$$\hat{H}_0 \equiv \int \hat{\psi}^\dagger(\xi_1)\left[\frac{\hat{\mathbf{p}}_1^2}{2m} + \mathscr{U}(\mathbf{r}_1)\right]\hat{\psi}(\xi_1)\,\mathrm{d}\xi_1, \tag{3.45}$$

where $\mathscr{U}(\mathbf{r})$ is a one-body potential. Suppose that the following one-particle eigenvalue problem has been solved:

$$\left[\frac{\hat{\mathbf{p}}_1^2}{2m} + \mathscr{U}(\mathbf{r}_1)\right]\varphi_q(\xi_1) = \varepsilon_q\varphi_q(\xi_1), \tag{3.46}$$

where q denotes a set of quantum numbers that specifies the one-particle eigenstate $\varphi_q \equiv |q\rangle$ and its eigenvalue ε_q. We assume that $\varphi_q(\xi) = \langle\xi|q\rangle$ satisfies orthonormality:

$$\langle q|q'\rangle \equiv \int \varphi_q^*(\xi_1)\varphi_{q'}(\xi_1)\mathrm{d}\xi_1 = \delta_{qq'}, \tag{3.47}$$

and completeness:

$$\langle\xi|\xi'\rangle \equiv \sum_q \varphi_q(\xi)\varphi_q^*(\xi') = \delta(\xi,\xi'). \tag{3.48}$$

Next, we expand the operators $(\hat{\psi}, \hat{\psi}^\dagger)$ in terms of $\varphi_q(\xi)$ as

$$\hat{\psi}(\xi) = \sum_q \hat{c}_q\varphi_q(\xi), \qquad \hat{\psi}^\dagger(\xi) = \sum_q \hat{c}_q^\dagger\varphi_q^*(\xi). \tag{3.49}$$

To obtain "coefficients" $\hat{c}_q$ and $\hat{c}_q^\dagger$, we multiply the two expansions by $\varphi_{q'}^*(\xi)$ and $\varphi_{q'}(\xi)$, respectively, perform integrations over ξ, and then use orthonormality (3.47). Replacing $q' \to q$ in the resulting expression, we obtain

$$\hat{c}_q = \int \varphi_q^*(\xi)\hat{\psi}(\xi)\,\mathrm{d}\xi, \qquad \hat{c}_q^\dagger = \int \varphi_q(\xi)\hat{\psi}^\dagger(\xi)\,\mathrm{d}\xi. \tag{3.50}$$

Using (3.17) and (3.47), they are shown to obey

$$\left[\hat{c}_q, \hat{c}_{q'}^\dagger\right]_\sigma = \delta_{qq'}, \qquad \left[\hat{c}_q, \hat{c}_{q'}\right]_\sigma = \left[\hat{c}_q^\dagger, \hat{c}_{q'}^\dagger\right]_\sigma = 0. \tag{3.51}$$

Let us substitute (3.49) into (3.45) and use (3.46) and (3.47). Hamiltonian (3.45) is thereby reduced to the diagonal form:

$$\hat{H}_0 = \sum_q \varepsilon_q\hat{c}_q^\dagger\hat{c}_q. \tag{3.52}$$

Next, we obtain the eigenstates of $\hat{H}_0$ in the same representation. To this end, we start with the product of N one-particle eigenfunctions:

$$\tilde{\Phi}_\nu(x_1, x_2, \cdots, x_N) \equiv \prod_{j=1}^{N} \langle \xi_j | q_j \rangle, \qquad \nu = (q_1, q_2, \cdots, q_N),$$

for which the eigenvalues are given by

$$\mathscr{E}_\nu = \sum_{j=1}^{N} \varepsilon_{q_j}. \tag{3.53}$$

Subsequently, we symmetrize or antisymmetrize the above wave function based on (3.32) to obtain

$$\Phi_\nu(\xi_1, \xi_2, \cdots, \xi_N) = \frac{A_N}{N!} \sum_{\hat{P}} \sigma^P \langle \xi_1 | q_{p_1} \rangle \langle \xi_2 | q_{p_2} \rangle \cdots \langle \xi_N | q_{p_N} \rangle. \tag{3.54}$$

It follows from (3.10) that its eigenenergy is still given by (3.53).

Let us focus on fermions. With $\sigma = -1$, the summation over $\hat{P}$ in (3.54) defines a determinant of a matrix of entries $\langle \xi_i | q_j \rangle$ [1, 8]

$$\Phi_\nu^{(\mathrm{F})}(\xi_1, \xi_2, \cdots, \xi_N) = \frac{A_N^{(\mathrm{F})}}{N!} \det \begin{bmatrix} \langle \xi_1 | q_1 \rangle & \cdots & \langle \xi_1 | q_N \rangle \\ \vdots & & \vdots \\ \langle \xi_N | q_1 \rangle & \cdots & \langle \xi_N | q_N \rangle \end{bmatrix}, \tag{3.55}$$

which is known as the *Slater determinant*. It follows from properties of the determinant [1, 8] that $\Phi_\nu^{(\mathrm{F})}$ vanishes when a pair of columns or rows are identical. The statement is precisely the *Pauli exclusion principle*; specifically, no pairs of identical fermions can simultaneously occupy the same one-particle state or coordinate (including spin). Thus, the Pauli exclusion principle naturally results from the permutation symmetry of the system of identical particles. In this context, it is not a *principle* but a natural *outcome* of the permutation symmetry. Nevertheless, we cannot overstate its historical importance; it provided a microscopic understanding of the periodic table and was also a precursor of the Fermi-Dirac statistics. Indeed, the latter may be regarded as a direct extension of the Pauli exclusion principle proposed specifically for electrons in atoms to other electronic systems at finite temperatures.

Noting that $\{q_j\}$ for $\sigma = -1$ are different from each other and using (3.47), we can express the normalization condition for (3.54) as

$$1 = \int d\xi_1 \cdots \int d\xi_N |\Phi_\nu^{(\mathrm{F})}(\xi_1, \cdots, \xi_N)|^2 = \frac{(A_N^{(\mathrm{F})})^2}{(N!)^2} \sum_{\hat{P}} \sum_{\hat{P}'} (-1)^{P'+P} \prod_{j=1}^{N} \langle q_{p'_j} | q_{p_j} \rangle$$

$$= \frac{\left(A_N^{(\mathrm{F})}\right)^2}{(N!)^2} \sum_{\hat{P}} \sum_{\hat{P}'} (-1)^{P'+P} \prod_{j=1}^{N} \delta_{p'_j p_j} = \frac{\left(A_N^{(\mathrm{F})}\right)^2}{(N!)^2} \sum_{\hat{P}} \sum_{\hat{P}'} (-1)^{P'+P} \delta_{\hat{P}'\hat{P}} = \frac{\left(A_N^{(\mathrm{F})}\right)^2}{N!}.$$

Hence, we obtain $A_N^{(\mathrm{F})}$ as

$$A_N^{(\mathrm{F})} = \sqrt{N!}. \tag{3.56}$$

For bosons, each state q_j can accommodate multiple particles. Consider specifically the case where there are n_j particles in the state q_j ($j = 1, 2, \cdots, \ell$; $\ell \le N$) as for

$$\nu = (\underbrace{q_1, \cdots, q_1}_{n_1}, \underbrace{q_2, \cdots, q_2}_{n_2}, \cdots\cdots, \underbrace{q_\ell, \cdots, q_\ell}_{n_\ell}), \qquad \sum_{j=1}^{\ell} n_j = N. \tag{3.57}$$

The corresponding wave function is given by (3.54) with $\sigma = 1$, where q_{p_j} denotes a permutation of ν in (3.57). The normalization condition becomes

$$1 = \int \mathrm{d}\xi_1 \cdots \int \mathrm{d}\xi_N |\Phi_\nu^{(\mathrm{B})}(\xi_1, \cdots, \xi_N)|^2 = \frac{\left(A_N^{(\mathrm{B})}\right)^2}{(N!)^2} \sum_{\hat{P}'} \sum_{\hat{P}} \prod_{j=1}^{N} \langle q_{p'_j} | q_{p_j} \rangle$$

$$= \frac{\left(A_N^{(\mathrm{B})}\right)^2}{(N!)^2} N! \sum_{\hat{P}} \prod_{j=1}^{N} \langle q_j | q_{p_j} \rangle = \frac{\left(A_N^{(\mathrm{B})}\right)^2}{N!} n_1! n_2! \cdots n_\ell!,$$

where the third equality obtains by multiplying the result for $\hat{P}' = \hat{1}$ by $N!$. Hence, we obtain the normalization constant as

$$A_N^{(\mathrm{B})} = \frac{\sqrt{N!}}{\sqrt{n_1! n_2! \cdots n_\ell!}}. \tag{3.58}$$

Now that the wave functions have been obtained, we can construct the corresponding bras. First, we consider fermions. Let us substitute $\sigma = -1$ and (3.56) into (3.54), insert the resulting wave function and (3.19) into (3.44), and use (3.50) and (3.51) for $\{\hat{c}_q^\dagger\}$. We thereby obtain $|\Phi_\nu^{(\mathrm{F})}\rangle$ as (**Problem** 3.4)

$$|\Phi_\nu^{(\mathrm{F})}\rangle = \hat{c}_{q_1}^\dagger \hat{c}_{q_2}^\dagger \cdots \hat{c}_{q_N}^\dagger |0\rangle. \tag{3.59}$$

This expression is much simpler than (3.55), and a transposition of $\hat{c}_{q_i}^\dagger$ and $\hat{c}_{q_j}^\dagger$ corresponds to a column exchange $q_i \leftrightarrow q_j$ in (3.55). The ground state is composed of N-lowest one-particle states, which is sometimes called the *Fermi vacuum* to emphasize the fact that there are no excitations; it is distinct from the vacuum state $|0\rangle$ of (3.18) with no particles.

For bosons, we substitute $\sigma = 1$ and (3.58) into (3.54), insert the resulting wave function and (3.19) into (3.44), and use (3.50) and (3.51) for $\{\hat{c}_q^\dagger\}$. We thereby obtain $|\Phi_\nu^{(B)}\rangle$ for the state ν in (3.57) as

$$|\Phi_\nu^{(B)}\rangle = \frac{(\hat{c}_{q_1}^\dagger)^{n_1}}{\sqrt{n_1!}} \cdots \frac{(\hat{c}_{q_\ell}^\dagger)^{n_\ell}}{\sqrt{n_\ell!}} |0\rangle. \tag{3.60}$$

This state is identical in form to that for the harmonic oscillator with multiple frequencies [10].

It follows from (3.51) that $(\hat{c}_q^\dagger)^2|0\rangle = 0$ holds for fermions of $\sigma = -1$. Noting this fact, we realize that (3.60) for bosons also includes (3.59) for fermions. We may also relax the condition $n_{q_j} \geq 1$ ($j = 1, 2, \cdots, \ell$) in (3.60) to $n_q \geq 0$ for all q's to remove the asymmetry in the notation between occupied and unoccupied states.

With these observations in mind, the main results of this section are summarized as follows. The non-interacting Hamiltonian (3.45) can be expressed alternatively in terms of the eigenvalues of (3.46) and operators of (3.50) as

$$\hat{H}_0 = \sum_q \varepsilon_q \hat{c}_q^\dagger \hat{c}_q, \tag{3.61}$$

where q denotes a set of one-particle quantum numbers, and $\hat{c}_q$ and $\hat{c}_q^\dagger$ satisfy

$$[\hat{c}_q, \hat{c}_{q'}^\dagger]_\sigma \equiv \hat{c}_q \hat{c}_{q'}^\dagger - \sigma \hat{c}_{q'}^\dagger \hat{c}_q = \delta_{qq'}, \qquad [\hat{c}_q, \hat{c}_{q'}]_\sigma = [\hat{c}_q^\dagger, \hat{c}_{q'}^\dagger]_\sigma = 0. \tag{3.62}$$

Every eigenstate ν of (3.61) is specified completely in terms of the number n_q of particles in each one-particle state q as

$$|\Phi_\nu\rangle \equiv |n_{q_1}, n_{q_2}, n_{q_3}, \cdots\rangle \equiv \frac{\left(\hat{c}_{q_1}^\dagger\right)^{n_{q_1}}}{\sqrt{n_{q_1}!}} \frac{\left(\hat{c}_{q_2}^\dagger\right)^{n_{q_2}}}{\sqrt{n_{q_2}!}} \frac{\left(\hat{c}_{q_3}^\dagger\right)^{n_{q_3}}}{\sqrt{n_{q_3}!}} \cdots \cdots |0\rangle. \tag{3.63}$$

The possible values for n_q differ between bosons ($\sigma = 1$) and fermions ($\sigma = -1$) as

$$n_q = \begin{cases} 0, 1, 2, \cdots & (\sigma = +1) \\ 0, 1 & (\sigma = -1) \end{cases}. \tag{3.64}$$

Particle number $\mathcal{N}_\nu$ and energy $\mathcal{E}_\nu$ of state ν are expressible as

$$\mathcal{N}_\nu = \sum_q n_q, \qquad \mathcal{E}_\nu = \sum_q n_q \varepsilon_q. \tag{3.65}$$

Figure 3.1 shows a diagrammatic representation of a non-interacting eigenstate for (a) bosons and (b) fermions. With no vertical energy scale, the one-particle

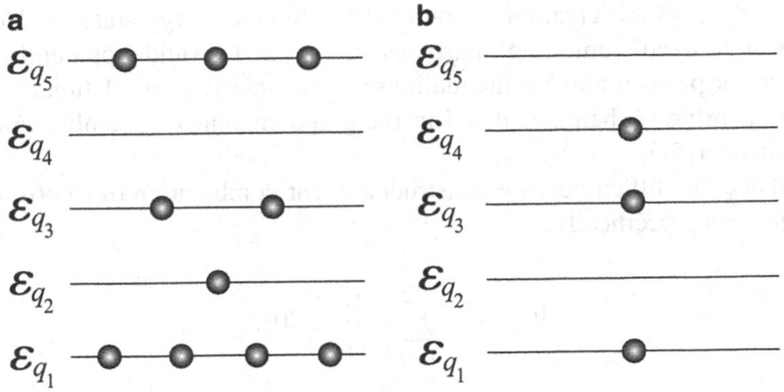

Fig. 3.1 Diagrammatic representation of a non-interacting eigenstate for **(a)** bosons and **(b)** fermions. The eigenstate corresponds to **(a)** $|\Phi_\nu\rangle = |4\ 1\ 2\ 0\ 3\ \cdots\rangle$ and **(b)** $|\Phi_\nu\rangle = |1\ 0\ 1\ 1\ 0\ \cdots\rangle$

energy levels are simply represented by horizontal lines; a filled circle on a level denotes a particle occupying the state. We remark that with degeneracies for the one-particle energy levels, the different occupancies are often distinguished using distinct symbols for particles occupying different states. For example, in the absence of a magnetic field, the spin states of an electron are two-fold degenerate and are marked using $\uparrow$ for $\alpha = \frac{1}{2}$ and $\downarrow$ for $\alpha = -\frac{1}{2}$.

A couple of comments are in order before closing the section. First, the second quantization is often described based on the occupancy representation of (3.63). However, there may be cases where an expansion of a many-particle wave function in terms of one-particle eigenstates is not appropriate. A typical example is superconductivity with the formation of coherent two-particle bound states, which is the main topic of this book. We shall see that (3.44) rather than (3.63) enables us to develop a theory of superconductivity so that the phase coherence is manifest. Second, (3.64) is sometimes called *Bose statistics* or *Fermi statistics*. As we have seen already, the statement has little to do with "statistics" or "probability" but is a direct consequence of the permutation symmetry inherent in systems of identical particles.

3.9 Coherent State

Here, we introduce the coherent state in terms of the ground state of non-interacting bosons.

Looking back at (3.63), one may express the ground state of n non-interacting bosons ($n \gg 1$) as

$$|\Phi_n\rangle = \frac{\left(\hat{c}^\dagger\right)^n}{\sqrt{n!}}|0\rangle, \tag{3.66}$$

where $\hat{c}^\dagger \equiv \hat{c}^\dagger_{q_1}$ is the creation operator of the lowest-energy state q_1. However, (3.66) with a fixed number of particles turns out to yield apparently wrong predictions for particle-number fluctuations and two-particle correlations of bosons at low temperatures when calculated in the grand canonical ensemble [6], as we shall see in Sect. 5.3.

To remove the difficulties, we construct a linear combination of (3.66) over the particle number; specifically,

$$|\Phi\rangle = A\sum_{n=0}^{\infty} a_n \frac{(\hat{c}^\dagger)^n}{\sqrt{n!}}|0\rangle, \tag{3.67}$$

where a_n is an expansion coefficient. The normalization constant is easily calculated using the orthonormality of $(\hat{c}^\dagger)^n|0\rangle/\sqrt{n!}$,

$$A = \left[\sum_{n=0}^{\infty}|a_n|^2\right]^{-1/2}. \tag{3.68}$$

One of the notable properties of (3.67) is that it yields a finite average of the annihilation operator,

$$\langle\Phi|\hat{c}|\Phi\rangle = |A|^2\sum_{n=0}^{\infty}\sqrt{n+1}\,a_n^* a_{n+1}. \tag{3.69}$$

Especially useful among the combinations of (3.67) is the *coherent state* introduced by Sudarshan [11] and Glauber [4] in the context of laser lights as an eigenstate of the annihilation operator:

$$\hat{c}|\Phi\rangle = \Theta|\Phi\rangle, \tag{3.70}$$

where Θ denotes the eigenvalue. To find its explicit expression, substitute (3.67) into (3.70), use the commutation relation $\left[\hat{c},(\hat{c}^\dagger)^n\right]_+ = n(\hat{c}^\dagger)^{n-1}$ (**Problem** 3.5) and $\hat{c}|0\rangle = 0$ to express $\hat{c}(\hat{c}^\dagger)^n|0\rangle = n(\hat{c}^\dagger)^{n-1}|0\rangle$ on the left-hand side, and compare the coefficients of $(\hat{c}^\dagger)^{n-1}|0\rangle$ on both sides. We thereby obtain the recursion relation $\sqrt{n}a_n = \Theta a_{n-1}$. It yields $a_n = a_0\Theta^n/\sqrt{n!}$, where a_0 can be chosen as a real positive number. Substitution of this result into (3.67) gives the coherent state in the form

$$|\Phi\rangle = \mathrm{e}^{-|\Theta|^2/2+\Theta\hat{c}^\dagger}|0\rangle. \tag{3.71}$$

Using expansion (3.49) and noting that $\hat{c} \equiv \hat{c}_{q_1}$ is the creation operator of the lowest-energy state, we can re-express (3.70) in terms of the field operator $\hat{\psi}(\xi)$ as

$$\hat{\psi}(\xi)|\Phi\rangle = \Psi(\xi)|\Phi\rangle, \tag{3.72}$$

where $\Psi(\xi) \equiv \Theta\varphi_{q_1}(\xi)$ denotes the *condensate wave function.*

Problems

3.1. Prove (3.26).

3.2. Prove (3.35).

3.3. Prove (3.37).

3.4. Prove (3.59).

3.5. Consider field operators $\hat{c}$ and $\hat{c}^\dagger$ that satisfy $\left[\hat{c}, \hat{c}^\dagger\right]_+ = 1$. Show that

$$\left[\hat{c}, \left(\hat{c}^\dagger\right)^n\right]_+ = n\left(\hat{c}^\dagger\right)^{n-1}$$

holds. Use it to prove

$$\left[\hat{c}, g(\hat{c}^\dagger)\right]_+ = g'(\hat{c}^\dagger),$$

where $g(x)$ is a function analytic at $x = 0$ and $g'(x) \equiv \mathrm{d}g(x)/\mathrm{d}x$.

References

1. A.C. Aitken, *Determinants and Matrices* (Oliver and Boyd, Edinburgh, 1956)
2. A.R. Edmonds, *Angular Momentum in Quantum Mechanics* (Princeton University Press, Princeton, 1957)
3. M. Fierz, Helv. Phys. Acta **12**, 3 (1939)
4. R.J. Glauber, Phys. Rev. **131**, 2766 (1963)
5. I.N. Herstein, I. Kaplansky, *Matters Mathematical* (Chelsea, New York, 1978)
6. J.R. Johnston, Am. J. Phys. **38**, 516 (1970)
7. L.D. Landau, E.M. Lifshitz, *Quantum Mechanics: Non-relativistic Theory*, 3rd edn. (Butterworth-Heinemann, Oxford, 1991)
8. S. Lang, *Linear Algebra* (Springer, New York, 1987)
9. W. Pauli, Phys. Rev. **58**, 716 (1940)
10. J.J. Sakurai, *Modern Quantum Mechanics*, rev. ed. (Addison-Wesley, Reading, 1994)
11. E.C.G. Sudarshan, Phys. Rev. Lett. **10**, 277 (1963)

Chapter 4
Statistical Mechanics of Ideal Gases

Abstract We clarify the thermodynamic properties of quantum ideal gases composed of identical monoatomic bosons or fermions. Aside from the standard content on the topic described in many textbooks, we specifically clarify the following: (i) Thermodynamic quantities of ideal gases can be expressed universally in terms of the single-particle density of states and either the Bose or Fermi distribution function. (ii) An appropriate choice of units enables us to study various homogeneous systems in a unified way. The result of (ii) is summarized in Fig. 4.1 below.

4.1 Bose and Fermi Distributions

As summarized in (3.61)–(3.65) of the previous chapter, the total particle number $\mathscr{N}_\nu$ and energy $\mathscr{E}_\nu$ for an ideal gas of identical particles can be written in terms of the one-particle energy ε_q and its occupation number n_q:

$$\mathscr{N}_\nu = \sum_q n_q, \qquad \mathscr{E}_\nu = \sum_q n_q \varepsilon_q, \tag{4.1}$$

where q distinguishes the single-particle eigenstates. A many-particle eigenstate ν of the ideal gas is specified by the set $\{n_q\}$ of occupation numbers for the one-particle eigenstates and written $|\nu\rangle = |n_{q1}, n_{q2}, n_{q3}, \cdots\rangle$. Moreover, the possible values of n_q differ between bosons and fermions in that

$$n_q = \begin{cases} 0, 1, 2, \cdots & : \text{bosons } (\sigma = 1) \\ 0, 1 & : \text{fermions } (\sigma = -1) \end{cases}, \tag{4.2}$$

where σ is the eigenvalue of transposition (3.3). We now apply the equilibrium statistical mechanics formulated in Sect. 2.2 to describe ideal gases, derive expressions for basic thermodynamic quantities of these systems, and obtain the mean occupation number given in (4.6) below.

The most mathematically suited for this purpose is the grand canonical ensemble given by (2.26) because it is free from the dual constraints of constant particle

© Springer Japan 2015

T. Kita, *Statistical Mechanics of Superconductivity*, Graduate Texts in Physics,
DOI 10.1007/978-4-431-55405-9_4

number $\mathscr{N}_\nu = N$ and energy $\mathscr{E}_\nu = U$. A fundamental quantity in the ensemble is the grand partition function (2.27). Its summation over ν can be performed using (4.1), giving

$$
\begin{aligned}
Z_{\rm G} &= \sum_\nu \mathrm{e}^{-\beta(\mathscr{E}_\nu - \mu\mathscr{N}_\nu)} = \sum_{\{n_q\}} \mathrm{e}^{-\beta\sum_q(\varepsilon_q-\mu)n_q} = \sum_{\{n_q\}}\prod_q \mathrm{e}^{-\beta(\varepsilon_q-\mu)n_q} \\
&= \prod_q \sum_{n_q} \mathrm{e}^{-\beta(\varepsilon_q-\mu)n_q} = \prod_q \begin{cases} \left[1+\mathrm{e}^{-\beta(\varepsilon_q-\mu)}+\mathrm{e}^{-2\beta(\varepsilon_q-\mu)}+\cdots\right] & (\sigma=1) \\ \left[1+\mathrm{e}^{-\beta(\varepsilon_q-\mu)}\right] & (\sigma=-1) \end{cases} \\
&= \prod_q \left[1-\sigma\mathrm{e}^{-\beta(\varepsilon_q-\mu)}\right]^{-\sigma} \qquad (\sigma=\pm 1). \qquad (4.3)
\end{aligned}
$$

Here, we have used $\mathrm{e}^{a+b+c+\cdots} = \mathrm{e}^a\mathrm{e}^b\mathrm{e}^c\cdots$ in the third equality. In the fourth equality, "the sum over possible sets of $\{n_q\}$ for $\prod_q \mathrm{e}^{-\beta(\varepsilon_q-\mu)n_q}$" is transformed into "a product over q of the sum over possible n_q's of $\mathrm{e}^{-\beta(\varepsilon_q-\mu)n_q}$." Its validity is illustrated for a system of fermions ($n_{q_j} = 0, 1$) with only two quantum states (q_1, q_2) as

$$
\sum_{\{n_1,n_2\}} \mathrm{e}^{-n_1x_1-n_2x_2} = 1+\mathrm{e}^{-x_1}+\mathrm{e}^{-x_2}+\mathrm{e}^{-x_1-x_2} = (1+\mathrm{e}^{-x_1})(1+\mathrm{e}^{-x_2}),
$$

with $n_j \equiv n_{q_j}$ and $x_j \equiv \beta(\varepsilon_{q_j}-\mu)$. The statement holds true irrespective of the upper limit of n_q or the number of states, as may be confirmed by changing them and checking the equality as above. In the fifth equality based on (4.2), we performed the summation over n_q of a geometric series.

Substituting (4.3) into (2.28), we obtain the grand potential $\Omega = \Omega(T,\mu)$ as

$$
\Omega = \frac{\sigma}{\beta}\sum_q \ln\left[1-\sigma\mathrm{e}^{-\beta(\varepsilon_q-\mu)}\right]. \qquad (4.4)
$$

This expression is effective even when the volume V is not an appropriate variable, as for a quantum dilute gas trapped in a harmonic potential. Next, we obtain the mean particle number $N = N(T,\mu)$ and internal energy $U = U(T,\mu)$ by substituting (4.4) and (4.3) into (2.29) and (2.30), respectively, to obtain

$$
N = \sum_q \bar{n}_q, \qquad U = \sum_q \varepsilon_q\bar{n}_q, \qquad (4.5)
$$

with

$$
\bar{n}_q \equiv \frac{1}{\mathrm{e}^{\beta(\varepsilon_q-\mu)}-\sigma} \qquad (\sigma=\pm 1). \qquad (4.6)
$$

Here $\bar{n}_q$ denotes the mean occupation number of state q. Indeed, a formal averaging of (4.1) yields (4.5). Equation (4.6) for $\sigma = 1$ (-1) is called the *Bose distribution*

(*Fermi distribution*). In the high-temperature limit of $e^{\beta(\varepsilon_q-\mu)} \gg 1$, they both approach the *Maxwell-Boltzmann distribution* $\bar{n}_q = \mathrm{e}^{-\beta(\varepsilon_q-\mu)}$.

The first of (4.5) may be regarded as an integral equation to determine $\mu = \mu(T, N)$ for a given N. Once the latter expression is obtained, we can make a change of variables for Ω from μ to N, specifically $\Omega(T, \mu(T, N))$.

As for entropy $S = S(T, \mu)$, we substitute (4.4) into the first expression of (2.29) and use $-\frac{\partial}{\partial T} = -\frac{d\beta}{dT}\frac{\partial}{\partial \beta} = k_B\beta^2\frac{\partial}{\partial \beta}$ and $\beta(\varepsilon_q - \mu) = \ln\left[(1 + \sigma\bar{n}_q)/\bar{n}_q\right]$ to obtain

$$
\begin{aligned}
S &= k_B \sum_q \left\{-\sigma \ln\left[1 - \sigma \mathrm{e}^{-\beta(\varepsilon_q-\mu)}\right] + \beta(\varepsilon_q - \mu)\bar{n}_q\right\} \\
&= k_B \sum_q \left[-\bar{n}_q \ln \bar{n}_q + \sigma(1 + \sigma\bar{n}_q)\ln(1 + \sigma\bar{n}_q)\right].
\end{aligned}
\tag{4.7}
$$

Heat capacity $C(T, N)$ is also obtained by substituting $S(T, \mu(T, N))$ into the thermodynamic relation $C = T(\partial S/\partial T)$ and using (4.6),

$$
\begin{aligned}
C(T, N) &= \sum_q (\varepsilon_q - \mu)\frac{\partial \bar{n}_q}{\partial T} \\
&= k_B \sum_q \left[x + \frac{1}{k_B}\frac{\partial \mu(T, N)}{\partial T}\right] \left.\frac{x\mathrm{e}^x}{(\mathrm{e}^x - \sigma)^2}\right|_{x=\beta(\varepsilon_q-\mu)}.
\end{aligned}
\tag{4.8}
$$

4.2 Single-Particle Density of States

Let us introduce the single-particle *density of states* by

$$
D(\epsilon) \equiv \sum_q \delta(\epsilon - \varepsilon_q),
\tag{4.9}
$$

where $\delta(x)$ is the *Dirac delta function* defined by

$$
\delta(x) \equiv \begin{cases} \infty & : x = 0 \\ 0 & : x \neq 0 \end{cases}, \qquad \int_{-\infty}^{\infty} \delta(x)\,\mathrm{d}x = 1.
\tag{4.10}
$$

It is the corresponding continuous-variable impulse function to the Kronecker delta (2.9) as well as being the derivative of the *Heaviside step function*

$$
\theta(x) \equiv \begin{cases} 1 & : x \geq 0 \\ 0 & : x < 0 \end{cases},
\tag{4.11}
$$

that is, $\delta(x) = \theta'(x)$. The latter fact may be realized by noting that $\delta(x) = \theta'(x)$ also satisfies (4.10).

The thermodynamic quantities of ideal gases such as (4.4), (4.5), (4.7), and (4.8) are all expressible in terms of the density of states (4.9) and the distribution function (4.6). For example, (4.5) can be written as

$$N = \int_{-\infty}^{\infty} \frac{D(\epsilon)}{e^{\beta(\epsilon-\mu)} - \sigma} d\epsilon, \tag{4.12}$$

$$U = \int_{-\infty}^{\infty} \frac{D(\epsilon)\epsilon}{e^{\beta(\epsilon-\mu)} - \sigma} d\epsilon. \tag{4.13}$$

These expressions show clearly what is necessary for the statistical-mechanical description of ideal gases. Rather than detailed expressions for ε_q, the important information is how single-particle energies $\{\varepsilon_q\}$ are distributed on the energy axis.

4.3 Monoatomic Gases in Three Dimensions

We now focus on a gaseous system composed of identical monoatomic molecules with mass m and spin s confined in a container of volume V in the absence of external magnetic fields and potentials. We analyze the thermodynamic properties of both Bose and Fermi gases, which will be shown to behave quite distinctly (Fig. 4.1).

4.3.1 *Single-Particle Density of States*

First, we obtain the single-particle density of states for free particles. Because the boundary conditions do not affect the bulk density of states [6], we adopt the most convenient set, that being the periodic boundary conditions. Specifically, consider a cubic container with edge length L and impose the periodic boundary conditions for solving the single-particle Schrödinger equation. In the absence of magnetic fields, the spin variable $\alpha = s, s-1, \cdots, -s$ is a good quantum number of the eigenstate as the spin operator commutes with the Hamiltonian, and hence is simultaneously compatible with the set of eigenenergies. Indeed, the eigenfunctions and corresponding eigenenergies are given explicitly by

$$\varphi_{\mathbf{k}\alpha'}(\mathbf{r}\alpha) = \langle \mathbf{r}\alpha|\mathbf{k}\alpha'\rangle = \delta_{\alpha'\alpha}\frac{1}{\sqrt{V}}e^{i\mathbf{k}\cdot\mathbf{r}}, \tag{4.14}$$

$$\varepsilon_k = \frac{\hbar^2 k^2}{2m}, \tag{4.15}$$

where $V \equiv L^3$ is the volume of the system, and the wave vector $\mathbf{k}$ is determined by the integers n_η ($\eta = x, y, z$) from

$$\mathbf{k} = \frac{2\pi}{L}(n_x, n_y, n_z). \tag{4.16}$$

The corresponding density of states is obtained from (4.9) by replacing $q \to \mathbf{k}\alpha$. Noting that the spacing of two adjacent quantum states in (4.16) is $\Delta k = 2\pi/L$, we can convert the sum into an integral in the limit $L \to \infty$ to obtain

$$\begin{aligned}
D(\epsilon) &= \sum_{\mathbf{k}\alpha} \delta(\epsilon - \varepsilon_k) && \text{insert } \frac{1}{(\Delta k)^3}(\Delta k)^3 \text{ and sum over } \alpha \\
&= (2s+1)\left(\frac{L}{2\pi}\right)^3 \sum_{\mathbf{k}} (\Delta k)^3 \delta(\epsilon - \varepsilon_k) && \text{sum} \to \text{integral} \\
&\approx (2s+1)\left(\frac{L}{2\pi}\right)^3 \int \mathrm{d}^3k\, \delta(\epsilon - \varepsilon_k) && \text{adopt polar coordinates} \\
&= \frac{(2s+1)L^3}{(2\pi)^3} 4\pi \int_0^\infty \mathrm{d}k k^2 \delta(\epsilon - \varepsilon_k) && \text{use } k = \left(\frac{2m\varepsilon_k}{\hbar^2}\right)^{1/2} \\
&= \frac{(2s+1)V}{2\pi^2}\left(\frac{2m}{\hbar^2}\right)^{3/2} \int_0^\infty \frac{\mathrm{d}\varepsilon_k}{2\varepsilon_k^{1/2}} \varepsilon_k \delta(\epsilon - \varepsilon_k) \\
&= \frac{(2s+1)V}{4\pi^2}\left(\frac{2m}{\hbar^2}\right)^{3/2} \epsilon^{1/2}\theta(\epsilon).
\end{aligned} \tag{4.17}$$

In the last equality, the step function $\theta(\epsilon)$ signifies that there is no state for $\epsilon < 0$. Hence, the density of states depends on the square root of the energy, $\epsilon^{1/2}$. In general, one can show that the density of states for free particles in d dimensions behaves as $D(\epsilon) \propto L^d \epsilon^{(d-2)/2}$ for $\epsilon \geq 0$ (**Problem** 4.1). This difference in energy dependence manifests itself in distinct physical properties among systems of different dimensions.

4.3.2 *Connection Between Internal Energy and Pressure*

Using (4.17), one obtains the relation,

$$PV = \frac{2}{3}U. \tag{4.18}$$

To derive this, we start from $PV = -\Omega$ of (1.28), substitute (4.4) for Ω, transform the sum over states into an integral over ϵ using the density of states (4.9), and substitute (4.17) for the density of states, expressing it more generally as

$$D(\epsilon) = A\epsilon^{\eta-1}\theta(\epsilon),$$

where A and $\eta > 0$ are constants. Quantity PV thereby becomes

$$\begin{aligned} PV = -\Omega &= -\frac{\sigma}{\beta}\int_0^\infty D(\epsilon)\ln\left[1-\sigma e^{-\beta(\epsilon-\mu)}\right]d\epsilon \\ &= -\frac{\sigma A}{\beta\eta}\epsilon^\eta \ln\left[1-\sigma e^{-\beta(\epsilon-\mu)}\right]\Big|_0^\infty + \frac{\sigma A}{\beta\eta}\int_0^\infty \epsilon^\eta \frac{\sigma\beta e^{-\beta(\epsilon-\mu)}}{1-\sigma e^{-\beta(\epsilon-\mu)}}d\epsilon \\ &= \frac{1}{\eta}\int_0^\infty \frac{D(\epsilon)\epsilon}{e^{\beta(\epsilon-\mu)}-\sigma}d\epsilon = \frac{1}{\eta}U, \end{aligned} \tag{4.19}$$

where we have used (4.13) in the last equality. Setting $\eta = 3/2$, we obtain (4.18). The above general derivation also implies that the energy dependence of the density of states manifests itself in the proportionality constant between PV and U. To be specific, (4.18) is replaced by $PL^d = (2/d)U$ for d dimensions.

4.3.3 *Introducing Dimensionless Variables*

Using a set of appropriate dimensionless variables often enables us to make physical arguments clearer and find similarities among apparently different systems. Let us introduce the following unit of length for an assembly of identical particles in three dimensions,

$$l_Q \equiv \left[\frac{V}{N/(2s+1)}\right]^{1/3}. \tag{4.20}$$

The order of l_Q is roughly the mean interparticle spacing of particles with spin component α. Using l_Q, we next define the units associated with wave number, energy, and temperature:

$$k_Q \equiv \frac{\pi}{l_Q}, \qquad \varepsilon_Q \equiv \frac{\hbar^2 k_Q^2}{2m}, \qquad T_Q \equiv \frac{\varepsilon_Q}{k_B}, \tag{4.21}$$

where π in k_Q is introduced for convenience. Each of these units represents the scale at which quantum effects are manifest. Table 4.1 presents values of T_Q for

Table 4.1 Estimates of T_Q below which quantum effects manifest themselves. The valence of Cu is taken to be 1

Systems	Conduction electrons in Cu	Liquid ^{4}He	Liquid ^{3}He
Magnitude of spin	1/2	0	1/2
Atomic mass (g/mol)	63.5	4.00	3.02
Density (g/cm^3 at 1 atm)	8.96 (298 K) [8]	0.125 (4.23 K) [7]	0.059 (3.19 K) [7]
T_Q (K)	5.31×10^4	4.21	2.58

three typical systems of identical particles. The temperature is of the order of 10^4 K for electrons in metals, whereas it is 4.21 and 2.58 K for liquid ^{4}He ($s = 0$) and ^{3}He ($s = 1/2$), respectively. Thus, values of T_Q differ considerably among systems. However, it is possible to find close similarities in different ensembles of the same species with integer or half-integer spins, e.g., between electrons and liquid ^{3}He with $s = 1/2$. For this reason, we make a change of variables for energy, temperature, chemical potential, internal energy, and heat capacity in units of (4.21) by defining 'tilde' quantities

$$\epsilon = \varepsilon_Q \tilde{\epsilon}, \qquad k_B T = \varepsilon_Q \tilde{T}, \qquad \mu = \varepsilon_Q \tilde{\mu}, \qquad U = N\varepsilon_Q \tilde{u}, \qquad C = N k_B \tilde{c}. \tag{4.22}$$

Using (4.17) and (4.20)–(4.22), we can transform (4.12) into a dimensionless form:

$$\begin{aligned} 1 &= \frac{(2s+1)V}{4\pi^2 N}\left(\frac{2m\varepsilon_Q}{\hbar^2}\right)^{3/2} \int_0^\infty \frac{\tilde{\epsilon}^{1/2}}{\mathrm{e}^{(\tilde{\epsilon}-\tilde{\mu})/\tilde{T}} - \sigma} \mathrm{d}\tilde{\epsilon} \\ &= \frac{\pi}{4} \int_0^\infty \frac{\tilde{\epsilon}^{1/2}}{\mathrm{e}^{(\tilde{\epsilon}-\tilde{\mu})/\tilde{T}} - \sigma} \mathrm{d}\tilde{\epsilon}, \end{aligned} \tag{4.23}$$

which forms an integral equation for $\tilde{\mu} = \tilde{\mu}(\tilde{T})$. Similarly, (4.13) for the internal energy simplifies to

$$\tilde{u} = \frac{\pi}{4} \int_0^\infty \frac{\tilde{\epsilon}^{3/2}}{\mathrm{e}^{(\tilde{\epsilon}-\tilde{\mu})/\tilde{T}} - \sigma} \mathrm{d}\tilde{\epsilon}. \tag{4.24}$$

We also express (4.8) for the heat capacity in terms of the density of states, and subsequently use (4.22) to write it as

$$\tilde{c} = \frac{\pi}{4} \int_0^\infty \tilde{\epsilon}^{1/2} \left(x + \frac{\partial \tilde{\mu}}{\partial \tilde{T}}\right) \frac{x\mathrm{e}^x}{(\mathrm{e}^x - \sigma)^2}\bigg|_{x=(\tilde{\epsilon}-\tilde{\mu})/\tilde{T}} \mathrm{d}\tilde{\epsilon}, \tag{4.25}$$

where $\partial\tilde{\mu}/\partial\tilde{T}$ is obtained by differentiating (4.23) with respect to $\tilde{T}$,

$$\frac{\partial \tilde{\mu}}{\partial \tilde{T}} = -\frac{\displaystyle\int_0^\infty \tilde{\epsilon}^{1/2} \frac{x\mathrm{e}^x}{(\mathrm{e}^x - \sigma)^2}\bigg|_{x=(\tilde{\epsilon}-\tilde{\mu})/\tilde{T}} \mathrm{d}\tilde{\epsilon}}{\displaystyle\int_0^\infty \tilde{\epsilon}^{1/2} \frac{\mathrm{e}^x}{(\mathrm{e}^x - \sigma)^2}\bigg|_{x=(\tilde{\epsilon}-\tilde{\mu})/\tilde{T}} \mathrm{d}\tilde{\epsilon}}. \tag{4.26}$$

From (4.7), entropy can also be expressed concisely in a dimensionless form.

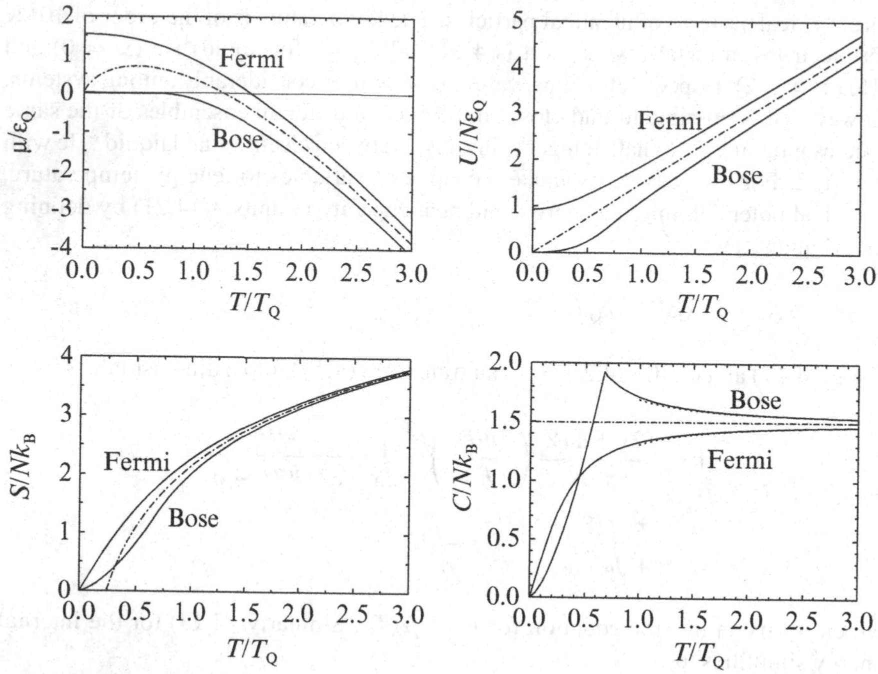

Fig. 4.1 Temperature dependences of the chemical potential μ, internal energy U, entropy S, and heat capacity C. *Solid lines* are exact results from numerical calculations, whereas *chain lines* are the classical Maxwell-Boltzmann results. *Dashed lines* in the heat capacity for $k_B T/\varepsilon_Q \geq 1$ show leading-order quantum corrections to the classical result

4.3.4 Temperature Dependences of Thermodynamic Quantities

Let us survey the temperature dependences of the thermodynamic quantities to capture their basic features. Figure 4.1 plots the chemical potential, internal energy, and heat capacity as a function of reduced temperature T/T_Q. They are obtained by solving (4.23)–(4.26) numerically for bosons ($\sigma = 1$) and fermions ($\sigma = -1$). In detail, we first solve (4.23) to obtain $\tilde{\mu}(\tilde{T})$, which is subsequently used in (4.24) and (4.26) to calculate $\tilde{u}$ and $\partial\tilde{\mu}/\partial\tilde{T}$. Finally, $\tilde{\mu}$ and $\partial\tilde{\mu}/\partial\tilde{T}$ are used to plot (4.25). Entropy has been calculated similarly.

The solid lines show the exact numerical results, whereas the chain lines are obtained using the classical Maxwell-Boltzmann distribution; the dashed lines for heat capacity plot the formula (4.31), which incorporates the leading quantum corrections to the classical result. Each solid line exhibits a considerable deviation from the classical result below around T_Q. We can also see a marked difference between bosons and fermions at low temperatures for each thermodynamic quantity. Noting (4.18), we may regard the lines of U as the quantum equations of state. Thus, we realize that the pressure for a system of bosons (fermions) decreases (increases) compared with the classical result. This implies that there is an effective attraction (repulsion) between each pair of bosons (fermions) with the same α due to the permutation symmetry. Whereas classical entropy shows unphysical behavior in falling below 0 and diverging to $-\infty$, entropy for bosons and fermions appropriately approaches 0 as $T \to 0$ in accordance with the third law (1.11) of thermodynamics; the same feature appears for heat capacity. The peak in heat capacity for bosons marks the onset of a phase transition called *Bose-Einstein condensation* (BEC), below which the chemical potential is pinned at the lowest one-particle energy $\varepsilon_0 = 0$. We shall elaborate on these features below.

4.4 High-Temperature Expansions

We first consider the high-temperature region of $\tilde{T} \equiv T/T_Q \gg 1$, where $e^{-(\tilde{\epsilon}-\tilde{\mu})/\tilde{T}} \ll 1$ holds in (4.23). Hence, we perform an expansion of its integrand in terms of $e^{-(\tilde{\epsilon}-\tilde{\mu})/\tilde{T}}$ and retain the leading two contributions to obtain the approximation

$$\begin{aligned} 1 &= \frac{\pi}{4}\int_0^\infty \tilde{\epsilon}^{1/2} e^{-(\tilde{\epsilon}-\tilde{\mu})/\tilde{T}} \left[1 + \sigma e^{-(\tilde{\epsilon}-\tilde{\mu})/\tilde{T}} + e^{-2(\tilde{\epsilon}-\tilde{\mu})/\tilde{T}} + \cdots \right] d\tilde{\epsilon} \\ &\approx \frac{\pi}{4}\tilde{T}^{3/2}\left(e^{\tilde{\mu}/\tilde{T}}\int_0^\infty x^{1/2}e^{-x}dx + \sigma e^{2\tilde{\mu}/\tilde{T}}\int_0^\infty x^{1/2}e^{-2x}dx\right) \\ &= \frac{\pi^{3/2}\tilde{T}^{3/2}}{8} e^{\tilde{\mu}/\tilde{T}}\left(1 + \sigma\frac{e^{\tilde{\mu}/\tilde{T}}}{2^{3/2}}\right). \end{aligned} \tag{4.27}$$

Here, we have written $\tilde{\epsilon} = \tilde{T}x$, expressed the integrals in terms of the Gamma function [1, 3]:

$$\Gamma(x) \equiv \int_0^\infty e^{-t} t^{x-1}\, dt \qquad (x > 0), \tag{4.28}$$

and used $\Gamma(3/2) = \sqrt{\pi}/2$. Taking the logarithm of (4.27) yields $\tilde{\mu}$,

$$\begin{aligned} \tilde{\mu} &\approx -\tilde{T}\left[\ln\left(\frac{\pi\tilde{T}}{4}\right)^{3/2} + \ln\left(1+\sigma\frac{e^{\tilde{\mu}/\tilde{T}}}{2^{3/2}}\right)\right] \\ &\approx -\frac{3}{2}\tilde{T}\ln\frac{\pi\tilde{T}}{4} - \sigma\left(\frac{2}{\pi}\right)^{3/2}\frac{1}{\tilde{T}^{1/2}}, \end{aligned} \tag{4.29}$$

where we have approximated $\ln(1+\sigma e^{\tilde{\mu}/\tilde{T}}/2^{3/2}) \approx \sigma e^{\tilde{\mu}/\tilde{T}}/2^{3/2}$ and subsequently replaced $e^{\tilde{\mu}/\tilde{T}}$ with its leading-order expression $e^{\tilde{\mu}/\tilde{T}} \approx (4/\pi\tilde{T})^{3/2}$. The first term in the last expression is the classical chemical potential, whereas the second one represents the leading quantum correction.

Similarly, internal energy (4.24) is estimated using (4.27),

$$\tilde{u} \approx \frac{3\pi^{3/2}\tilde{T}^{5/2}}{16} e^{\tilde{\mu}/\tilde{T}} \left(1 + \sigma \frac{e^{\tilde{\mu}/\tilde{T}}}{2^{5/2}}\right) \approx \frac{3}{2}\tilde{T}\left(1 - \frac{\sqrt{2}\sigma}{\pi^{3/2}\tilde{T}^{3/2}}\right). \tag{4.30}$$

Differentiating with respect to $\tilde{T}$ yields the dimensionless heat capacity,

$$\tilde{c} = \frac{3}{2}\left(1 + \frac{\sigma}{\sqrt{2}\pi^{3/2}\tilde{T}^{3/2}}\right). \tag{4.31}$$

The first term in the round brackets of (4.31) gives the classical heat capacity $3/2$, whereas the second term incorporates the leading quantum correction with opposite signs for bosons and fermions. As shown by the dashed lines in Fig. 4.1, (4.31) reproduces precisely the behaviors of heat capacity for $\tilde{T} \gtrsim 1$.

4.5 Fermions at Low Temperatures

Next, we consider fermions ($\sigma = -1$) at low temperatures. The results here were obtained by Sommerfeld in his theory of electrons in metals in 1927, which clarified that the heat capacity of electrons should vanish as $T \to 0$.

4.5.1 *Fermi Energy and Fermi Wave Number*

As we have already seen, solving (4.23) yields $\tilde{\mu} = \tilde{\mu}(\tilde{T})$. We specifically consider the case of $\tilde{T} \to 0$, where the Fermi distribution function in the integrand behaves as

$$\bar{n}_{\mathrm{F}}(\tilde{\epsilon}) = \frac{1}{e^{(\tilde{\epsilon}-\tilde{\mu})/\tilde{T}} + 1} \xrightarrow{\tilde{T}\to 0} \theta(\tilde{\mu} - \tilde{\epsilon}), \tag{4.32}$$

with θ denoting the step function (4.11); see also Fig. 4.2. Its derivative with respect to $\tilde{\epsilon}$ can be expressed in the limit $\tilde{T} \to 0$ as

$$\frac{\partial \bar{n}_{\mathrm{F}}(\tilde{\epsilon})}{\partial \tilde{\epsilon}} = -\frac{1}{T}\frac{e^{(\tilde{\epsilon}-\tilde{\mu})/\tilde{T}}}{[e^{(\tilde{\epsilon}-\tilde{\mu})/\tilde{T}} + 1]^2} \xrightarrow{\tilde{T}\to 0} -\delta(\tilde{\epsilon} - \tilde{\mu}), \tag{4.33}$$

where δ is the delta function (4.10).

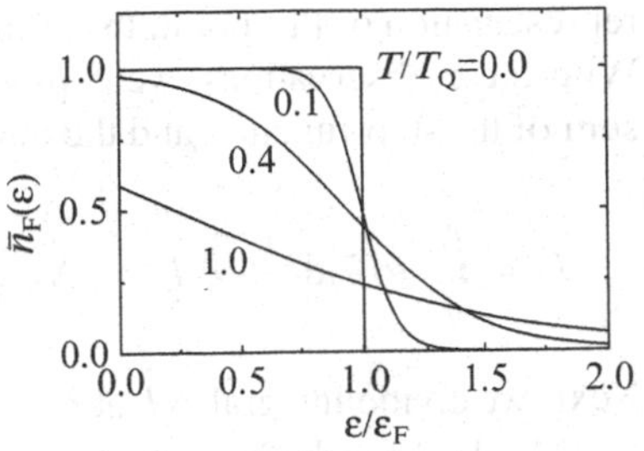

Fig. 4.2 Fermi distribution function defined by (4.32) at $T/T_Q = 0.0, 0.1, 0.4$, and 1.0. The chemical potential has been calculated by solving (4.23) numerically

Substituting the limiting form of (4.32) into (4.23) and performing the integration, we obtain $1 = \frac{\pi}{6}\tilde{\mu}^{3/2}$. Hence, we find the chemical potential at zero temperature in reduced units, $\tilde{\mu}(0) \equiv \tilde{\varepsilon}_F$, and in standard units, $\varepsilon_F \equiv \varepsilon_Q\tilde{\varepsilon}_F$, using (4.20) and (4.21),

$$\tilde{\varepsilon}_F = (6/\pi)^{2/3} = 1.54, \qquad \varepsilon_F = \frac{\hbar^2}{2m}\left[\frac{6\pi^2 N}{(2s+1)V}\right]^{2/3}. \tag{4.34}$$

The quantity ε_F introduced here is called the *Fermi energy*, which is of the same order as ε_Q in (4.21). The corresponding wave number $k_F \equiv (2m\varepsilon_F/\hbar^2)^{1/2}$, i.e.,

$$k_F = \left[\frac{6\pi^2 N}{(2s+1)V}\right]^{1/3} \tag{4.35}$$

is called the *Fermi wave number*, which depends only on the density of a single spin component α. It follows from (4.32) that a Fermi gas at zero temperature realizes the lowest-energy state where the single-particle states for $\varepsilon_k \le \varepsilon_F$ ($k \le k_F$) are all occupied with no vacancies. This ground state is sometimes called the *Fermi sea* or *Fermi vacuum*.

4.5.2 Sommerfeld Expansion

In general, the thermodynamic quantities for fermionic systems are expressible in terms of the integral,

$$I \equiv \int_0^\infty \frac{g(\tilde{\epsilon})}{e^{(\tilde{\epsilon}-\tilde{\mu})/\tilde{T}} + 1} d\tilde{\epsilon}. \tag{4.36}$$

For example, (4.23) for the chemical potential has this form with $g(\tilde{\epsilon}) = \frac{\pi}{4}\tilde{\epsilon}^{1/2}$. Let us evaluate the above integral analytically for $\tilde{T} \ll 1$.

The Fermi distribution function in (4.36) approaches the singular step function $\theta(\tilde{\mu} - \tilde{\epsilon})$ as $\tilde{T} \to 0$ with a discontinuity at $\tilde{\epsilon} = \tilde{\mu}$; see also Fig. 4.2 for a graphical representation of this point. In contrast, function $g(\tilde{\epsilon})$ is in general smooth at $\tilde{\epsilon} = \tilde{\mu}$. With these observations, we express the Fermi distribution function in (4.36) as a sum of the step function and the deviation from it:

$$I = \int_0^{\tilde{\mu}} g(\tilde{\epsilon})\mathrm{d}\tilde{\epsilon} + \Delta I, \quad \Delta I \equiv \int_0^{\infty} g(\tilde{\epsilon}) \left[\frac{1}{\mathrm{e}^{(\tilde{\epsilon}-\tilde{\mu})/\tilde{T}} + 1} - \theta(\tilde{\mu} - \tilde{\epsilon}) \right] \mathrm{d}\tilde{\epsilon}. \tag{4.37}$$

Next, we divide integral ΔI at $\tilde{\epsilon} = \tilde{\mu}$ into two contributions, use $(\mathrm{e}^x + 1)^{-1} - 1 = -(\mathrm{e}^{-x}+1)^{-1}$ for the low-energy part, and with a change of variable $x \equiv \pm(\tilde{\epsilon}-\tilde{\mu})/\tilde{T}$ for $\tilde{\varepsilon} \gtrless \tilde{\mu}$ obtain

$$\Delta I = -\int_0^{\tilde{\mu}} \frac{g(\tilde{\epsilon})}{\mathrm{e}^{-(\tilde{\epsilon}-\tilde{\mu})/\tilde{T}} + 1}\mathrm{d}\tilde{\epsilon} + \int_{\tilde{\mu}}^{\infty} \frac{g(\tilde{\epsilon})}{\mathrm{e}^{(\tilde{\epsilon}-\tilde{\mu})/\tilde{T}} + 1}\mathrm{d}\tilde{\epsilon}$$
$$= -\tilde{T} \int_0^{\tilde{\mu}/\tilde{T}} \frac{g(\tilde{\mu} - \tilde{T}x)}{\mathrm{e}^x + 1}\mathrm{d}x + \tilde{T} \int_0^{\infty} \frac{g(\tilde{\mu} + \tilde{T}x)}{\mathrm{e}^x + 1}\mathrm{d}x.$$

The integrands decrease exponentially for $x \gg 1$ because of the factor e^x in the denominator. Also, noting $\tilde{\mu}/\tilde{T} \gg 1$ at low temperatures, we can replace the upper limit of the integration in the first term by ∞ to an excellent approximation. Thus, ΔI becomes

$$\Delta I \approx \tilde{T} \int_0^{\infty} \frac{g(\tilde{\mu} + \tilde{T}x) - g(\tilde{\mu} - \tilde{T}x)}{\mathrm{e}^x + 1}\mathrm{d}x$$
$$= 2g'(\tilde{\mu})\tilde{T}^2 \int_0^{\infty} \frac{x}{\mathrm{e}^x + 1}\mathrm{d}x + \frac{2g^{(3)}(\tilde{\mu})}{3!}\tilde{T}^4 \int_0^{\infty} \frac{x^3}{\mathrm{e}^x + 1}\mathrm{d}x + \cdots. \tag{4.38}$$

The integrals in the second line can be evaluated exactly,

$$J_n^{\mathrm{F}} \equiv \int_0^{\infty} \frac{x^{n-1}}{\mathrm{e}^x + 1}\mathrm{d}x = \int_0^{\infty} x^{n-1} \frac{\mathrm{e}^{-x}}{1 + \mathrm{e}^{-x}}\mathrm{d}x = \sum_{m=1}^{\infty} (-1)^{m-1} \int_0^{\infty} x^{n-1}\mathrm{e}^{-mx}\mathrm{d}x$$
$$= \sum_{m=1}^{\infty} \frac{(-1)^{m-1}}{m^n}\Gamma(n) = \left(1 - \frac{1}{2^{n-1}}\right)\zeta(n)\Gamma(n), \tag{4.39}$$

where Γ is defined by (4.28), and ζ denotes the Riemann zeta function [1]:

$$\zeta(x) \equiv \sum_{m=1}^{\infty} \frac{1}{m^x}. \tag{4.40}$$

The relevant values are $\Gamma(2) = 1$, $\zeta(2) = \pi^2/6$, $\Gamma(4) = 3!$, $\zeta(4) = \pi^4/90$; these then yield $J_2^{\mathrm{F}} = \pi^2/12$ and $J_4^{\mathrm{F}} = 7\pi^4/120$ for the integral (4.39). Substituting the values into (4.38), we obtain for the low-temperature expansion of (4.37),

$$I \equiv \int_0^\infty \frac{g(\tilde{\epsilon})}{e^{(\tilde{\epsilon}-\tilde{\mu})/\tilde{T}}+1}d\tilde{\epsilon} \approx \int_0^{\tilde{\mu}} g(\tilde{\epsilon})d\tilde{\epsilon} + \frac{\pi^2}{6}g'(\tilde{\mu})\tilde{T}^2 + \frac{7\pi^4}{360}g^{(3)}(\tilde{\mu})\tilde{T}^4 + \cdots, \tag{4.41}$$

which is known as the *Sommerfeld expansion.*

4.5.3 Chemical Potential and Heat Capacity

Using (4.41) with $g(\tilde{\epsilon}) = \frac{\pi}{4}\tilde{\epsilon}^{1/2}$ in (4.23) and retaining terms up to order $\tilde{T}^2$, we hence obtain an equation that determines the chemical potential for $\tilde{T} \ll 1$,

$$1 = \frac{\pi}{6}\tilde{\mu}^{3/2}\left[1 + \frac{\pi^2}{8}\left(\frac{\tilde{T}}{\tilde{\mu}}\right)^2\right].$$

Noting $(6/\pi)^{2/3} = \tilde{\varepsilon}_F$ in (4.34), we obtain $\tilde{\mu}$ as

$$\tilde{\mu} = \tilde{\varepsilon}_F\left[1 + \frac{\pi^2}{8}\left(\frac{\tilde{T}}{\tilde{\mu}}\right)^2\right]^{-2/3} \approx \tilde{\varepsilon}_F\left[1 - \frac{\pi^2}{12}\left(\frac{\tilde{T}}{\tilde{\varepsilon}_F}\right)^2\right], \tag{4.42}$$

where the second expression has been derived by expanding $(1+x)^{-2/3} \approx 1 - \frac{2}{3}x$ in the first expression and subsequently approximating $\mu \approx \tilde{\varepsilon}_F$.

Internal energy (4.24) corresponds to $g(\tilde{\epsilon}) = \frac{\pi}{4}\tilde{\epsilon}^{3/2}$ in (4.41). Retaining terms up to order $\tilde{T}^2$ in the resulting expression, we obtain

$$\tilde{u} \approx \frac{\pi}{10}\tilde{\mu}^{5/2}\left[1 + \frac{5\pi^2}{8}\left(\frac{\tilde{T}}{\tilde{\mu}}\right)^2\right] \approx \frac{\pi}{10}\tilde{\varepsilon}_F^{5/2}\left[1 + \frac{5\pi^2}{12}\left(\frac{\tilde{T}}{\tilde{\varepsilon}_F}\right)^2\right], \tag{4.43}$$

where we have substituted (4.42) into $\tilde{\mu}$ to obtain the second expression. Differentiation of this expression with respect to $\tilde{T}$ yields the heat capacity at low temperatures,

$$\tilde{c} = \frac{\pi^3}{12}\tilde{\varepsilon}_F^{1/2}\tilde{T}, \quad C = Nk_B\tilde{c} = \frac{\pi^2}{3}D(\varepsilon_F)k_B^2 T. \tag{4.44}$$

Hence, the heat capacity of low-temperature fermions is proportional to T, and the density of states at the Fermi energy is relevant in its presence in the prefactor. See also Fig. 4.1.

4.6 Bosons at Low Temperatures

Next, we consider bosons at low temperatures. The most peculiar feature of the Bose distribution function (4.6) with $\sigma = 1$ is the singularity at $\varepsilon_q = \mu$ that causes the function to diverge. This singularity has no physical relevance at high temperatures where $\mu < 0$ so that $\varepsilon_q > \mu$ for any q. As the temperature is lowered, however, μ increases gradually to approach the lowest eigenenergy ε_0 from below. Depending on the density of states, it may finally reach ε_0 at a certain temperature T_0 where the singularity in the Bose distribution function manifests itself as a phase transition called *Bose-Einstein condensation* (BEC). This phase places a macroscopic number of particles into the lowest energy level.

We present a preliminary survey of the BEC based on Fig. 4.1 for free bosons in three dimensions. With decreasing temperature, the chemical potential approaches $\varepsilon_0 = 0$ from below to eventually become zero at the temperature $T_0 = 0.671 T_{\mathrm{Q}}$. For $T \leq T_0$, the chemical potential stays constant at $\mu = 0$, while more and more particles occupy the lowest energy level as $T \to 0$. We give a detailed description of this BEC state below.

Bose-Einstein condensation was predicted by Einstein in 1925 in his attempt to extend the statistical-mechanical theory of photons by Bose in 1924 to massive particles, hence his connection to BEC. It is well known that before his discovery, Einstein not only realized the importance of Bose's preprint sent to him, but also showed kindness and sincerity in translating it into German and getting it published. No BEC systems were known for a long time except for the strongly correlated ^{4}He liquid. Finally, in 1995, systems that could be described quantitatively by the theory of Einstein were realized using atomic gases trapped in harmonic potentials [2, 5].

4.6.1 *Critical Temperature of Condensation*

The equation to determine the critical temperature $\tilde{T}_0$ is obtained from (4.23) with $\sigma = 1$ by setting $\tilde{T} = \tilde{T}_0$ and $\tilde{\mu} = 0$ as

$$1 = \frac{\pi}{4} \int_0^\infty \frac{\tilde{\epsilon}^{1/2}}{\mathrm{e}^{\tilde{\epsilon}/\tilde{T}_0} - 1} \mathrm{d}\tilde{\epsilon} = \frac{\pi}{4} \tilde{T}_0^{3/2} \int_0^\infty \frac{x^{1/2}}{\mathrm{e}^x - 1} \mathrm{d}x. \tag{4.45}$$

This integral can be evaluated quite generally in the same manner as (4.39) for fermions,

$$J_n^{\mathrm{B}} \equiv \int_0^\infty \frac{x^{n-1}}{\mathrm{e}^x - 1} \mathrm{d}x = \sum_{m=1}^\infty \int_0^\infty x^{n-1} \mathrm{e}^{-mx} \mathrm{d}x = \zeta(n) \Gamma(n), \tag{4.46}$$

where $\Gamma(n)$ and $\zeta(n)$ are defined by (4.28) and (4.40), respectively. The relevant values are $\zeta(3/2) = 2.612\cdots$, $\Gamma(3/2) = \sqrt{\pi}/2$, $\zeta(5/2) = 1.341\cdots$, and $\Gamma(5/2) = 3\sqrt{\pi}/4$. Using them in (4.46) to evaluate (4.45) and noting (4.20) and (4.21), we obtain $\tilde{T}_0$ and $T_0 = T_Q\tilde{T}_0$:

$$\tilde{T}_0 = \frac{4}{\pi\zeta(3/2)^{2/3}} = 0.671, \qquad T_0 = \frac{\hbar^2}{2mk_B}\frac{4\pi(N/V)^{2/3}}{[(2s+1)\zeta(3/2)]^{2/3}}. \tag{4.47}$$

4.6.2 Thermodynamic Quantities of $T < T_0$

For $T < T_0$, a macroscopic number N_0 of particles occupies the lowest energy level with $\varepsilon_0 = 0$. The ratio N_0/N can be calculated by subtracting the fraction of excited particles with $\varepsilon_k > 0$ from 1 and using (4.45),

$$\begin{aligned}\frac{N_0}{N} &= 1 - \frac{\pi}{4}\int_0^\infty \frac{\tilde{\epsilon}^{1/2}}{e^{\tilde{\epsilon}/\tilde{T}}-1}d\tilde{\epsilon} = 1 - \frac{\pi}{4}\tilde{T}^{3/2}\int_0^\infty \frac{x^{1/2}}{e^x-1}dx = 1 - \left(\frac{\tilde{T}}{\tilde{T}_0}\right)^{3/2} \\ &= 1 - \left(\frac{T}{T_0}\right)^{3/2}.\end{aligned} \tag{4.48}$$

The internal energy $\tilde{u}$ for $T < T_0$ is obtained from (4.24) by setting $\tilde{\mu} = 0$ and using (4.46),

$$\tilde{u} = \frac{\pi}{4}\int_0^\infty \frac{\tilde{\epsilon}^{3/2}}{e^{\tilde{\epsilon}/\tilde{T}}-1}d\tilde{\epsilon} = \frac{\pi}{4}\tilde{T}^{5/2}\int_0^\infty \frac{x^{3/2}}{e^x-1}dx = \frac{3\pi^{3/2}\zeta(5/2)}{16}\tilde{T}^{5/2}. \tag{4.49}$$

Using $[8/\pi^{3/2}\zeta(3/2)]\tilde{T}_0^{-3/2} = 1$ from (4.47), we can transform the heat capacity $\tilde{c} = d\tilde{u}/d\tilde{T}$ as

$$\tilde{c} = \frac{15\pi^{3/2}\zeta(5/2)}{32}\tilde{T}^{3/2} = \frac{15\zeta(5/2)}{4\zeta(3/2)}\left(\frac{T}{T_0}\right)^{3/2} = 1.93\left(\frac{T}{T_0}\right)^{3/2}. \tag{4.50}$$

Note that value 1.93 is larger than the value 1.5 for the classical Maxwell-Boltzmann distribution. See also Fig. 4.1 on this point.

4.6.3 Chemical Potential and Heat Capacity for $T \gtrsim T_0$

To find the explicit temperature dependence of the chemical potential for $\tilde{T} \gtrsim \tilde{T}_0$ where $\tilde{\mu} \lesssim 0$, we transform the integral in (4.23) to obtain

$$
\begin{aligned}
1 &= \frac{\pi}{4}\int_0^\infty \frac{\tilde{\epsilon}^{1/2}}{e^{(\tilde{\epsilon}-\tilde{\mu})/\tilde{T}}-1}d\tilde{\epsilon} \\
&= \frac{\pi}{4}\int_0^\infty \frac{\tilde{\epsilon}^{1/2}}{e^{\tilde{\epsilon}/\tilde{T}}-1}d\tilde{\epsilon} + \frac{\pi}{4}\int_0^\infty \left[\frac{1}{e^{(\tilde{\epsilon}-\tilde{\mu})/\tilde{T}}-1} - \frac{1}{e^{\tilde{\epsilon}/\tilde{T}}-1}\right]\tilde{\epsilon}^{1/2}d\tilde{\epsilon}
\end{aligned}
$$

expand denominators in the second term in terms of $(\tilde{\epsilon}-\tilde{\mu})/\tilde{T}$ and $\tilde{\epsilon}/\tilde{T}$

$$
\approx \frac{\pi}{4}\tilde{T}^{3/2}\int_0^\infty \frac{x^{1/2}}{e^x-1}dx + \frac{\pi}{4}\int_0^\infty \left(\frac{\tilde{T}}{\tilde{\epsilon}-\tilde{\mu}} - \frac{\tilde{T}}{\tilde{\epsilon}}\right)\tilde{\epsilon}^{1/2}d\tilde{\epsilon}
$$

use (4.45) for the first term and set $\tilde{\epsilon} = |\tilde{\mu}|x^2$ in the second term

$$
= \left(\frac{\tilde{T}}{\tilde{T}_0}\right)^{3/2} - \frac{\pi}{2}\tilde{T}|\tilde{\mu}|^{1/2}\int_0^\infty \frac{1}{x^2+1}dx
$$

set $x = \tan\theta$ in the second term

$$
\begin{aligned}
&= \left(1+\frac{\tilde{T}-\tilde{T}_0}{\tilde{T}_0}\right)^{3/2} - \frac{\pi}{2}\tilde{T}|\tilde{\mu}|^{1/2}\int_0^{\pi/2} d\theta \\
&\approx 1 + \frac{3}{2}\frac{\tilde{T}-\tilde{T}_0}{\tilde{T}_0} - \frac{\pi^2}{4}\tilde{T}_0|\tilde{\mu}|^{1/2}.
\end{aligned}
$$

We thereby obtain $\tilde{\mu}$ for $\tilde{T} \gtrsim \tilde{T}_0$ as

$$
\tilde{\mu} \approx -\frac{36}{\pi^4\tilde{T}_0^2}\left(\frac{\tilde{T}-\tilde{T}_0}{\tilde{T}_0}\right)^2. \tag{4.51}
$$

This expression tells us that both $\tilde{\mu}$ and $\partial\tilde{\mu}/\partial\tilde{T}$ are continuous at $\tilde{T} = \tilde{T}_0$. Accordingly, heat capacity is also continuous at $\tilde{T} = \tilde{T}_0$, as seen from (4.25).

4.7 Bose-Einstein Condensation and Density of States

The thermodynamic quantities of ideal gases can be expressed as integrals in terms of the density of states $D(\epsilon)$ and distribution function $[e^{\beta(\epsilon-\mu)} - \sigma]^{-1}$, where all the effects of external potentials and system dimensions are contained in $D(\epsilon)$. For example, we have seen earlier that the density of states $D(\epsilon)$ for free particles in d dimensions is expressible as $D(\epsilon) \propto \epsilon^{(d-2)/2}\theta(\epsilon)$ in terms of a d-dependent exponent. With these observations, it is worth considering a general model with the density of states:

$$
D(\epsilon) = A(\epsilon-\varepsilon_0)^{\eta-1}\theta(\epsilon-\varepsilon_0), \tag{4.52}
$$

where $A > 0$ is a constant, ε_0 is the lowest one-particle energy, and $\theta(x)$ is the step function (4.11). Indeed, setting $\eta = 3/2$ ($\eta = 1$) and $\varepsilon_0 = 0$ yields the free Bose gas in three (two) dimensions. We shall clarify how the critical temperature T_0 for the BEC transition depends on the exponent η.

The equation to determine T_0 is obtained by substituting (4.52) into (4.12) with $\mu = \varepsilon_0$ and $T = T_0$. The resulting equation is further transformed as

$$N = \int_{\varepsilon_0}^{\infty} \frac{A(\epsilon - \varepsilon_0)^{\eta-1}}{e^{(\epsilon-\varepsilon_0)/k_B T_0} - 1} d\epsilon = A(k_B T_0)^{\eta} \int_0^{\infty} \frac{x^{\eta-1}}{e^x - 1} dx = A(k_B T_0)^{\eta} \zeta(\eta)\Gamma(\eta),$$

where we have used (4.46). Thus, we obtain T_0 as

$$T_0 = \frac{1}{k_B} \left[\frac{N}{A\zeta(\eta)\Gamma(\eta)} \right]^{1/\eta}. \tag{4.53}$$

Specifically, we conclude $T_0 \to 0$ as $\eta \to 1$ from above, because $\zeta(\eta \to 1) \to \infty$. This also implies that there is no BEC transition for $\eta < 1$. A critical case with $\eta = 1$ is the free Bose gas in two dimensions, where the BEC transition occurs at $T = 0$.

As for the thermodynamic quantities, we only need to replace the results of Sect. 4.6 with, for example, $\zeta(3/2) \to \zeta(\eta)$, $\zeta(5/2) \to \zeta(\eta + 1)$, and $(T/T_0)^{3/2} \to (T/T_0)^{\eta}$ (**Problem** 4.3).

Problems

4.1. Consider a free particle with mass m and spin s that moves in a rectangular area of d dimensions with edge length L. For $d = 1, 2$, show that the density of states is given by

$$D(\epsilon) = \begin{cases} \dfrac{(2s+1)L}{2\pi} \left(\dfrac{2m}{\hbar^2}\right)^{1/2} \epsilon^{-1/2}\theta(\epsilon) & : d = 1 \\ \dfrac{(2s+1)L^2}{4\pi} \left(\dfrac{2m}{\hbar^2}\right) \theta(\epsilon) & : d = 2 \end{cases}. \tag{4.54}$$

4.2. N identical monoatomic molecules with mass m and spin $1/2$ are confined in a two-dimensional square area with edge length L.

(a) Obtain expressions for the Fermi energy and Fermi wave number.
(b) Find the expression for μ at low temperatures up to order T^2 using (4.41).

4.3. N identical monoatomic molecules with mass m and spin 0 are confined in a harmonic potential $\mathscr{U}(\mathbf{r}) = \frac{1}{2}m\omega^2\mathbf{r}^2$ in three dimensions without mutual interactions. The one-particle eigenenergy is given by [4]

$$\varepsilon_{n_x n_y n_z} \equiv \left(n_x + n_y + n_z + \frac{3}{2}\right)\hbar\omega,$$

where n_η ($\eta = x, y, z$) are non-negative integers.

(a) Evaluate (4.9) for $q \equiv (n_x, n_y, n_z)$ to show that the density of states is given to an excellent approximation for $T \gtrsim \hbar\omega/k_B$ by

$$D(\epsilon) = \frac{(\epsilon - \varepsilon_0)^2}{2(\hbar\omega)^3}\theta(\epsilon - \varepsilon_0)$$

with $\varepsilon_0 \equiv 3\hbar\omega/2$.

(b) Show that the BEC transition temperature T_0 is given by

$$T_0 = \frac{1}{k_B}\left[\frac{2(\hbar\omega)^3 N}{\zeta(3)\Gamma(3)}\right]^{1/3} = \left[\frac{N}{\zeta(3)}\right]^{1/3}\frac{\hbar\omega}{k_B}.$$

(c) Show that the number N_0 of condensed particles for $T < T_0$ is given by

$$\frac{N_0}{N} = 1 - \left(\frac{T}{T_0}\right)^3.$$

(d) Show that internal energy U and heat capacity C for $T < T_0$ are given by

$$U = N\left[\varepsilon_0 + \frac{3\zeta(4)}{\zeta(3)}k_B T_0\left(\frac{T}{T_0}\right)^4\right], \qquad C = Nk_B\frac{12\zeta(4)}{\zeta(3)}\left(\frac{T}{T_0}\right)^3.$$

References

1. M. Abramowitz, I.A. Stegun, (eds.), *Handbook of Mathematical Functions: With Formulas, Graphs, and Mathematical Tables* (Dover, New York, 1965)
2. M.H. Anderson, J.R. Ensher, M.R. Matthews, C.E. Wieman, E.A. Cornell, Science **269**, 198 (1995)
3. G.B. Arfken, H.J. Weber, *Mathematical Methods for Physicists* (Academic, New York, 2012)
4. L.D. Landau, E.M. Lifshitz, *Quantum Mechanics: Non-relativistic Theory*, 3rd edn. (Butterworth-Heinemann, Oxford, 1991)
5. C.J. Pethick, H. Smith, *Bose-Einstein Condensation in Dilute Gases* (Cambridge University Press, Cambridge, 2002)
6. F. Reif, *Fundamentals of Statistical and Thermal Physics* (McGraw-Hill, New York, 1965)
7. Teragon's Summary of Cryogen Properties
8. R.C. Weast (ed.), *Handbook of Chemistry and Physics* (Chemical Rubber, Cleveland, 1965–)

Chapter 5
Density Matrices and Two-Particle Correlations

Abstract In this chapter, we first introduce two new concepts named *density matrix* and *reduced density matrices* as (5.1) and (5.3). The reduced density matrices are closely connected with the n-particle correlations in equilibrium ($n = 1, 2, \cdots$). Next, we give a proof of the *Bloch–De Dominicis theorem*, i.e., a thermodynamic extension of *Wick's theorem*, which enables us to express the n-particle correlations of ideal gases in terms of one-particle correlations as (5.11). Finally, the theorem is applied to obtain the two-particle correlations of homogeneous ideal Bose and Fermi gases in three dimensions. The results are summarized in Fig. 5.1 below. It clearly shows that there exists a special quantum-mechanical correlation between each pair of identical particles due to the permutation symmetry, which is completely different in nature between Bose and Fermi gases.

5.1 Density Matrices

The density matrix is defined in terms of eigenstates $|\Phi_\nu\rangle$ of the Schrödinger equation (3.43) and its probability w_ν of realization,

$$\hat{\rho} \equiv \sum_\nu |\Phi_\nu\rangle w_\nu \langle\Phi_\nu|. \tag{5.1}$$

The distributions frequently used for $\{w_\nu\}$ are the microcanonical distribution (2.12), the canonical distribution (2.18), and the grand canonical distribution (2.26). Once $\hat{\rho}$ is obtained, we can calculate the expectation of an arbitrary Hermitian operator $\hat{\mathcal{O}}$ by

$$\langle\hat{\mathcal{O}}\rangle \equiv \mathrm{Tr}\,\hat{\rho}\,\hat{\mathcal{O}} = \sum_\nu w_\nu \langle\Phi_\nu|\hat{\mathcal{O}}|\Phi_\nu\rangle, \tag{5.2}$$

where Tr denotes trace.

Next, we introduce reduced density matrices in terms of (5.2) for the expectation by

$$\rho^{(n)}(\xi_1, \cdots, \xi_n; \xi_1', \cdots, \xi_n') \equiv \langle\hat{\psi}^\dagger(\xi_1')\cdots\hat{\psi}^\dagger(\xi_n')\hat{\psi}(\xi_n)\cdots\hat{\psi}(\xi_1)\rangle$$

© Springer Japan 2015
T. Kita, *Statistical Mechanics of Superconductivity*, Graduate Texts in Physics,
DOI 10.1007/978-4-431-55405-9_5

$$= \sum_{\nu} w_{\nu} \frac{\mathscr{N}_{\nu}!}{(\mathscr{N}_{\nu}-n)!} \int d\xi_{n+1} \cdots \int d\xi_{\mathscr{N}_{\nu}} \Phi_{\nu}(\xi_1, \cdots, \xi_n, \xi_{n+1} \cdots \xi_{\mathscr{N}_{\nu}})$$
$$\times \Phi_{\nu}^{*}(\xi_1', \cdots, \xi_n', \xi_{n+1} \cdots \xi_{\mathscr{N}_{\nu}}) \tag{5.3}$$

with $n = 1, 2, \cdots, \mathscr{N}_{\nu}$, where we have successively used (3.26)–(3.28), (3.22), and (3.12) to give the concise expression in terms of the wave functions. It should be noted that the particle number in the microcanonical and canonical ensembles is a constant that does not depend on ν as $\mathscr{N}_{\nu} = N$. Quantity $\rho^{(n)}$ is also called the *n-particle density matrix*. Its physical meaning may be realized by looking at the case $n = 2$ with $\xi_1' = \xi_1$ and $\xi_2' = \xi_2$:

$$\rho^{(2)}(\xi_1, \xi_2; \xi_1, \xi_2) = \langle \hat{\psi}^{\dagger}(\xi_1) \hat{\psi}^{\dagger}(\xi_2) \hat{\psi}(\xi_2) \hat{\psi}(\xi_1) \rangle$$
$$= \sum_{\nu} w_{\nu} \mathscr{N}_{\nu}(\mathscr{N}_{\nu}-1) \int d\xi_3 \cdots \int d\xi_{\mathscr{N}_{\nu}} |\Psi_{\nu}(\xi_1, \xi_2, \xi_3, \cdots, \xi_{\mathscr{N}_{\nu}})|^2, \tag{5.4}$$

which is proportional to the probability that a pair of particles are simultaneously at ξ_1 and ξ_2. Thus, we can clarify many-particle correlations once reduced density matrices have been obtained. We also realize from Hamiltonian (3.42) that we can evaluate the n-particle operators using $\rho^{(n)}$.

Definition (5.2) of the density matrix contains bras and kets, which are indispensable for describing systems with spontaneous symmetry breaking such as ferromagnets. To be specific, the free energy for an isotropic ferromagnet is degenerate with respect to the direction of the macroscopic moment. However, this rotational symmetry is broken spontaneously to realize a macroscopic moment that is directed along some specific direction. Because a huge number of magnetic moments are aligned cooperatively along a single direction, there is essentially no chance for the macroscopic moment to change its direction at a time. Thus, the direction of the magnetic moment is essentially fixed with possible fluctuations around it, and the brackets in (5.2) should also describe this situation by excluding those possibilities where the moment is aligned along other directions.

5.2 Bloch–De Dominicis Theorem

Equation (5.3) is given as an expectation of a product of $2n$ field operators. How can we evaluate it? For the special cases of ideal gases in the grand canonical distribution, the theorem of Bloch and De Dominicis [2] enables us to perform this concisely. It is a thermodynamic extension of Wick's theorem for the evaluation of the S-matrix in relativistic field theory [7]. The present proof follows that given by Gaudin [3], which is more elementary and easier to understand.

Consider a system without interactions described by Hamiltonian $\hat{H}_0$ in (3.45). As we are considering a grand canonical ensemble, we subtract the product term involving the chemical potential μ and number operator $\hat{\mathscr{N}}$ from $\hat{H}_0$ to introduce

$$\hat{\mathscr{H}}_0 \equiv \hat{H}_0 - \mu\hat{\mathscr{N}} = \int \mathrm{d}\xi\, \hat{\psi}^\dagger(\xi)\left[\frac{\hat{\mathbf{p}}^2}{2m} + \mathscr{U}(\mathbf{r}) - \mu\right]\hat{\psi}(\xi)$$
$$= \sum_q (\varepsilon_q - \mu)\hat{c}_q^\dagger \hat{c}_q. \tag{5.5}$$

Here ε_q is a single-particle eigenenergy determined by (3.46), and $\hat{c}_q$ obeys the commutation relation (3.51). Substituting (2.26) into (5.1) and using (3.35), we obtain the corresponding density matrix,

$$\hat{\rho} = \sum_\nu |\Phi_\nu\rangle \mathrm{e}^{\beta\Omega - \beta(\mathscr{E}_\nu - \mu\mathscr{N}_\nu)}\langle\Phi_\nu| = \sum_\nu \mathrm{e}^{\beta\Omega - \beta\hat{\mathscr{H}}_0}|\Phi_\nu\rangle\langle\Phi_\nu|$$
$$= \mathrm{e}^{\beta(\Omega - \hat{\mathscr{H}}_0)}. \tag{5.6}$$

As preliminaries for the theorem, let us prove the identities:

$$\hat{c}_q \mathrm{e}^{-\beta\hat{\mathscr{H}}_0} = a_q \mathrm{e}^{-\beta\hat{\mathscr{H}}_0}\hat{c}_q, \qquad a_q \equiv \mathrm{e}^{-\beta(\varepsilon_q - \mu)}, \tag{5.7}$$
$$\hat{c}_q^\dagger \mathrm{e}^{-\beta\hat{\mathscr{H}}_0} = \frac{1}{a_q}\mathrm{e}^{-\beta\hat{\mathscr{H}}_0}\hat{c}_q^\dagger. \tag{5.8}$$

Equation (5.8) is the Hermitian conjugate of (5.7) so that it suffices to show the latter. For this purpose, we introduce

$$\hat{c}_q(\beta) \equiv \mathrm{e}^{\beta\hat{\mathscr{H}}_0}\hat{c}_q \mathrm{e}^{-\beta\hat{\mathscr{H}}_0}. \tag{5.9}$$

We differentiate this equation with respect to β, substituting (5.5), and using (3.51) to bring the derivative to the form

$$\frac{\mathrm{d}\hat{c}_q(\beta)}{\mathrm{d}\beta} = \mathrm{e}^{\beta\hat{\mathscr{H}}_0}(\hat{\mathscr{H}}_0\hat{c}_q - \hat{c}_q\hat{\mathscr{H}}_0)\mathrm{e}^{-\beta\hat{\mathscr{H}}_0}$$
$$= \mathrm{e}^{\beta\hat{\mathscr{H}}_0}\sum_{q'}(\varepsilon_{q'} - \mu)(\hat{c}_{q'}^\dagger\hat{c}_{q'}\hat{c}_q - \hat{c}_q\hat{c}_{q'}^\dagger\hat{c}_{q'})\mathrm{e}^{-\beta\hat{\mathscr{H}}_0}$$
$$= \mathrm{e}^{\beta\hat{\mathscr{H}}_0}\sum_{q'}(\varepsilon_{q'} - \mu)[\hat{c}_{q'}^\dagger\sigma\hat{c}_q\hat{c}_{q'} - (\delta_{qq'} + \sigma\hat{c}_{q'}^\dagger\hat{c}_q)\hat{c}_{q'}]\mathrm{e}^{-\beta\hat{\mathscr{H}}_0}$$
$$= -\mathrm{e}^{\beta\hat{\mathscr{H}}_0}(\varepsilon_q - \mu)\hat{c}_q\mathrm{e}^{-\beta\hat{\mathscr{H}}_0}$$
$$= -(\varepsilon_q - \mu)\hat{c}_q(\beta).$$

This first-order differential equation can be solved easily using the initial condition $\hat{c}_q(0) = \hat{c}_q$ from (5.9) as

$$\hat{c}_q(\beta) = \mathrm{e}^{-\beta(\varepsilon_q - \mu)}\hat{c}_q. \tag{5.10}$$

Finally, we equate the right-hand sides of (5.9) and (5.10) and multiply the resulting equation by $e^{-\beta\hat{\mathscr{H}}_0}$ from the left. We thereby obtain (5.7). This completes the proof.

We proceed to the main theorem, the statement of which is

Bloch–De Dominicis Theorem

$$\begin{aligned}\langle \hat{C}_1\hat{C}_2\cdots\hat{C}_{2n}\rangle &\equiv \mathrm{Tr}\, e^{\beta(\Omega-\hat{\mathscr{H}}_0)}\hat{C}_1\hat{C}_2\cdots\hat{C}_{2n} \\ &= \sum_{\hat{P}}{}' \sigma^P \langle \hat{C}_{p_1}\hat{C}_{p_2}\rangle\langle \hat{C}_{p_3}\hat{C}_{p_4}\rangle\cdots\langle \hat{C}_{p_{2n-1}}\hat{C}_{p_{2n}}\rangle. \quad (5.11)\end{aligned}$$

Here $\hat{C}_j$ represents either $\hat{c}_{q_j}$ or $\hat{c}^\dagger_{q_j}$, and symbol $\sum_{\hat{P}}{}'$ with the prime denotes a restricted sum that, of the permutations given by

$$\hat{P} = \begin{pmatrix} 1 & 2 & 3 & \cdots & 2n-1 & 2n \\ p_1 & p_2 & p_3 & \cdots & p_{2n-1} & p_{2n} \end{pmatrix}, \tag{5.12}$$

only those obeying the conditions

$$p_1 < p_2,\ p_3 < p_4,\ \cdots,\ p_{2n-1} < p_{2n}, \qquad p_1 < p_3 < \cdots < p_{2n-1}, \tag{5.13}$$

enter in the summation. The transformation (5.11) is called the *Wick decomposition*. The first condition of (5.13) implies that the alignment order in each expectation remains the same before and after the Wick decomposition; the second condition excludes the double counting of the same decomposition. See (5.18) below for a specific example.

To prove (5.11), it is convenient to express the commutation relations of (3.51) in a unified way with the notation

$$[\hat{C}_i,\hat{C}_j]_\sigma = (i,j), \qquad (i,j) \equiv \begin{cases} \delta_{ij} & : \hat{C}_i = \hat{c}_{q_i},\ \hat{C}_j = \hat{c}^\dagger_{q_j} \\ -\sigma\delta_{ij} & : \hat{C}_i = \hat{c}^\dagger_{q_i},\ \hat{C}_j = \hat{c}_{q_j} \\ 0 & : \text{otherwise} \end{cases}. \tag{5.14}$$

By noting that (i,j) is a constant, we move $\hat{C}_1$ on the left-hand side of (5.11) to the rightmost position,

$$\begin{aligned}\langle \hat{C}_1\hat{C}_2\cdots\hat{C}_{2n}\rangle &= \langle[(1,2)+\sigma\hat{C}_2\hat{C}_1]\hat{C}_3\cdots\hat{C}_{2n}\rangle \\ &= (1,2)\langle\hat{C}_3\cdots\hat{C}_{2n}\rangle + \sigma\langle\hat{C}_2[(1,3)+\sigma\hat{C}_3\hat{C}_1]\hat{C}_4\cdots\hat{C}_{2n}\rangle \\ &= (1,2)\langle\hat{C}_3\cdots\hat{C}_{2n}\rangle + \sigma(1,3)\langle\hat{C}_2\hat{C}_4\cdots\hat{C}_{2n}\rangle + \cdots \\ &\quad + \sigma^{2n-2}(1,2n)\langle\hat{C}_2\hat{C}_3\cdots\hat{C}_{2n-1}\rangle + \sigma^{2n-1}\langle\hat{C}_2\cdots\hat{C}_{2n}\hat{C}_1\rangle.\end{aligned}$$

The last term can be transformed by noting the definition of the expectation in (5.11) and using (5.7) and (5.8),

$$\langle \hat{C}_2 \cdots \hat{C}_{2n} \hat{C}_1 \rangle = a_1^\eta \langle \hat{C}_1 \hat{C}_2 \cdots \hat{C}_{2n} \rangle, \qquad \eta = \begin{cases} 1 & : \hat{C}_1 = \hat{c}_{q_1} \\ -1 & : \hat{C}_1 = \hat{c}_{q_1}^\dagger \end{cases} .$$

Combining the two equations above, we can express the original expectation of the $2n$ product of operators in terms of those with a $2n-2$ product of operators,

$$\begin{aligned} \langle \hat{C}_1 \hat{C}_2 \cdots \hat{C}_{2n} \rangle = & \frac{(1,2)}{1-\sigma a_1^\eta} \langle \hat{C}_3 \hat{C}_4 \cdots \hat{C}_{2n} \rangle + \sigma \frac{(1,3)}{1-\sigma a_1^\eta} \langle \hat{C}_2 \hat{C}_4 \cdots \hat{C}_{2n} \rangle \\ & + \cdots + \sigma^{2n-2} \frac{(1,2n)}{1-\sigma a_1^\eta} \langle \hat{C}_2 \hat{C}_3 \cdots \hat{C}_{2n-1} \rangle . \end{aligned} \tag{5.15}$$

In particular, the case $n=1$ yields

$$\langle \hat{C}_1 \hat{C}_2 \rangle = \frac{(1,2)}{1-\sigma a_1^\eta}, \qquad \eta = \begin{cases} 1 & : \hat{C}_1 = \hat{c}_{q_1} \\ -1 & : \hat{C}_1 = \hat{c}_{q_1}^\dagger \end{cases} . \tag{5.16}$$

Substituting this back into (5.15), we obtain

$$\begin{aligned} \langle \hat{C}_1 \hat{C}_2 \cdots \hat{C}_{2n} \rangle = & \langle \hat{C}_1 \hat{C}_2 \rangle \langle \hat{C}_3 \hat{C}_4 \cdots \hat{C}_{2n} \rangle + \sigma \langle \hat{C}_1 \hat{C}_3 \rangle \langle \hat{C}_2 \hat{C}_4 \cdots \hat{C}_{2n} \rangle \\ & + \cdots + \sigma^{2n-2} \langle \hat{C}_1 \hat{C}_{2n} \rangle \langle \hat{C}_2 \hat{C}_3 \cdots \hat{C}_{2n-1} \rangle . \end{aligned}$$

This is a recursion formula that enables us to reduce the number of products in the expectation by 2. With repeated use, we obtain (5.11). Note that condition (5.13) is satisfied here.

Three comments on the Bloch–De Dominicis theorem are in order. First, C_j in (5.11) is originally specified as an eigenoperator corresponding to an eigenstate of the single-particle Schrödinger equation (3.46). However, the theorem holds true directly for an arbitrary linear combination of $\hat{C}_j$ given by $\hat{B}_k \equiv \sum_k U_{kj} \hat{C}_j$; that is,

$$\langle \hat{B}_1 \hat{B}_2 \cdots \hat{B}_{2n} \rangle = \sum_{\hat{P}}{}' \sigma^P \langle \hat{B}_{p_1} \hat{B}_{p_2} \rangle \langle \hat{B}_{p_3} \hat{B}_{p_4} \rangle \cdots \langle \hat{B}_{p_{2n-1}} \hat{B}_{p_{2n}} \rangle . \tag{5.17}$$

In proof, we only need to express the left-hand side of (5.17) in terms of $\hat{C}$, use (5.11), and finally put $\sum_k U_{kj}$ back into the appropriate pair of brackets. Note especially that $\hat{B}_k$ above can be a linear combination of $\hat{c}_{q_j}$ and $\hat{c}_{q_j}^\dagger$. We shall encounter this situation when we consider superconductivity.

As a simple application of (5.17), consider the two-particle density matrix (5.4) for a normal system without interactions. Its Wick decomposition can be performed concisely as follows. First, we enumerate distinct decompositions by marking each pair of operators with a common symbol on top;

$$
\begin{aligned}
\rho^{(2)}(\xi_1,\xi_2;\xi_1,\xi_2) &\equiv \langle \hat{\psi}^\dagger(\xi_1)\hat{\psi}^\dagger(\xi_2)\hat{\psi}(\xi_2)\hat{\psi}(\xi_1)\rangle \\
&= \langle \dot{\hat{\psi}}^\dagger(\xi_1)\ddot{\hat{\psi}}^\dagger(\xi_2)\ddot{\hat{\psi}}(\xi_2)\dot{\hat{\psi}}(\xi_1)\rangle + \langle \dot{\hat{\psi}}^\dagger(\xi_1)\ddot{\hat{\psi}}^\dagger(\xi_2)\dot{\hat{\psi}}(\xi_2)\ddot{\hat{\psi}}(\xi_1)\rangle,
\end{aligned}
$$

where we used the fact that a pair of annihilation operators have null expectation in the normal state. Next, we focus successively on a single operator from the left in each term and move its partner to its right, multiplying by $\sigma = \pm 1$ upon each exchange of operators until all the pairs are coupled. We then obtain

$$
\begin{aligned}
&\rho^{(2)}(\xi_1,\xi_2;\xi_1,\xi_2) \\
&\quad = \sigma^2\langle \dot{\hat{\psi}}^\dagger(\xi_1)\dot{\hat{\psi}}(\xi_1)\ddot{\hat{\psi}}^\dagger(\xi_2)\ddot{\hat{\psi}}(\xi_2)\rangle + \sigma\langle \dot{\hat{\psi}}^\dagger(\xi_1)\dot{\hat{\psi}}(\xi_2)\ddot{\hat{\psi}}^\dagger(\xi_2)\ddot{\hat{\psi}}(\xi_1)\rangle.
\end{aligned}
$$

Finally, we place around each coupled pair angle brackets and simultaneously remove common symbols on top of them,

$$
\begin{aligned}
&\rho^{(2)}(\xi_1,\xi_2;\xi_1,\xi_2) \\
&\quad = \langle \hat{\psi}^\dagger(\xi_1)\hat{\psi}(\xi_1)\rangle\langle \hat{\psi}^\dagger(\xi_2)\hat{\psi}(\xi_2)\rangle + \sigma\langle \hat{\psi}^\dagger(\xi_1)\hat{\psi}(\xi_2)\rangle\langle \hat{\psi}^\dagger(\xi_2)\hat{\psi}(\xi_1)\rangle \\
&\quad = \rho^{(1)}(\xi_1,\xi_1)\rho^{(1)}(\xi_2,\xi_2) + \sigma\rho^{(1)}(\xi_2,\xi_1)\rho^{(1)}(\xi_1,\xi_2). \qquad (5.18)
\end{aligned}
$$

Thus, the two-particle density matrix has been expressed successfully in terms of one-particle density matrices $\rho^{(1)}(\xi_1,\xi_2) \equiv \langle \hat{\psi}^\dagger(\xi_2)\hat{\psi}(\xi_1)\rangle$.

Second, we extend (5.11) to the BEC phases with $\sigma = 1$, where a macroscopic number of particles occupies the lowest-energy state (Sect. 4.6). In this case, we should express $\hat{\psi}(\xi)$ in (3.49) as a sum of the *condensate wave function*:

$$
\Psi(\xi) \equiv \langle \hat{\psi}(\xi)\rangle \qquad (5.19)
$$

and the rest as

$$
\hat{\psi}(\xi) \equiv \Psi(\xi) + \sum_q{}' \hat{c}_q\varphi_q(\xi) \equiv \Psi(\xi) + \hat{\varphi}(\xi), \qquad (5.20)
$$

and apply the theorem only to the field $\hat{\varphi}(\xi)$ without the lowest-energy state. Note that the finite average $\langle \hat{\psi}(\xi)\rangle \neq 0$ is possible only when we consider a superposition over the occupation number for the lowest-energy state as (3.67), which in turn will be shown to yield a two-particle correlation that is physically reasonable, as seen below. See also [1] for a justification in setting $\langle \hat{\psi}(\xi)\rangle \neq 0$. The one-particle density matrix with this procedure is given by

$$\begin{aligned}\rho^{(1)}(\xi_1,\xi_2) &\equiv \langle\hat{\psi}^\dagger(\xi_2)\hat{\psi}(\xi_1)\rangle \\ &= \Psi(\xi_1)\Psi^*(\xi_2) + \langle\hat{\varphi}^\dagger(\xi_2)\hat{\varphi}(\xi_1)\rangle \\ &= \Psi(\xi_1)\Psi^*(\xi_2) + \sum_q{}' \varphi_q(\xi_1)\varphi_q^*(\xi_2)\bar{n}_q,\end{aligned} \tag{5.21}$$

where we have used (5.20), $\langle\hat{\varphi}(\xi)\rangle = 0$, and $\langle\hat{c}_{q'}^\dagger\hat{c}_q\rangle = \delta_{q'q}\bar{n}_q$. As for the two-particle density matrix, we substitute (5.20) into (5.4) and apply the theorem (5.17) only to the field $\hat{\varphi}$, noting that $\langle\hat{\varphi}\rangle = 0$ and $\langle\hat{\varphi}\hat{\varphi}\rangle = 0$. We thereby obtain

$$\begin{aligned}&\rho^{(2)}(\xi_1,\xi_2;\xi_1,\xi_2) \\ &\quad= \langle[\Psi^*(\xi_1)+\hat{\varphi}^\dagger(\xi_1)][\Psi^*(\xi_2)+\hat{\varphi}^\dagger(\xi_2)][\Psi(\xi_2)+\hat{\varphi}(\xi_2)][\Psi(\xi_1)+\hat{\varphi}(\xi_1)]\rangle \\ &\quad= |\Psi(\xi_1)|^2|\Psi(\xi_2)|^2 + |\Psi(\xi_1)|^2\langle\hat{\varphi}^\dagger(\xi_2)\hat{\varphi}(\xi_2)\rangle + |\Psi(\xi_2)|^2\langle\hat{\varphi}^\dagger(\xi_1)\hat{\varphi}(\xi_1)\rangle \\ &\qquad+\Psi(\xi_1)\Psi^*(\xi_2)\langle\hat{\varphi}^\dagger(\xi_1)\hat{\varphi}(\xi_2)\rangle + \Psi(\xi_2)\Psi^*(\xi_1)\langle\hat{\varphi}^\dagger(\xi_2)\hat{\varphi}(\xi_1)\rangle \\ &\qquad+\langle\hat{\varphi}^\dagger(\xi_1)\hat{\varphi}(\xi_1)\rangle\langle\hat{\varphi}^\dagger(\xi_2)\hat{\varphi}(\xi_2)\rangle + \langle\hat{\varphi}^\dagger(\xi_1)\hat{\varphi}(\xi_2)\rangle\langle\hat{\varphi}^\dagger(\xi_2)\hat{\varphi}(\xi_1)\rangle \\ &\quad= \rho^{(1)}(\xi_1,\xi_1)\rho^{(1)}(\xi_2,\xi_2) + \sigma\rho^{(1)}(\xi_2,\xi_1)\rho^{(1)}(\xi_1,\xi_2) - |\Psi(\xi_1)|^2|\Psi(\xi_2)|^2,\end{aligned} \tag{5.22}$$

where in the last equality, we have inserted the factor $\sigma = 1$ for later convenience and expressed $\rho^{(2)}$ concisely in terms of $\rho^{(1)}$, (5.21).

Third, theorem (5.11) does not apply to the canonical and microcanonical ensembles. This may be realized by inspecting (5.16) in the proof. For the grand canonical ensemble, this one-particle expectation (5.16) for each of $\hat{C}_j = \hat{c}_{q_j}$ and $\hat{c}_{q_j}^\dagger$ adequately yields the single-particle occupation number in equilibrium as $\langle\hat{c}_{q_1}^\dagger\hat{c}_{q_2}\rangle = \delta_{q_1q_2}\bar{n}_{q_1}$, $\langle\hat{c}_{q_1}\hat{c}_{q_2}^\dagger\rangle = \delta_{q_1q_2}(1+\sigma\bar{n}_{q_1})$, $\langle\hat{c}_{q_1}\hat{c}_{q_2}\rangle = \langle\hat{c}_{q_1}^\dagger\hat{c}_{q_2}^\dagger\rangle = 0$ with $\bar{n}_{q_1}$ given by (4.6). The above proof may seem applicable to the canonical ensemble. However, if we set $\mu \to 0$, (5.16) does not yield the correct one-particle expectation for the canonical ensemble. The reason for this is that (5.11) makes use of processes where the particle number changes. In the canonical or microcanonical ensembles, this is not permitted.

5.3 Two-Particle Correlations of Monoatomic Ideal Gases

Expressions (5.21) and (5.22) for the reduced density matrices of ideal gases can be used not only for the BEC phases, but also for normal states of both bosons ($\sigma = 1$) and fermions ($\sigma = -1$) by setting $\Psi \to 0$. Here, we use (5.22) to clarify the two-particle correlations of free bosons and fermions as a supplement to the thermodynamic considerations of Sects. 4.3–4.6.

Because $\rho^{(2)}$ in (5.22) is expressed in terms of $\rho^{(1)}$, we first consider (5.21). Its quantum number q is specified in the present case by a combination of wave vector

k and spin index $\alpha = s, s-1, \cdots, -s$. The corresponding wave function is given by (4.14); that is,

$$\varphi_{\mathbf{k}\alpha}(\xi') = \langle \mathbf{r}'\alpha'|\mathbf{k}\alpha\rangle = \delta_{\alpha\alpha'}\frac{1}{\sqrt{V}}\mathrm{e}^{\mathrm{i}\mathbf{k}\cdot\mathbf{r}'}. \tag{5.23}$$

In contrast, the wave function of a homogeneous BEC carries wave vector **0** and can be written in terms of volume V, spin s, and condensate particle number N_0 of (4.48) as

$$\Psi(\xi) = \sqrt{\frac{N_0}{(2s+1)V}}, \tag{5.24}$$

where the phase of $\Psi(\xi)$ has been put equal to 0 irrespective of spin element α as it does not affect the energy at all for ideal gases. Note that integrating $|\Psi(\xi)|^2$ as (3.16) yields N_0.

Substituting (5.23) and (5.24) into (5.21), we obtain the one-particle density matrix,

$$\begin{aligned}\rho^{(1)}(\xi_1,\xi_2) &= \frac{N_0}{(2s+1)V} + \frac{\delta_{\alpha_1\alpha_2}}{V}\sum_{\mathbf{k}}{}' \frac{\mathrm{e}^{\mathrm{i}\mathbf{k}\cdot(\mathbf{r}_1-\mathbf{r}_2)}}{\mathrm{e}^{\beta(\varepsilon_k-\mu)}-\sigma} \\ &= \frac{N_0}{(2s+1)V} + \delta_{\alpha_1\alpha_2}\int\frac{\mathrm{d}^3k}{(2\pi)^3}\frac{\mathrm{e}^{\mathrm{i}\mathbf{k}\cdot(\mathbf{r}_1-\mathbf{r}_2)}}{\mathrm{e}^{\beta(\varepsilon_k-\mu)}-\sigma},\end{aligned} \tag{5.25}$$

where we have replaced the sum over **k** into an integral as in (4.17) by noting that the point $\mathbf{k} = \mathbf{0}$ measures zero in the integral. Let us choose $\mathbf{r} \equiv \mathbf{r}_1 - \mathbf{r}_2$ along the z axis in (5.25) and express the wave vector in polar coordinates as $\mathbf{k} = (k\sin\theta\cos\varphi, k\sin\theta\sin\varphi, k\cos\theta)$. We subsequently perform the integration over the solid angle,

$$\int_0^{2\pi}\mathrm{d}\varphi\int_0^{\pi}\mathrm{d}\theta\,\sin\theta\,\mathrm{e}^{ikr\cos\theta} = 2\pi\int_{-1}^{1}\mathrm{e}^{ikrt}\mathrm{d}t = 4\pi\frac{\sin kr}{kr}.$$

Next, we express 4π above as an integral over the solid angle, substitute it back into (5.25), and rewrite the resulting expression as a sum over **k**. We thereby obtain

$$\begin{aligned}\rho^{(1)}(\xi_1,\xi_2) &= \frac{N_0}{(2s+1)V} + \frac{\delta_{\alpha_1\alpha_2}}{V}\sum_{\mathbf{k}}{}' \frac{1}{\mathrm{e}^{\beta(\varepsilon_k-\mu)}-\sigma}\frac{\sin kr}{kr} \\ &= \frac{N_0}{(2s+1)V} + \frac{\delta_{\alpha_1\alpha_2}}{(2s+1)V}\int_{-\infty}^{\infty}\mathrm{d}\epsilon\frac{D(\epsilon)}{\mathrm{e}^{\beta(\epsilon-\mu)}-\sigma}\frac{\sin\sqrt{2m\epsilon/\hbar^2}r}{\sqrt{2m\epsilon/\hbar^2}r},\end{aligned}$$

where we have used the first expression of (4.17) in the second equality. Using the last expression of (4.17) and making a change of variables specified in (4.22), we obtain

$$\begin{aligned}\rho^{(1)}(\xi_1,\xi_2) &= \frac{N_0}{(2s+1)V} + \frac{\delta_{\alpha_1\alpha_2}N}{(2s+1)V}\frac{\pi}{4}\int_0^\infty d\tilde{\epsilon}\frac{\tilde{\epsilon}^{1/2}}{e^{(\tilde{\epsilon}-\tilde{\mu})/\tilde{T}}-\sigma}\frac{\sin\left(k_Q r\tilde{\epsilon}^{1/2}\right)}{k_Q r\tilde{\epsilon}^{1/2}} \\ &= \frac{N}{(2s+1)V}\left[\frac{N_0}{N} + \delta_{\alpha_1\alpha_2}\ell(k_Q|\mathbf{r}_1-\mathbf{r}_2|)\right], \qquad (5.26)\end{aligned}$$

where k_Q is given in (4.21), and function $\ell(x)$ is defined by

$$\ell(x) \equiv \frac{\pi}{4}\int_0^\infty \frac{1}{e^{(\tilde{\epsilon}-\tilde{\mu})/\tilde{T}}-\sigma}\frac{\sin(\tilde{\epsilon}^{1/2}x)}{x}d\tilde{\epsilon}. \qquad (5.27)$$

Recalling (4.23) for normal states ($N_0 = 0$) and (4.48) for BEC phases ($N_0 > 0$; $\tilde{\mu} = 0$), we can write limiting behaviors of $\ell(x)$ in a unified way as

$$\ell(x) \longrightarrow \begin{cases} 1 - N_0/N & : x \to 0 \\ 0 & : x \to \infty \end{cases}. \qquad (5.28)$$

Equation (5.27) can be evaluated numerically in the same way as (4.23).

It follows from (5.26) and (5.28) that

$$\rho^{(1)}(\xi,\xi) = \frac{N}{(2s+1)V}, \qquad (5.29)$$

which denotes the particle density per single spin component α. We also conclude that, in BEC phases with $N_0 > 0$, the one-particle density matrix remains finite even for $|\mathbf{r}_1-\mathbf{r}_2| \to \infty$ as

$$\rho^{(1)}(\xi_1,\xi_2) \;\to\; \frac{N_0}{(2s+1)V} \qquad (|\mathbf{r}_1-\mathbf{r}_2| \to \infty). \qquad (5.30)$$

This property characteristic of BEC phases is named *off-diagonal long-range order* [5, 6, 8].

Now, we focus on $\rho^{(2)}$ to clarify two-particle correlations. Let us substitute (5.24), (5.26), and (5.29) into (5.22). We then obtain an expression for the two-particle density matrix

$$\rho^{(2)}(\xi_1,\xi_2;\xi_1,\xi_2) = \left[\frac{N}{(2s+1)V}\right]^2 g_{\alpha_1\alpha_2}(|\mathbf{r}_1-\mathbf{r}_2|), \qquad (5.31)$$

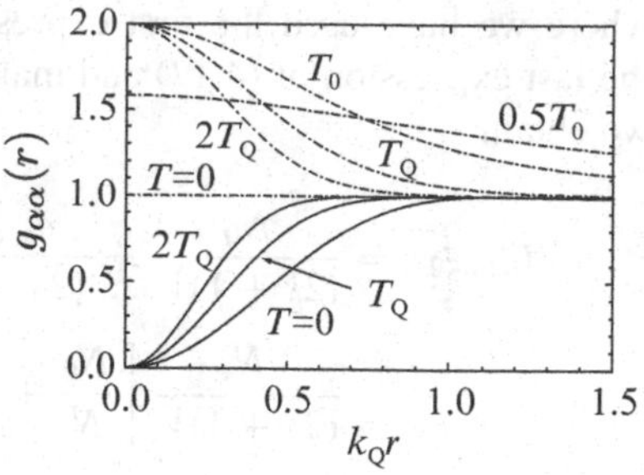

Fig. 5.1 Pair distribution function $g_{\alpha\alpha}(r)$ for fermions (*solid lines*) at $T = 0, T_Q, 2T_Q$, and for bosons (*chain lines*) at $T = 0, 0.5T_0, T_0, T_Q, 2T_Q$, where T_Q and T_0 are given in (4.21) and (4.47), respectively

with

$$g_{\alpha_1\alpha_2}(r) \equiv 1 + \delta_{\alpha_1\alpha_2}\sigma \left\{ [\ell(k_Q r)]^2 + 2\frac{N_0}{N}\ell(k_Q r) \right\}. \tag{5.32}$$

Function $g_{\alpha_1\alpha_2}(r)$ introduced here is called the *pair distribution function*, which represents the probability that one finds a pair of particles simultaneously in states $\xi_1 = \mathbf{r}_1\alpha_1$ and $\xi_2 = \mathbf{r}_2\alpha_2$. When $\alpha_1 \neq \alpha_2$, $g_{\alpha_1\alpha_2}(r) = 1$, independent of r, implying that there are no correlations between a pair of particles with different spin components. For $\alpha_1 = \alpha_2$, the second term of (5.32) is finite; hence, in contrast, we expect finite correlations between pairs of particles.

Figure 5.1 plots the pair distribution function $g_{\alpha\alpha}(r)$ for a pair of particles with the same spin component α obtained by calculating (5.32) numerically. We observe a clear difference between bosons ($\sigma = 1$) and fermions ($\sigma = -1$). For fermions, $g_{\alpha\alpha}(r)$ approaches 0 rapidly for $k_Q r \lesssim 1$, indicating the presence of an effective repulsive force due to the antisymmetry of the corresponding state. Therefore, the probability is zero for a pair of particles in the same spin state to be simultaneously at the same position, in accordance with the Pauli exclusion principle. The region near $r = 0$ where $g_{\alpha\alpha}(r)$ reduces considerably from 1 is called the *exchange hole*. Conversely, for normal bosons, $g_{\alpha\alpha}(r)$ approaches 2 for $r \to 0$, implying an effective attraction between a pair of particles due to the symmetry of the corresponding state. The value of $g_{\alpha\alpha}(0)$ starts to decrease below the BEC transition temperature T_0 and finally reaches 1 at $T = 0$. The latter fact indicates the absence of correlations between any pair of particles at $T = 0$. Hence, each particle loses its individual characteristics following its complete assimilation to the same quantum state in the condensate.

Finally, it is worth mentioning what would result without the finite off-diagonal expectation (5.19) for the BEC phase. If this were the case, the pair distribution function at $T = 0$ would become $g_{\alpha\alpha}(r) = 2$ irrespective of r instead of $g_{\alpha\alpha}(r) = 1$ above, implying a huge fluctuation in the particle number [4]. This suggests that the superposition over the particle number for the condensate is an integrable part of BEC [1].

Problems

5.1. Perform the Wick decomposition for the three-particle density matrix of normal ideal gases given by

$$\rho^{(3)}(\xi_1, \xi_2, \xi_3; \xi_1', \xi_2', \xi_3') \equiv \langle \hat{\psi}^\dagger(\xi_1')\hat{\psi}^\dagger(\xi_2')\hat{\psi}^\dagger(\xi_3')\hat{\psi}(\xi_3)\hat{\psi}(\xi_2)\hat{\psi}(\xi_1)\rangle.$$

5.2. Use (4.32) and (4.34) to show that function (5.27) for ideal Fermi gases ($\sigma = -1$) can be expressed analytically at $T = 0$ in terms of the Fermi wave number $k_\mathrm{F} = (6/\pi)^{1/3} k_\mathrm{Q}$ as

$$\ell(k_\mathrm{Q} r) \xrightarrow{T\to 0} \frac{3\,(-k_\mathrm{F} r \cos k_\mathrm{F} r + \sin k_\mathrm{F} r)}{(k_\mathrm{F} r)^3}.$$

Show also that near $r = 0$ this function behaves as

$$\ell(k_\mathrm{Q} r) \xrightarrow{r\to 0} 1 - \frac{(k_\mathrm{F} r)^2}{10}.$$

References

1. P.W. Anderson, Rev. Mod. Phys. **38**, 298 (1966)
2. C. Bloch, C. De Donimicis, Nucl. Phys. **7**, 459 (1958)
3. M. Gaudin, Nucl. Phys. **15**, 89 (1960)
4. J.R. Johnston, Am. J. Phys. **38**, 516 (1970)
5. O. Penrose, Philos. Mag. **42**, 1373 (1951)
6. O. Penrose, L. Onsager, Phys. Rev. **104**, 576 (1956)
7. G.C. Wick, Phys. Rev. **80**, 268 (1950)
8. C.N. Yang, Rev. Mod. Phys. **34**, 694 (1962)

Chapter 6
Hartree–Fock Equations and Landau's Fermi-Liquid Theory

Abstract In the previous two chapters, we have considered only non-interacting quantum many-particle systems in obtaining exact results for basic thermodynamic quantities and two-particle correlations. However, in real systems, particles interact, making exact statistical-mechanical calculations impossible except for some low-dimensional solvable models. Thus, we are almost always obliged to introduce some approximation when studying interacting systems. Here, we derive the *Hartree–Fock equations*, i.e., one of the simplest approximation schemes for studying interaction effects at finite temperatures, based on a variational principle for the grand potential. They are most effective when interactions are weak and repulsive and are crucial in describing molecular-field effects, but may not be applicable to systems with attractive potentials, as will be seen in later chapters. Next, we apply the Hartree–Fock equations to fermions at low temperatures to clarify how the interaction affects thermodynamic properties along the lines of *Landau's Fermi-liquid theory*.

6.1 Variational Principle in Statistical Mechanics

The thermodynamic equilibrium of a system exchanging heat and particles with a reservoir is given by the grand canonical distribution (2.26), which has been obtained by minimizing functional (2.25). Thus, (2.25) forms a variational principle in statistical mechanics. Here, we express it in terms of the density matrix of (5.1) to make it convenient for subsequent discussions.

First, we transform (2.25) into a functional of density matrix $\hat{\rho}$. Noting $w_\nu \geq 0$ and $\sum_\nu w_\nu = 1$, we may state the definition of $\hat{\rho}$ in (5.1) alternatively as follows: *The density matrix $\hat{\rho}$ is a Hermitian operator that is positive semidefinite (i.e., all eigenvalues are nonnegative) and satisfies*

$$\mathrm{Tr}\,\hat{\rho} = 1. \tag{6.1}$$

Using $\hat{\rho}$, we can express (2.25) as

$$\Omega[\hat{\rho}] \equiv \mathrm{Tr}\,\hat{\rho}\,(\hat{\mathscr{H}} + \beta^{-1}\ln\hat{\rho}), \qquad \hat{\mathscr{H}} \equiv \hat{H} - \mu\hat{\mathscr{N}}. \tag{6.2}$$

© Springer Japan 2015
T. Kita, *Statistical Mechanics of Superconductivity*, Graduate Texts in Physics,
DOI 10.1007/978-4-431-55405-9_6

This may be confirmed by substituting (5.1) and $\ln\hat{\rho} = \sum_\nu |\Phi_\nu\rangle \ln w_\nu \langle\Phi_\nu|$ into (6.2) and then using (3.35), (3.43), and $\hat{\mathscr{N}}|\Phi_\nu\rangle = \mathscr{N}_\nu|\Phi_\nu\rangle$.

The density matrix in equilibrium is given by (5.6) with replacement $\hat{\mathscr{H}}_0 \to \hat{\mathscr{H}}$. Thus, operator

$$\hat{\rho}_{\rm eq} \equiv {\rm e}^{\beta(\Omega_{\rm eq}-\hat{\mathscr{H}})}$$

with $\Omega_{\rm eq} \equiv -\beta^{-1}\ln{\rm Tr}\,{\rm e}^{-\beta\hat{\mathscr{H}}}$ minimizes $\Omega[\hat{\rho}]$, i.e., inequality

$$\Omega[\hat{\rho}] \geq \Omega[\hat{\rho}_{\rm eq}] \tag{6.3}$$

holds. The variational principle (6.3) also enables us to obtain $\hat{\rho}$ approximately. To be specific, one may construct an approximate $\hat{\rho}$ by incorporating variational parameters in it and choosing them to minimize $\Omega[\hat{\rho}]$. The smaller the value, the closer we expect $\hat{\rho}$ is to the real density matrix.

6.2 Hartree–Fock Equations

We consider a system of normal bosons or fermions described by the Hamiltonian[1]:

$$\begin{aligned}\hat{\mathscr{H}} \equiv& \int d\xi_1 \hat{\psi}^\dagger(\xi_1)\left[\frac{\hat{\mathbf{p}}_1^2}{2m} + \mathscr{U}(\mathbf{r}_1) - \mu\right]\hat{\psi}(\xi_1) \\ &+ \frac{1}{2}\int d\xi_1 \int d\xi_2 \mathscr{V}(|\mathbf{r}_1-\mathbf{r}_2|)\hat{\psi}^\dagger(\xi_1)\hat{\psi}^\dagger(\xi_2)\hat{\psi}(\xi_2)\hat{\psi}(\xi_1),\end{aligned} \tag{6.4}$$

which has an additional interaction term in comparison with (5.5) of the previous chapter. We derive the Hartree–Fock equations for this system using two different methods. One is based on the variational principle (6.3), whereas the other relies on a self-consistent Wick decomposition technique. Both will be shown to give an identical set of equations, (6.12) and (6.13), below.

6.2.1 Derivation Based on the Variational Principle

To incorporate interaction effects approximately, we make use of (6.3) and choose $\hat{\rho}$ in an ideal-gas form given by (4.4), (5.5), and (5.6) as

[1]We use symbol $\hat{\mathscr{H}}$ to denote $\hat{\mathscr{H}} \equiv \hat{H} - \mu\hat{\mathscr{N}}$ from now on, which is distinct from those in Chap. 3 like one in (3.39).

$$\hat{\rho} = \exp\left\{\beta\left[\Omega_0 - \sum_q (\varepsilon_q - \mu)\hat{c}_q^\dagger \hat{c}_q\right]\right\}, \qquad \Omega_0 \equiv \frac{\sigma}{\beta}\sum_q \ln\left[1 - \sigma e^{-\beta(\varepsilon_q - \mu)}\right], \tag{6.5}$$

where $\hat{c}_q^\dagger$ and $\hat{c}_q$ are supposed to obey the commutation relation (3.51). A key point here is that ε_q is a variational parameter, to be determined appropriately below, which has the physical meaning of a renormalized one-particle energy with interaction effects. Thus, we no longer use (3.46) for the one-particle eigenstate φ_q and eigenenergy ε_q, but derive new equations for them in such a form as to include interaction effects.

Let us substitute (6.4) and (6.5) into (6.2) and evaluate the grand potential by noting the following: (i) The Bloch-De Dominicis theorem holds true for the variational density matrix so that the interaction term in (6.4) can be evaluated with the Wick decomposition of (5.18); (ii) Entropy $S \equiv -k_B \mathrm{Tr}\, \hat{\rho} \ln \hat{\rho}$ for (6.5) can be expressed in terms of the mean occupation number:

$$\bar{n}_q \equiv \langle \hat{c}_q^\dagger \hat{c}_q \rangle = \frac{1}{e^{\beta(\varepsilon_q - \mu)} - \sigma}, \tag{6.6}$$

as

$$\begin{aligned} S &= -k_B \beta \left[\Omega_0 - \sum_q (\varepsilon_q - \mu)\bar{n}_q\right] \\ &= k_B \sum_q \left[-\bar{n}_q \ln \bar{n}_q + \sigma(1 + \sigma\bar{n}_q)\ln(1 + \sigma\bar{n}_q)\right]. \end{aligned} \tag{6.7}$$

Thus, it is identical in form to (4.7) for ideal gases. Using (i) and (ii) above, we can express $\Omega[\hat{\rho}]$ as

$$\begin{aligned} \Omega[\hat{\rho}] = &\int d\xi_1 \left[\frac{\hat{\mathbf{p}}_1^2}{2m} + \mathscr{U}(\mathbf{r}_1) - \mu\right] \rho^{(1)}(\xi_1, \xi_2)\Big|_{\xi_2 = \xi_1} + \frac{1}{2}\int d\xi_1 \int d\xi_2 \\ &\times \mathscr{V}(|\mathbf{r}_1 - \mathbf{r}_2|)\left[\rho^{(1)}(\xi_1, \xi_1)\rho^{(1)}(\xi_2, \xi_2) + \sigma\rho^{(1)}(\xi_2, \xi_1)\rho^{(1)}(\xi_1, \xi_2)\right] \\ &- \frac{1}{\beta}\sum_q \left[-\bar{n}_q \ln \bar{n}_q + \sigma(1 + \sigma\bar{n}_q)\ln(1 + \sigma\bar{n}_q)\right]. \end{aligned} \tag{6.8}$$

The one-particle density matrix $\rho^{(1)}(\xi_1, \xi_2) \equiv \langle \hat{\psi}^\dagger(\xi_2)\hat{\psi}(\xi_1)\rangle$ is transformed by expanding the field operators formally as (3.49) and using $\langle \hat{c}_{q'}^\dagger \hat{c}_q \rangle = \delta_{q'q}\bar{n}_q$ into

$$\rho^{(1)}(\xi_1, \xi_2) = \sum_q \varphi_q(\xi_1)\varphi_q^*(\xi_2)\bar{n}_q. \tag{6.9}$$

In addition, (6.2) tells us that we obtain the expectation $\langle \hat{\mathscr{N}} \rangle \equiv N$ for the particle number from $N = -\partial\Omega[\hat{\rho}]/\partial\mu$. This differentiation can be performed easily for (6.8),

$$N = \int \rho^{(1)}(\xi_1, \xi_1)\, d\xi_1 = \sum_q \bar{n}_q. \tag{6.10}$$

Next, we minimize $\Omega[\hat{\rho}]$ in terms of variational parameters $\{\varepsilon_q\}$ in (6.5). A necessary condition for this is that (6.8) is stationary with respect to ε_q [4]. Noting (6.6) and (6.9), however, we realize that (6.8) depends on ε_q only through $\bar{n}_q$. Thus, condition $\delta\Omega[\hat{\rho}]/\delta\varepsilon_q = 0$ is equivalent to $\delta\Omega[\hat{\rho}]/\delta\bar{n}_q = 0$, which is expanded to give

$$\begin{aligned}
0 &= \frac{\delta\Omega[\hat{\rho}]}{\delta\bar{n}_q} \\
&= \int d\xi_1 \varphi_q^*(\xi_1)\left[\frac{\hat{\mathbf{p}}_1^2}{2m} + \mathscr{U}(\mathbf{r}_1) - \mu\right]\varphi_q(\xi_1) + 2 \times \frac{1}{2}\int d\xi_1 \int d\xi_2 \mathscr{V}(|\mathbf{r}_1 - \mathbf{r}_2|) \\
&\quad \times \left[\varphi_q^*(\xi_1)\varphi_q(\xi_1)\rho^{(1)}(\xi_2, \xi_2) + \sigma\varphi_q^*(\xi_1)\varphi_q(\xi_2)\rho^{(1)}(\xi_1, \xi_2)\right] - \frac{1}{\beta}\ln\frac{1+\sigma\bar{n}_q}{\bar{n}_q} \\
&= \int d\xi_1 \varphi_q^*(\xi_1)\left\{\left[\frac{\hat{\mathbf{p}}_1^2}{2m} + \mathscr{U}(\mathbf{r}_1)\right]\varphi_q(\xi_1) + \int d\xi_2\, \mathscr{U}_{\rm HF}(\xi_1, \xi_2)\varphi_q(\xi_2)\right\} \\
&\quad - \varepsilon_q - \mu\left[\int d\xi_1 |\varphi_q(\xi_1)|^2 - 1\right],
\end{aligned} \tag{6.11}$$

where $\mathscr{U}_{\rm HF}(\xi_1, \xi_2)$ denotes the *Hartree–Fock potential*:

$$\mathscr{U}_{\rm HF}(\xi_1, \xi_2) \equiv \delta(\xi_1, \xi_2)\int d\xi_3 \mathscr{V}(|\mathbf{r}_1 - \mathbf{r}_3|)\rho^{(1)}(\xi_3, \xi_3) + \sigma\mathscr{V}(|\mathbf{r}_1 - \mathbf{r}_2|)\rho^{(1)}(\xi_1, \xi_2). \tag{6.12}$$

The last equality in (6.11) may be confirmed by substituting (6.12) and $\varepsilon_q - \mu = \beta^{-1}\ln[(1+\sigma\bar{n}_q)/\bar{n}_q]$ into the final expression. Condition (6.11) can be satisfied by solving the eigenvalue problem:

$$\left[\frac{\hat{\mathbf{p}}_1^2}{2m} + \mathscr{U}(\mathbf{r}_1)\right]\varphi_q(\xi_1) + \int d\xi_2\, \mathscr{U}_{\rm HF}(\xi_1, \xi_2)\varphi_q(\xi_2) = \varepsilon_q\varphi_q(\xi_1). \tag{6.13}$$

This is seen as follows. First, symmetry $\mathscr{U}_{\rm HF}^*(\xi_1, \xi_2) = \mathscr{U}_{\rm HF}(\xi_2, \xi_1)$ of (6.12) tells us that (6.13) is a Hermitian eigenvalue problem. Hence, ε_q is real and $\{\varphi_q\}_q$ can be constructed to form a complete orthonormal set that satisfies (3.47) and (3.48). Next, we can use (6.13) and (3.47) to show that the last expression of (6.11) is certainly zero.

Equations (6.12) and (6.13) are called the *Hartree–Fock equations*, which form a closed set of self-consistency equations. To be specific, (6.13) can be interpreted as an eigenvalue equation for ε_q and $\varphi_q(x)$ for a given $\mathscr{U}_{\rm HF}(\xi_1, \xi_2)$, However, as seen from (6.9) and (6.12), $\mathscr{U}_{\rm HF}(\xi_1, \xi_2)$ itself includes ε_q and $\varphi_q(x)$ that are to be determined. Hence, (6.12) and (6.13) form a set of self-consistency (i.e., nonlinear) equations for ε_q and $\varphi_q(x)$. In this circumstance, we need to make an appropriate guess about the solutions to start solving the nonlinear equations, as they may have multiple solutions including unphysical ones. For the Hartree–Fock equations, one may initially adopt the eigenvalues and eigenfunctions for $\mathscr{U}_{\rm HF} = 0$ and improve on them iteratively using (6.12) and (6.13) until a convergence is reached.

Let us denote the minimum of (6.8) by $\Omega_{\rm HF}$. Its concise expression is obtained from (6.8) using (6.7), (6.9), (6.12) and (6.13) and finally substituting Ω_0 of (6.5). This gives

$$\begin{aligned}\Omega_{\rm HF} &= \sum_q (\varepsilon_q - \mu)\bar{n}_q - \frac{1}{2}\int d\xi_1 \int d\xi_2 \mathscr{U}_{\rm HF}(\xi_1, \xi_2)\rho^{(1)}(\xi_2, \xi_1) + \Omega_0 - \sum_q (\varepsilon_q - \mu)\bar{n}_q \\ &= \frac{\sigma}{\beta}\sum_q \ln[1 - \sigma e^{-\beta(\varepsilon_q - \mu)}] - \frac{1}{2}\int d\xi_1 \int d\xi_2\, \mathscr{U}_{\rm HF}(\xi_1, \xi_2)\rho^{(1)}(\xi_2, \xi_1). \qquad (6.14)\end{aligned}$$

The second term in the last expression is an additional term from the interaction, which removes the double counting of the interaction energy in ε_q obtained by (6.12) and (6.13).

6.2.2 Derivation Based on Wick Decomposition

There is an alternative concise method for deriving the Hartree–Fock equations, which proceeds as follows: (i) Consider all the possible Wick decompositions for the interaction in (6.4); (ii) Express each pair for the decomposition as a sum of their average and the deviation from it; and (iii) Neglect terms that are second order in the deviations. Following this procedure and using abbreviations $\hat{\psi}(\xi_i) \to \hat{\psi}_i$, the product of four field operators in the interaction transforms to

$$\begin{aligned}&\hat{\psi}_1^\dagger \hat{\psi}_2^\dagger \hat{\psi}_2 \hat{\psi}_1 \\ &\to \big(\langle \hat{\psi}_1^\dagger \hat{\psi}_1\rangle + \hat{\psi}_1^\dagger \hat{\psi}_1 - \langle \hat{\psi}_1^\dagger \hat{\psi}_1\rangle\big)\big(\langle \hat{\psi}_2^\dagger \hat{\psi}_2\rangle + \hat{\psi}_2^\dagger \hat{\psi}_2 - \langle \hat{\psi}_2^\dagger \hat{\psi}_2\rangle\big) \\ &\quad + \sigma\big(\langle \hat{\psi}_1^\dagger \hat{\psi}_2\rangle + \hat{\psi}_1^\dagger \hat{\psi}_2 - \langle \hat{\psi}_1^\dagger \hat{\psi}_2\rangle\big)\big(\langle \hat{\psi}_2^\dagger \hat{\psi}_1\rangle + \hat{\psi}_2^\dagger \hat{\psi}_1 - \langle \hat{\psi}_2^\dagger \hat{\psi}_1\rangle\big) \\ &\approx \langle \hat{\psi}_1^\dagger \hat{\psi}_1\rangle\langle \hat{\psi}_2^\dagger \hat{\psi}_2\rangle + \big(\hat{\psi}_1^\dagger \hat{\psi}_1 - \langle \hat{\psi}_1^\dagger \hat{\psi}_1\rangle\big)\langle \hat{\psi}_2^\dagger \hat{\psi}_2\rangle + \langle \hat{\psi}_1^\dagger \hat{\psi}_1\rangle\big(\hat{\psi}_2^\dagger \hat{\psi}_2 - \langle \hat{\psi}_2^\dagger \hat{\psi}_2\rangle\big) \\ &\quad + \sigma\big[\langle \hat{\psi}_1^\dagger \hat{\psi}_2\rangle\langle \hat{\psi}_2^\dagger \hat{\psi}_1\rangle + \big(\hat{\psi}_1^\dagger \hat{\psi}_2 - \langle \hat{\psi}_1^\dagger \hat{\psi}_2\rangle\big)\langle \hat{\psi}_2^\dagger \hat{\psi}_1\rangle \\ &\quad + \langle \hat{\psi}_1^\dagger \hat{\psi}_2\rangle\big(\hat{\psi}_2^\dagger \hat{\psi}_1 - \langle \hat{\psi}_2^\dagger \hat{\psi}_1\rangle\big)\big]\end{aligned}$$

$$= \hat{\psi}_1^\dagger\hat{\psi}_1\langle\hat{\psi}_2^\dagger\hat{\psi}_2\rangle + \langle\hat{\psi}_1^\dagger\hat{\psi}_1\rangle\hat{\psi}_2^\dagger\hat{\psi}_2 + \sigma\left[\hat{\psi}_1^\dagger\hat{\psi}_2\langle\hat{\psi}_2^\dagger\hat{\psi}_1\rangle + \langle\hat{\psi}_1^\dagger\hat{\psi}_2\rangle\hat{\psi}_2^\dagger\hat{\psi}_1\right]$$
$$- \langle\hat{\psi}_1^\dagger\hat{\psi}_1\rangle\langle\hat{\psi}_2^\dagger\hat{\psi}_2\rangle - \sigma\langle\hat{\psi}_1^\dagger\hat{\psi}_2\rangle\langle\hat{\psi}_2^\dagger\hat{\psi}_1\rangle. \tag{6.15}$$

The Hamiltonian (6.4) is thereby replaced by an expression given in terms of (6.12) as

$$\hat{\mathscr{H}}_{\rm HF} \equiv \int d\xi_1 \hat{\psi}^\dagger(\xi_1)\left\{\left[\frac{\hat{\mathbf{p}}_1^2}{2m} + \mathscr{U}(\mathbf{r}_1) - \mu\right]\hat{\psi}(\xi_1) + \int d\xi_2 \mathscr{U}_{\rm HF}(\xi_1,\xi_2)\hat{\psi}(\xi_2)\right\}$$
$$- \frac{1}{2}\int d\xi_1 \int d\xi_2\, \mathscr{U}_{\rm HF}(\xi_1,\xi_2)\rho^{(1)}(\xi_2,\xi_1). \tag{6.16}$$

The second term on the right-hand side is a constant, whereas the first term has a quadratic form with respect to $\hat{\psi}^\dagger$ and $\hat{\psi}$. Expanding them formally as (3.49) and determining eigenstate $\varphi_q(\xi)$ by (6.13), we can put the first term into a diagonal form,

$$\hat{\mathscr{H}}_{\rm HF} = \sum_q (\varepsilon_q - \mu)\hat{c}_q^\dagger\hat{c}_q - \frac{1}{2}\int d\xi_1 \int d\xi_2\, \mathscr{U}_{\rm HF}(\xi_1,\xi_2)\rho^{(1)}(\xi_2,\xi_1).$$

Let us use this expression and (6.7) to estimate $\Omega_{\rm HF} \equiv \langle\hat{\mathscr{H}}_{\rm HF}\rangle - TS$. We thereby reproduce (6.14) for the grand potential. Thus, the Hartree–Fock formalism has been derived also by the Wick decomposition procedure.

6.2.3 *Homogeneous Cases*

We apply the Hartree–Fock formalism to homogeneous systems to derive a simplified equation. This helps in understanding clearly how the one-particle energy is determined self-consistently while accounting for interaction effects. Linearization of the equation with respect to external perturbations gives the basic equation of Landau's Fermi-liquid theory with molecular-field effects, as will be seen shortly.

With no external potential ($\mathscr{U} = 0$), the eigenfunctions of (6.12) and (6.13) are plane waves given by (5.23). This is shown self-consistently as follows. Let us assume that (5.23) is indeed an eigenfunction and substitute it into (6.9) with $q \to \mathbf{k}\alpha$. We then obtain

$$\rho^{(1)}(\xi_1,\xi_2) = \frac{\delta_{\alpha_1\alpha_2}}{V}\sum_{\mathbf{k}} e^{i\mathbf{k}\cdot(\mathbf{r}_1-\mathbf{r}_2)}\bar{n}_{\mathbf{k}\alpha_1}. \tag{6.17}$$

Accordingly, we expand the delta function and the interaction potential of (6.12) as

$$\delta(\xi_1,\xi_2) = \frac{\delta_{\alpha_1\alpha_2}}{V}\sum_{\mathbf{k}} e^{i\mathbf{k}\cdot(\mathbf{r}_1-\mathbf{r}_2)}, \qquad \mathscr{V}(|\mathbf{r}_1-\mathbf{r}_2|) = \frac{1}{V}\sum_{\mathbf{k}} \mathscr{V}_k\, e^{i\mathbf{k}\cdot(\mathbf{r}_1-\mathbf{r}_2)}. \tag{6.18}$$

Using (6.17) and (6.18), we can express the Hartree–Fock potential (6.12) in the form

$$\mathscr{U}_{\text{HF}}(\xi_1, \xi_2) = \frac{\delta_{\alpha_1\alpha_2}}{V}\sum_{\mathbf{k}} \mathrm{e}^{\mathrm{i}\mathbf{k}\cdot(\mathbf{r}_1-\mathbf{r}_2)}\frac{1}{V}\sum_{\mathbf{k}'\alpha'}\left(\mathscr{V}_0 + \sigma\delta_{\alpha'\alpha_1}\mathscr{V}_{|\mathbf{k}-\mathbf{k}'|}\right)\bar{n}_{\mathbf{k}'\alpha'}. \tag{6.19}$$

Substitution of (6.19) into (6.13) with $\mathscr{U} = 0$ confirms that (5.23) is certainly an eigenfunction of the Hartree–Fock equations with eigenvalue

$$\varepsilon_{\mathbf{k}\alpha} = \varepsilon_k^0 + \frac{1}{V}\sum_{\mathbf{k}'\alpha'}(\mathscr{V}_0 + \sigma\delta_{\alpha\alpha'}\mathscr{V}_{|\mathbf{k}-\mathbf{k}'|})\bar{n}_{\mathbf{k}'\alpha'}, \tag{6.20}$$

where

$$\varepsilon_k^0 \equiv \frac{\hbar^2 k^2}{2m} \tag{6.21}$$

denotes the one-particle energy assuming no interaction. Note that $\bar{n}_{\mathbf{k}'\alpha'}$ on the right-hand side of (6.20) is a function of $\varepsilon_{\mathbf{k}'\alpha'}$. Hence, (6.20) forms a set of self-consistency (or nonlinear) equations for $\{\varepsilon_{\mathbf{k}\alpha}\}$.

Equation (6.20) also determines changes of $\varepsilon_{\mathbf{k}\alpha}$ when external perturbations are applied. The corresponding first-order variations $\{\delta\varepsilon_{\mathbf{k}\alpha}\}$ obey coupled equations obtained by linearizing (6.20),

$$\delta\varepsilon_{\mathbf{k}\alpha} = \delta\varepsilon_{\mathbf{k}\alpha}^0 + \frac{1}{V}\sum_{\mathbf{k}'\alpha'} f_{\alpha\alpha'}(\mathbf{k}, \mathbf{k}')\frac{\partial\bar{n}_{k'}}{\partial\varepsilon_{k'}}\delta\varepsilon_{\mathbf{k}'\alpha'}. \tag{6.22}$$

Here, we have replaced the subscript of ε^0, $k \to \mathbf{k}\alpha$, by considering the possibility of some $\mathbf{k}\alpha$ dependence because of perturbations; function $f_{\alpha\alpha'}(\mathbf{k}, \mathbf{k}')$ is defined by

$$f_{\alpha\alpha'}(\mathbf{k}, \mathbf{k}') \equiv \mathscr{V}_0 + \sigma\delta_{\alpha\alpha'}\mathscr{V}_{|\mathbf{k}-\mathbf{k}'|}. \tag{6.23}$$

Further, we have replaced the derivative of the mean occupation number by an isotropic term without perturbations as justified in the first-order approximation. The second term on the right-hand side of (6.22) represents molecular-field (or mean-field) effects. It indicates that a perturbation to the system causes a change in the molecular field originating from the interaction, which then produces a feedback effect on the one-particle energy.

Equation (6.22) for low-temperature identical fermions ($\sigma = -1$) forms the starting point of Landau's Fermi-liquid theory [5], and function (6.23) may be regarded as the Hartree–Fock approximation to the Landau f function. This Landau theory describes how low-temperature fermions respond to external perturbations; details will be given below.

6.3 Application to Low-Temperature Fermions

Among systems of identical fermions are those with spin 1/2 particles, which include typical targets of statistical mechanics such as electrons in metals and liquid ^{3}He. Thus, we apply (6.22) to low-temperature fermions ($\sigma = -1$) with spin 1/2 to clarify how the interaction affects their thermodynamic properties.

6.3.1 *Fermi Wave Number and Fermi Energy*

First, we focus on the zero-temperature case in the absence of a magnetic field to clarify how the Fermi wave number and Fermi energy are affected by the interaction.

As the system under consideration is isotropic, the energies $\varepsilon_{\mathbf{k}\alpha}$ for the zero magnetic field will depend only on the wave number k. Thus, we may set $\varepsilon_{\mathbf{k}\alpha} \to \varepsilon_k$ in (6.20). To determine the chemical potential $\mu(0) \equiv \varepsilon_{\mathrm{F}}$, we substitute (6.6) into (6.10), set $q \to \mathbf{k}\alpha$ and $T \to 0$, and use (4.32). The resulting equation is given by

$$N = \sum_{\mathbf{k}\alpha} \theta(\varepsilon_{\mathrm{F}} - \varepsilon_k) = \sum_{\mathbf{k}\alpha} \theta(k_{\mathrm{F}} - k), \tag{6.24}$$

where we have transformed the condition for ε_k into that for k. The latter expression is identical to that for the non-interacting case in terms of the wave vector (4.16). Hence, we conclude that the Fermi wave number k_{F} is still given by (4.35) with $s = 1/2$, that is,

$$k_{\mathrm{F}} = \left(\frac{3\pi^2 N}{V}\right)^{1/3}. \tag{6.25}$$

Thus, k_{F} does not depend on the interaction. This is because the interaction does not change the density of particles confined in a fixed volume. The theorem, which is known as the *Fermi-surface sum rule*, also holds true for electrons in anisotropic metals where the volume enclosed by the surface $\varepsilon_{\mathbf{k}} = \varepsilon_{\mathrm{F}}$ remains invariant as a function of the interaction strength [6]. In contrast, the first expression of (6.24), given in terms of ε_k, is affected by the interaction. Hence, the value of the Fermi energy ε_{F} changes because of the interaction.

6.3.2 *Effective Mass, Density of States, and Heat Capacity*

We now introduce the concept of effective mass m^*, which will be shown to describe how the density of states at the Fermi energy and low-temperature heat capacity are modified by the interaction.

The one-particle energy ε_k of isotropic systems depends only on $k = |\mathbf{k}|$ and may be expanded for $k \sim k_\mathrm{F}$ as

$$\varepsilon_k \approx \varepsilon_\mathrm{F} + \left.\frac{\mathrm{d}\varepsilon_k}{\mathrm{d}k}\right|_{k=k_\mathrm{F}}(k-k_\mathrm{F}) = \varepsilon_\mathrm{F} + \frac{\hbar^2 k_\mathrm{F}}{m^*}(k-k_\mathrm{F}), \tag{6.26}$$

where m^* denotes the *effective mass* defined by

$$\frac{1}{m^*} \equiv \frac{1}{\hbar^2 k_\mathrm{F}}\left.\frac{\mathrm{d}\varepsilon_k}{\mathrm{d}k}\right|_{k=k_\mathrm{F}}. \tag{6.27}$$

It follows from (6.21) that m^* is identical to m for ideal gases.

The effective mass manifests itself in the single-particle density of states defined by

$$D(\epsilon) \equiv \sum_{\mathbf{k}\alpha} \delta(\epsilon - \varepsilon_k). \tag{6.28}$$

Indeed, $D(\epsilon)$ at the Fermi energy $\epsilon = \varepsilon_\mathrm{F}$ can be transformed in the same way as (4.17) for $s = 1/2$ as

$$\begin{aligned} D(\varepsilon_\mathrm{F}) &= \frac{2V}{(2\pi)^3}4\pi\int_0^\infty \mathrm{d}k k^2 \delta(\varepsilon_\mathrm{F}-\varepsilon_k) = \frac{V}{\pi^2}\int_{\varepsilon_0}^\infty \frac{\mathrm{d}\varepsilon_k}{\mathrm{d}\varepsilon_k/\mathrm{d}k} k^2 \delta(\varepsilon_\mathrm{F}-\varepsilon_k) \\ &= \frac{V}{\pi^2}\frac{k_\mathrm{F}^2}{\mathrm{d}\varepsilon_k/\mathrm{d}k\big|_{k=k_\mathrm{F}}} = \frac{V k_\mathrm{F} m^*}{\pi^2\hbar^2}, \end{aligned} \tag{6.29}$$

where we have used (6.27) in the last equality. Noting that k_F does not depend on the interaction, we realize that the density of states at the Fermi energy is modified by factor m^*/m because of the interaction.

This change in the density of states is observable in the low-temperature heat capacity. To see this, we start from (6.7) for entropy in the Hartree–Fock approximation. It has the same expression as (4.7) for ideal gases with the difference being only in the one-particle energy in $\bar{n}_q$. Hence, the heat capacity $C = T(\partial S/\partial T)$ in the Hartree–Fock approximation is identical in form to (4.8) for ideal gases; in terms of the density of states (6.28), we have

$$C = k_\mathrm{B}\int_{-\infty}^{\infty} D(\epsilon)\left(x + \frac{1}{k_\mathrm{B}}\frac{\partial\mu}{\partial T}\right)\left.\frac{x\mathrm{e}^x}{(\mathrm{e}^x+1)^2}\right|_{x=\beta(\epsilon-\mu)} \mathrm{d}\epsilon. \tag{6.30}$$

In particular, near $T = 0$, we may approximate $D(\epsilon) \approx D(\varepsilon_\mathrm{F})$ and make a change of variables, $x \equiv \beta(\epsilon - \varepsilon_\mathrm{F})$. Consequently, the term with $\partial\mu/\partial T$ in the integrand becomes odd in x to yield a null contribution. Let us transform the remaining integral over $-\infty < x < \infty$ into that over $0 \le x < \infty$ with a change of variables for $x < 0$ and perform an integration by parts to evaluate it with (4.39). We thereby obtain

$$C \approx 2D(\varepsilon_F)k_B^2 T \int_0^\infty \frac{x^2 e^x}{(e^x+1)^2}dx = 2D(\varepsilon_F)k_B^2 T\left(-\frac{x^2}{e^x+1}\bigg|_0^\infty + 2\int_0^\infty \frac{x}{e^x+1}dx\right)$$
$$= \frac{\pi^2}{3}D(\varepsilon_F)k_B^2 T. \tag{6.31}$$

The last result is identical in form to (4.44) for ideal gases, but $D(\varepsilon_F)$ is modified by factor m^*/m.

6.3.3 *Effective Mass and Landau Parameter*

In the present homogeneous case, because the mass change $m \to m^*$ is caused solely by the interaction, it is reasonable to expect that ratio m^*/m is expressible by the Landau f function on the Fermi surface. We show that this is indeed the case and write m^*/m in terms of the *Landau parameters*.

First, let us adopt a coordinate system that moves with velocity $\mathbf{u}$ ($u \ll v_F$) relative to the original system, where $v_F \equiv \hbar k_F/m^*$ denotes the Fermi velocity. The Hamiltonian in the new coordinate system is given by (6.4) with $\hat{\mathbf{p}}_1 \to \hat{\mathbf{p}}_1 - m\mathbf{u}$ and $\mathscr{U} = 0$. Thus, it is only the kinetic-energy operator that is affected by the change in coordinate system. The corresponding Hartree–Fock equation for the homogeneous system is obtained from (6.20) by replacing every wave vector with $\mathbf{k} \to \mathbf{k} - m\mathbf{u}/\hbar$. Thus, the wave vector is modified by

$$\delta\mathbf{k} = -m\mathbf{u}/\hbar$$

from the change in coordinate system. Accordingly, the one-particle energy near the Fermi surface is shifted to first order in $\delta\mathbf{k}$ by

$$\delta\varepsilon_{\mathbf{k}} = \frac{\partial\varepsilon_k}{\partial\mathbf{k}}\cdot\delta\mathbf{k} = \frac{d\varepsilon_k}{dk}\frac{\partial k}{\partial\mathbf{k}}\cdot\delta\mathbf{k} \approx \frac{\hbar^2 k_F}{m^*}\frac{\mathbf{k}}{k}\cdot\left(-\frac{m\mathbf{u}}{\hbar}\right) \approx -\frac{m}{m^*}\hbar\mathbf{k}\cdot\mathbf{u},$$

where we have used (6.26), $k = (k_x^2 + k_y^2 + k_z^2)^{1/2}$, and $k \approx k_F$. Similarly, (6.21) is also changed by $\delta\varepsilon_{\mathbf{k}}^0 = -\hbar\mathbf{k}\cdot\mathbf{u}$. In contrast, function $f_{\alpha\alpha'}(\mathbf{k},\mathbf{k}')$ given by (6.23) remains invariant as it depends only on $\mathbf{k} - \mathbf{k}'$. Let us substitute these expressions into (6.22) with $\mathbf{k} \to \mathbf{k}_F$ and note that $\mathbf{u}$ can be chosen arbitrarily. We thereby obtain

$$\mathbf{k}_F\frac{m}{m^*} = \mathbf{k}_F + \frac{1}{V}\sum_{\mathbf{k}'\alpha'} f_{\alpha\alpha'}(\mathbf{k}_F,\mathbf{k}')\frac{\partial\bar{n}_{k'}}{\partial\varepsilon_{k'}}\mathbf{k}'\frac{m}{m^*}.$$

Further, we form the scalar product of the equation with $\mathbf{k}_\mathrm{F}/k_\mathrm{F}^2$, set $T \to 0$, and use (4.33). This yields

$$\frac{m}{m^*} = 1 - \frac{1}{V}\sum_{\mathbf{k}'\alpha'} f_{\alpha\alpha'}(\mathbf{k}_\mathrm{F},\mathbf{k}')\delta(\varepsilon_{k'}-\varepsilon_\mathrm{F})\frac{\mathbf{k}_\mathrm{F}\cdot\mathbf{k}'}{k_\mathrm{F}^2}\frac{m}{m^*}.$$

Subsequently, we transform the sum over $\mathbf{k}'$ into an integral as (4.17), adopt polar coordinates $\mathbf{k}' = (k'\sin\theta'\cos\varphi', k'\sin\theta'\sin\varphi', k'\cos\theta')$, and express the angular integrations as

$$\int_0^\pi \mathrm{d}\theta'\sin\theta'\int_0^{2\pi}\mathrm{d}\varphi' \equiv \int \mathrm{d}\Omega', \tag{6.32}$$

to obtain

$$\begin{aligned}\frac{m}{m^*} &= 1 - \sum_{\alpha'}\int_0^\infty \frac{\mathrm{d}k'}{(2\pi)^3}k'^2\delta(\varepsilon_{k'}-\varepsilon_\mathrm{F})\int \mathrm{d}\Omega' f_{\alpha\alpha'}(\mathbf{k}_\mathrm{F},\mathbf{k}'_\mathrm{F})\frac{\mathbf{k}_\mathrm{F}\cdot\mathbf{k}'_\mathrm{F}}{k_\mathrm{F}^2}\frac{m}{m^*}\\ &= 1 - \frac{D(\varepsilon_\mathrm{F})}{2V}\sum_{\alpha'}\int\frac{\mathrm{d}\Omega'}{4\pi}f_{\alpha\alpha'}(\mathbf{k}_\mathrm{F},\mathbf{k}'_\mathrm{F})\frac{\mathbf{k}_\mathrm{F}\cdot\mathbf{k}'_\mathrm{F}}{k_\mathrm{F}^2}\frac{m}{m^*},\end{aligned} \tag{6.33}$$

where $D(\varepsilon_\mathrm{F})$ is the density of states at the Fermi energy given by (6.29).

It follows from (6.23) that function $f_{\alpha\alpha'}(\mathbf{k}_\mathrm{F},\mathbf{k}'_\mathrm{F})$ on the Fermi surface depends only on the scalar product $\mathbf{k}_\mathrm{F}\cdot\mathbf{k}'_\mathrm{F}/k_\mathrm{F}^2 \equiv \cos\theta'$. Let us multiply $f_{\alpha\alpha'}(\mathbf{k}_\mathrm{F},\mathbf{k}'_\mathrm{F})$ by $D(\varepsilon_\mathrm{F})/2V$ and expand it in terms of Legendre polynomials $\{P_\ell(\cos\theta')\}$,

$$\frac{D(\varepsilon_\mathrm{F})}{2V}f_{\alpha\alpha'}(\mathbf{k}_\mathrm{F},\mathbf{k}'_\mathrm{F}) = \sum_{\ell=0}^\infty\left(\frac{F_\ell^\mathrm{s}-F_\ell^\mathrm{a}}{2}+\delta_{\alpha\alpha'}F_\ell^\mathrm{a}\right)P_\ell(\cos\theta'), \tag{6.34}$$

where F_ℓ^s and F_ℓ^a are dimensionless parameters called *Landau parameters*. Specifically, $P_\ell(x)$ is defined by [1, 2]

$$P_\ell(x) \equiv \frac{1}{2^\ell\ell!}\frac{\mathrm{d}^\ell}{\mathrm{d}x^\ell}(x^2-1)^\ell, \tag{6.35}$$

and satisfies

$$\int_{-1}^1 P_\ell(x)P_{\ell'}(x)\mathrm{d}x = \frac{2}{2\ell+1}\delta_{\ell\ell'}. \tag{6.36}$$

The first few low-order polynomials are $P_0(x) = 1$, $P_1(x) = x$, $P_2(x) = \frac{1}{2}(3x^2-1)$. Let us substitute (6.34) into (6.33), write $\mathbf{k}_\mathrm{F}\cdot\mathbf{k}'_\mathrm{F}/k_\mathrm{F}^2 = P_1(\cos\theta')$, and use (6.36). Equation (6.33) thereby becomes

$$
\begin{aligned}
\frac{m}{m^*} &= 1 - \sum_{\ell=0}^{\infty} \sum_{\alpha'=\pm\frac{1}{2}} \left(\frac{F_\ell^{\rm s} - F_\ell^{\rm a}}{2} + \delta_{\alpha\alpha'} F_\ell^{\rm a} \right) \int \frac{{\rm d}\Omega'}{4\pi} P_\ell(\cos\theta') P_1(\cos\theta') \frac{m}{m^*} \\
&= 1 - \sum_{\ell=0}^{\infty} F_\ell^{\rm s} \int_{-1}^{1} \frac{{\rm d}x'}{2} P_\ell(x') P_1(x') \frac{m}{m^*} = 1 - \frac{F_1^{\rm s}}{3} \frac{m}{m^*}.
\end{aligned}
$$

That is, ratio m^*/m is given solely in terms of $F_1^{\rm s}$ as

$$
\frac{m^*}{m} = 1 + \frac{F_1^{\rm s}}{3}. \tag{6.37}
$$

6.3.4 Spin Susceptibility

Next, we study the spin susceptibility χ at $T = 0$. We shall see that it is also modified from the ideal-gas value by a couple of factors, which are expressible in terms of Landau parameters.

Suppose that there is a weak magnetic flux density of magnitude B along the z axis. In this situation, the non-interacting one-particle energy is shifted because of the spin degrees of freedom by

$$
\delta\varepsilon_{\mathbf{k}\alpha}^0 = -\mu_{\rm m}^0 \alpha B. \tag{6.38}
$$

This energy splitting between $\alpha = \pm 1/2$ is the *Zeeman effect*. The quantity $\mu_{\rm m}^0$ denotes the magnetic moment whose magnitude varies from particle to particle; for electrons, for example, it is given in terms of the *Bohr magneton*

$$
\mu_{\rm B} \equiv \frac{|e|\hbar}{2m} = 9.27 \times 10^{-24}\,{\rm J \cdot T^{-1}} \tag{6.39}
$$

as $\mu_{\rm m}^0 = -2\mu_{\rm B}$. We also expect that the one-particle energy with interactions is expressible as

$$
\delta\varepsilon_{\mathbf{k}\alpha} = -\mu_{\rm m} \alpha B, \tag{6.40}
$$

where $\mu_{\rm m}$ is an unknown constant having the physical meaning of an effective magnetic moment.

Let us substitute (6.38) and (6.40) into (6.22), set $\mathbf{k} = \mathbf{k}_{\rm F}$, divide both sides by $-B$, and take the limit $T \to 0$. We thereby obtain

$$
\mu_{\rm m}\alpha = \mu_{\rm m}^0 \alpha - \frac{1}{V} \sum_{\mathbf{k}'\alpha'} f_{\alpha\alpha'}(\mathbf{k}_{\rm F}, \mathbf{k}') \delta(\varepsilon_{\rm F} - \varepsilon_{k'}) \mu_{\rm m} \alpha'.
$$

Subsequently, we express the sum over $\mathbf{k}'$ into an integral over $\mathbf{k}'$ as (6.33), substitute (6.34), and use (6.36) and $P_0(x) = 1$. The above equation thereby becomes

$$\begin{aligned}\mu_\mathrm{m}\alpha &= \mu_\mathrm{m}^0\alpha - \frac{D(\varepsilon_\mathrm{F})}{2V}\sum_{\alpha'=\pm\frac{1}{2}}\int\frac{\mathrm{d}\Omega'}{4\pi}f_{\alpha\alpha'}(\mathbf{k}_\mathrm{F},\mathbf{k}_\mathrm{F}')\mu_\mathrm{m}\alpha' \\ &= \mu_\mathrm{m}^0\alpha - \sum_{\alpha'=\pm\frac{1}{2}}\mu_\mathrm{m}\alpha'\sum_{\ell=0}^{\infty}\left(\frac{F_\ell^\mathrm{s}-F_\ell^\mathrm{a}}{2}+\delta_{\alpha\alpha'}F_\ell^\mathrm{a}\right)\frac{1}{2}\int_{-1}^{1}P_\ell(x')P_0(x')\mathrm{d}x' \\ &= \mu_\mathrm{m}^0\alpha - \mu_\mathrm{m}\alpha F_0^\mathrm{a}.\end{aligned}$$

Hence, we obtain μ_m in terms of μ_m^0 as

$$\mu_\mathrm{m} = \frac{\mu_\mathrm{m}^0}{1+F_0^\mathrm{a}}. \tag{6.41}$$

The total moment M due to spin is expressible generally in terms of the spin magnetic moment $\mu_\mathrm{m}^0\alpha$ and one-particle density matrix as

$$M = \int \mathrm{d}\xi\,\mu_\mathrm{m}^0\alpha\,\rho^{(1)}(\xi,\xi), \tag{6.42}$$

where $\mathrm{d}\xi$ is defined by (3.16). Let us substitute (6.17) into the above expression, expand $\bar{n}_{\mathbf{k}\alpha} \approx \bar{n}_k + (\partial\bar{n}_k/\partial\varepsilon_k)\delta\varepsilon_{\mathbf{k}\alpha}$ with $\partial\bar{n}_k/\partial\varepsilon_k = -\delta(\varepsilon_\mathrm{F}-\varepsilon_k)$, and successively use (6.40), (6.28), and (6.41) to arrange it in the form

$$\begin{aligned}M &= \sum_{\mathbf{k}\alpha}\mu_\mathrm{m}^0\alpha\bar{n}_{\mathbf{k}\alpha} \approx \sum_{\mathbf{k}\alpha}\mu_\mathrm{m}^0\alpha\frac{\partial\bar{n}_k}{\partial\varepsilon_k}\delta\varepsilon_{\mathbf{k}\alpha} = \mu_\mathrm{m}^0\mu_\mathrm{m}B\sum_{\mathbf{k}}\sum_{\alpha=\pm\frac{1}{2}}\alpha^2\delta(\varepsilon_\mathrm{F}-\varepsilon_k) \\ &= \frac{(\mu_\mathrm{m}^0/2)^2D(\varepsilon_\mathrm{F})}{1+F_0^\mathrm{a}}B.\end{aligned} \tag{6.43}$$

Hence, the spin susceptibility is obtained from $\chi \equiv \partial M/\partial B$, giving

$$\chi = \frac{(\mu_\mathrm{m}^0/2)^2D(\varepsilon_\mathrm{F})}{1+F_0^\mathrm{a}}. \tag{6.44}$$

The numerator $(\mu_\mathrm{m}^0/2)^2D(\varepsilon_\mathrm{F})$ is identical to that for ideal gases, but the density of states here is modified from that for the non-interacting system by the factor $m^*/m = 1 + F_1^\mathrm{s}/3$, as seen from (6.29) and (6.37). Moreover, there is another factor $(1+F_0^\mathrm{a})^{-1}$ that originates from the spin-dependent part of the interaction, (6.34).

6.3.5 Compressibility

Compressibility κ is defined in terms of volume V and pressure P by

$$\kappa \equiv -\frac{1}{V}\frac{\partial V}{\partial P}. \tag{6.45}$$

For $T = 0$, we clarify how this is affected by the interaction.

To begin, we transform κ at $T = 0$ into an alternative convenient expression. An infinitesimal variation of the grand potential Ω can be expressed as in (1.27). It also follows from (1.28) that $\mathrm{d}\Omega = -P\mathrm{d}V - V\mathrm{d}P$ holds. Equating the two expressions yields the *Gibbs-Duhem relation*:

$$\mathrm{d}P = \frac{S}{V}\mathrm{d}T + \frac{N}{V}\mathrm{d}\mu, \tag{6.46}$$

variables (P, T, μ) of which are all intensive as well as the coefficients $(S/V, N/V)$. Using (6.46) with $\mathrm{d}T = 0$ at $T = 0$ and noting that μ depends on extensive variables (V, N) only through ratio V/N, we write the inverse of (6.45) as

$$\frac{1}{\kappa} = -V\frac{\partial P}{\partial V} = -V\frac{N}{V}\frac{\partial \mu}{\partial V} = -\frac{\partial \mu}{\partial (V/N)} = -\frac{1}{V}\frac{\partial \mu}{\partial (1/N)} = \frac{N^2}{V}\frac{\partial \mu}{\partial N}. \tag{6.47}$$

Thus, compressibility at $T = 0$ takes the alternative form with $\partial N/\partial \mu$.

To find $\partial N/\partial \mu$ at $T = 0$, we start from (6.10) with $q \to \mathbf{k}\alpha$ and rearrange its small variation due to $\delta\mu$, which should accompany no spin polarizations, using (4.33) and (6.6) to obtain

$$\delta N = \sum_{\mathbf{k}\alpha} \delta\bar{n}_{\mathbf{k}} = \sum_{\mathbf{k}\alpha}\left(\frac{\partial \bar{n}_k}{\partial \mu}\delta\mu + \frac{\partial \bar{n}_k}{\partial \varepsilon_k}\delta\varepsilon_{\mathbf{k}}\right) = \sum_{\mathbf{k}\alpha} \delta(\varepsilon_k - \varepsilon_{\mathrm{F}})(\delta\mu - \delta\varepsilon_{\mathbf{k}}). \tag{6.48}$$

The contribution proportional to $\delta\varepsilon_{\mathbf{k}}$ determines an indirect effect that $\delta\mu$ brings about through the interaction. Noting (6.22), we may express this $\delta\varepsilon_{\mathbf{k}}$ as

$$\delta\varepsilon_{\mathbf{k}} = \frac{1}{V}\sum_{\mathbf{k}'\alpha'} f_{\alpha\alpha'}(\mathbf{k}, \mathbf{k}')\delta\bar{n}_{\mathbf{k}'}.$$

As $\delta\mu$ is infinitesimal and $T = 0$, we only need to consider those $\mathbf{k}$ and $\mathbf{k}'$ that lie on the Fermi surface in the above expression. In addition, $\delta\bar{n}_{\mathbf{k}'}$ due to $\delta\mu$ should be isotropic. Keeping these points in mind, we transform the sum over $\mathbf{k}'$ above into an integral in the same way as in (6.33), and then use $P_0(x) = 1$, (6.34), and (6.36) to obtain

$$\delta\varepsilon_{\mathbf{k}} = \int_0^\infty \frac{dk'}{(2\pi)^3} k'^2 \delta\bar{n}_{\mathbf{k}'} \sum_{\alpha'=\pm\frac{1}{2}} \int d\Omega' f_{\alpha\alpha'}(\mathbf{k},\mathbf{k}')$$

$$= \int_0^\infty \frac{dk'}{(2\pi)^3} k'^2 \delta\bar{n}_{\mathbf{k}'} \frac{2V}{D(\varepsilon_F)} \sum_{\ell=0}^{\infty} \sum_{\alpha'=\pm\frac{1}{2}} \left(\frac{F_\ell^s - F_\ell^a}{2} + \delta_{\alpha\alpha'} F_\ell^a \right)$$

$$\times 2\pi \int_{-1}^{1} P_\ell(x') P_0(x') dx'$$

$$= \int_0^\infty \frac{dk'}{(2\pi)^3} k'^2 \delta\bar{n}_{\mathbf{k}'} \frac{2V}{D(\varepsilon_F)} 4\pi F_0^s = \frac{F_0^s}{D(\varepsilon_F)} \sum_{\mathbf{k}'\alpha'} \delta\bar{n}_{\mathbf{k}'} = \frac{F_0^s}{D(\varepsilon_F)} \delta N.$$

Substituting this expression into (6.48) and using (6.28), we obtain $\delta N = D(\varepsilon_F)[\delta\mu - F_0^s \delta N / D(\varepsilon_F)]$, i.e.,

$$\frac{\partial N}{\partial \mu} = \frac{D(\varepsilon_F)}{1 + F_0^s}. \tag{6.49}$$

A further substitution into (6.47) yields the compressibility at $T = 0$ as

$$\kappa = \frac{V}{N^2} \frac{D(\varepsilon_F)}{1 + F_0^s}. \tag{6.50}$$

Thus, κ is also affected by the interaction.

6.3.6 Landau Parameters

We have considered low-temperature fermions with $s = 1/2$ based on the Hartree–Fock approximation to clarify how the interaction affects thermodynamic properties. The main results are (6.25), (6.29), (6.31), (6.37), (6.44), and (6.50), which are all given in terms of the Landau parameters $F_\ell^{s,a}$. Here, we derive microscopic expressions of the parameters within the Hartree–Fock theory, and discuss a possible extension of the theory to systems with strong interactions where the Hartree–Fock formalism is no longer effective.

First, let us derive microscopic expressions for the Landau parameters within the Hartree–Fock theory. To this end, we note that $\mathscr{V}_{|\mathbf{k}-\mathbf{k}'|}$ in (6.23) depends only on $|\mathbf{k}-\mathbf{k}'| = (k^2 + k'^2 - 2kk' \cos\theta_{\mathbf{k}\mathbf{k}'})^{1/2}$, where $\theta_{\mathbf{k}\mathbf{k}'}$ denotes the angle between $\mathbf{k}$ and $\mathbf{k}'$. Thus, we can expand $\mathscr{V}_{|\mathbf{k}-\mathbf{k}'|}$ generally in terms of Legendre polynomials (6.35) as

$$\mathscr{V}_{|\mathbf{k}-\mathbf{k}'|} = \sum_{\ell=0}^{\infty} (2\ell+1) \mathscr{V}_\ell(k,k') P_\ell(\cos\theta_{\mathbf{k}\mathbf{k}'}). \tag{6.51}$$

Table 6.1 Landau parameters of liquid ^{3}He [7]

P (bar)	F_0^{s}	F_1^{s}	F_0^{a}
0	9.30	5.39	−0.695
3	15.99	6.49	−0.723
6	22.49	7.45	−0.733

Let us substitute (6.51) into (6.23) with $\sigma = -1$, choose $\mathbf{k}$ and $\mathbf{k}'$ on the Fermi surface, and compare the resulting expression with (6.34). We thereby obtain F_ℓ^{s} and F_ℓ^{a},

$$F_\ell^{\mathrm{s}} = \frac{D(\varepsilon_{\mathrm{F}})}{2V}\left[2\mathcal{V}_0\delta_{\ell 0} - (2\ell+1)\mathcal{V}_\ell(k_{\mathrm{F}},k_{\mathrm{F}})\right], \qquad F_\ell^{\mathrm{a}} = -\frac{D(\varepsilon_{\mathrm{F}})}{2V}(2\ell+1)\mathcal{V}_\ell(k_{\mathrm{F}},k_{\mathrm{F}}). \tag{6.52}$$

These expressions indicate that, for a repulsive interaction with $\mathcal{V}_k > 0$, we may expect $F_0^{\mathrm{s}} > 0$ and $F_0^{\mathrm{a}} < 0$ generally (**Problem** 6.1). In particular, inequality $F_0^{\mathrm{a}} < 0$ in (6.34) indicates that the repulsive interaction favors a pair of particles with the same spin alignment ($\alpha_1 = \alpha_2$) rather than the opposing alignment ($\alpha_1 \neq \alpha_2$); the state associated with the former can naturally suppress the repulsive force because of the Pauli exclusion principle. Thus, the Hartree–Fock equations enable an explanation of how the interaction affects thermodynamic properties qualitatively. However, because they have been derived based on the variational density matrix (6.5) of an ideal-gas form, (6.52) cannot be used quantitatively for describing strongly interacting systems.

In contrast, Landau's Fermi-liquid theory [5] starts from (6.22) and (6.34) and treats F_ℓ^{s} and F_ℓ^{a} as phenomenological parameters. It is also applicable to strongly interacting systems so that (6.25), (6.29), (6.31), (6.37), (6.44), and (6.50) can be used as they are. This is because scatterings between *quasiparticles* (i.e., renormalized entities with interaction effects that work like "particles") are suppressed at low temperatures because of the Fermi degeneracy so that they behave like real particles with an infinite lifetime to form an ideal gas [3, 5].

Landau's Fermi-liquid theory has been quite successful in describing liquid ^{3}He ($s = 1/2$) at low temperatures. Table 4.1 shows that quantum effects in ^{3}He should be substantial below around $T_{\mathrm{Q}} = 2.58$K, and Landau's Fermi-liquid theory is applicable for extremely low temperatures of $T \ll T_{\mathrm{Q}}$. Table 6.1 presents values of Landau parameters extracted from various experiments. Both F_0^{s} and F_1^{s} are large and positive, indicating that the interaction between particles in ^{3}He is mainly repulsive and strong. Thus, according to (6.50) and (6.37), its compressibility is small and the effective mass is enhanced substantially, making particle motion difficult. In addition, F_0^{a} is negative and close to -1 to give a large spin susceptibility according to (6.44). This enhancement of χ indicates that the system is close to the instability of the ferromagnetic transition for $F_0^{\mathrm{a}} \to -1$.

Problems

6.1. Suppose that the interaction potential in (6.4) is given by

$$\mathscr{V}(r) = U_0\, \mathrm{e}^{-r/r_0}, \tag{6.53}$$

where $r_0 > 0$ and U_0 are constants.

(a) Show that coefficient $\mathscr{V}_k$ of the Fourier expansion in (6.18) is given by

$$\mathscr{V}_k = \frac{8\pi U_0 r_0^3}{(1 + r_0^2 k^2)^2}. \tag{6.54}$$

(b) Replace k with $|\mathbf{k} - \mathbf{k}'|$ in (6.54) and expand it as in (6.51). Show that $V_0(k, k')$ is given by

$$V_0(k, k') = \frac{8\pi U_0 r_0^3}{\left(1 + r_0^2 k^2 + r_0^2 k'^2\right)^2 - 4 r_0^4 k^2 k'^2}. \tag{6.55}$$

(c) The Landau parameters in the Hartree–Fock approximation can be written generally as in (6.52). Express F_0^{s} and F_0^{a} of the present model in terms of U_0, r_0, and the density of states $D(\varepsilon_{\mathrm{F}})$ at the Fermi energy.

References

1. M. Abramowitz, I.A. Stegun (eds.), *Handbook of Mathematical Functions: With Formulas, Graphs, and Mathematical Tables* (Dover, New York, 1965)
2. G.B. Arfken, H.J. Weber, *Mathematical Methods for Physicists* (Academic, New York, 2012)
3. G. Baym, C. Pethick, *Landau Fermi-Liquid Theory: Concepts and Applications* (Wiley, New York, 1991)
4. I.M. Gelfand, S.V. Fomin, *Calculus of Variations* (Prentice-Hall, Englewood Cliffs, 1963)
5. L.D. Landau, J. Exp. Theor. Phys. **30**, 1058 (1956). (Sov. Phys. JETP **3**, 920 (1957))
6. J.M. Luttinger, Phys. Rev. **119**, 1153 (1960)
7. D. Vollhardt, P. Wölfle, *The Superfluid Phases of Helium 3* (Taylor & Francis, London, 1990), p. 31

Chapter 7
Attractive Interaction and Bound States

Abstract Struggling to find a way to theoretically explain the phenomenon of superconductivity, in 1956 Cooper eventually reached a simplified version of the problem of two particles on the Fermi surface under a mutual attraction. *Cooper's problem*, which represented a breakthrough in constructing a microscopic theory of superconductivity, is essentially identical to a one-particle problem with an attractive potential in two dimensions. In this chapter, we consider attractive potentials to clarify under what conditions a bound state is formed. First, we consider one-particle problems with an attractive potential in two and three dimensions to show that an infinitesimal attraction suffices in two dimensions to form a bound state whereas a finite threshold is requisite in three dimensions. Next, we shall see that this qualitative difference between two and three dimensions is caused by whether the one-particle density of states is finite at zero energy. Finally, the presence of the Fermi surface in Cooper's problem will be shown to make the density of states at the excitation threshold finite even in three dimensions, resulting in the formation of a bound state from only an infinitesimal attraction.

7.1 Attractive Potential in Two and Three Dimensions

We consider a particle in a central potential $\mathscr{V}(r)$ that obeys the Schrödinger equation:

$$\left[\frac{\hat{\mathbf{p}}^2}{2m} + \mathscr{V}(r)\right]\phi(\mathbf{r}) = \epsilon\phi(\mathbf{r}), \tag{7.1}$$

where $\hat{\mathbf{p}} \equiv -i\hbar\nabla$ is the momentum operator, m is the particle mass, $\phi(\mathbf{r})$ denotes the wave function, and ϵ is the eigenenergy. Here, we adopt a square-well potential given by (see also Fig. 7.1)

$$\mathscr{V}(r) = \begin{cases} -U_0 & : r < a \\ 0 & : r \geq a \end{cases} \tag{7.2}$$

to clarify under what conditions a bound state is formed in two and three dimensions.

© Springer Japan 2015
T. Kita, *Statistical Mechanics of Superconductivity*, Graduate Texts in Physics,
DOI 10.1007/978-4-431-55405-9_7

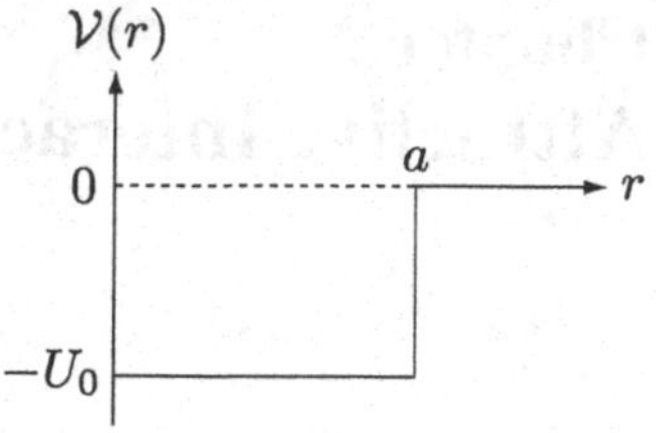

Fig. 7.1 A square-well potential

7.1.1 Bound State in Three Dimensions

First, we consider the three-dimensional case to show that there is a finite threshold for U_0 that determines whether a bound state can form; see (7.9) below.

The lowest-energy eigenstate is expected to have an isotropic s-wave symmetry. Its wave function may be written in terms of the spherical harmonic function [2, 5] $Y_{00}(\theta,\varphi) = (4\pi)^{-1/2}$ as $\phi(\mathbf{r}) = R(r)Y_{00}(\theta,\varphi)$ $(0 \le r \le \infty, 0 \le \theta \le \pi, 0 \le \varphi \le 2\pi)$. Substituting it into (7.1) and restating the resulting equation in polar coordinates [2, 7], we thereby obtain the radial Schrödinger equation,

$$\frac{1}{r}\frac{\mathrm{d}^2}{\mathrm{d}r^2} rR(r) + \frac{2m[\epsilon - \mathcal{V}(r)]}{\hbar^2} R(r) = 0. \tag{7.3}$$

By noting (7.2), setting $-U_0 \le \epsilon < 0$, and imposing boundary conditions $|R(0)| < \infty$ and $R(\infty) = 0$, we can easily solve (7.3) separately for $r < a$ and $r > a$ in terms of $rR(r)$ as

$$rR(r) = \begin{cases} A \sin kr & : r < a \\ B\,\mathrm{e}^{-\kappa r} & : r \ge a \end{cases}, \tag{7.4}$$

where A and B are constants, and k and κ are defined by

$$k \equiv \frac{\sqrt{2m(U_0 + \epsilon)}}{\hbar}, \qquad \kappa \equiv \frac{\sqrt{-2m\epsilon}}{\hbar}. \tag{7.5}$$

Subsequently, we match these solutions so that $[rR(r)]'/rR(r)$ is continuous at $r = a$. The condition can be expressed concisely in terms of the dimensionless quantities:

$$\xi \equiv ka > 0, \qquad \eta \equiv \kappa a > 0 \tag{7.6}$$

as

$$\eta = -\xi \cot \xi. \tag{7.7}$$

Also, from (7.5) and (7.6), variables (ξ, η) satisfy

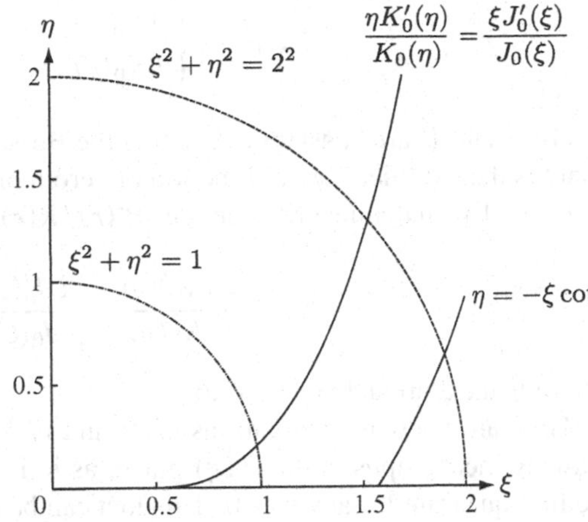

Fig. 7.2 A plot of the three constraints in the (ξ, η) plane, i.e., $\xi^2 + \eta^2 = r^2$ for $r = 1, 2$, $\eta = -\xi \cot \xi$, and $\eta K_0'(\eta)/K_0(\eta) = \xi J_0'(\xi)/J_0(\xi)$

$$\xi^2 + \eta^2 = \frac{2ma^2}{\hbar^2} U_0. \tag{7.8}$$

In Fig. 7.2, plots of (7.7) and (7.8) are drawn by which we can see graphically whether a solution of the coupled equations exists. An intersection point is present in the first quadrant of the (ξ, η) plane if $\left[(2ma^2/\hbar^2)U_0\right]^{1/2} \geq \pi/2$ is met. This condition is expressed more concisely as

$$U_0 \geq \frac{\hbar^2}{2m}\left(\frac{\pi}{2a}\right)^2. \tag{7.9}$$

That is, no bound state exists in three dimensions unless U_0 exceeds a finite value.

7.1.2 Bound State in Two Dimensions

Next, we consider the two-dimensional case to show that an infinitesimal attraction suffices in forming a bound state, as revealed in (7.14) below.

The isotropic wave function in two dimensions is similarly expressed in polar coordinates $\mathbf{r} = (r\cos\varphi, r\sin\varphi)$ as $\phi(\mathbf{r}) = R(r)/\sqrt{2\pi}$. Substitution into (7.1) and a change to the two-dimensional polar coordinates [2, 7] yields the radial Schrödinger equation

$$\frac{1}{r}\frac{\mathrm{d}}{\mathrm{d}r}r\frac{\mathrm{d}}{\mathrm{d}r}R(r) + \frac{2m[\epsilon - \mathscr{V}(r)]}{\hbar^2}R(r) = 0. \tag{7.10}$$

Substituting (7.2), we find that (7.10) reduces to Bessel's (modified Bessel's) differential equation of zeroth order for $r < a$ $(r > a)$ [1, 2]. Imposing the same boundary conditions $|R(0)| < \infty$ and $R(\infty) = 0$, we obtain the solution

$$R(r) = \begin{cases} AJ_0(kr) & : r < a \\ BK_0(\kappa r) & : r \geq a \end{cases}, \tag{7.11}$$

where A and B are constants, $J_0(kr)$ is the Bessel function of zeroth order, $K_0(\kappa r)$ denotes the modified Bessel function of zeroth order [1, 2], and k and κ are defined by (7.5). The matching criterion, i.e. $R'(r)/R(r)$ is continuous at $r = a$, yields

$$\frac{\eta K_0'(\eta)}{K_0(\eta)} = \frac{\xi J_0'(\xi)}{J_0(\xi)}, \tag{7.12}$$

where η and ξ are defined by (7.6).

The pair of coupled equations (7.8) and (7.12) may be solved qualitatively by drawing their graphs in the (ξ, η) plane, as in Fig. 7.2. The two curves intersect in the first quadrant for any $\eta > 0$. This fact can be stated quantitatively by expanding $J_0(\xi) \approx 1 - \xi^2/4 + O(\xi^4)$ and $K_0(\eta) = -\ln(\eta e^\gamma/2) + O(\eta^2 \ln \eta)$ in the weak-coupling region of $0 < \xi, \eta \ll 1$ [1, 2], where $\gamma = 0.57721\cdots$ is Euler's constant. Substitution of both into (7.12) yields $[\ln(\eta e^\gamma/2)]^{-1} = -\xi^2/2$, i.e.,

$$\eta = 2e^{-\gamma} \exp\left(-\frac{2}{\xi^2}\right) \approx 2e^{-\gamma} \exp\left(-\frac{\hbar^2}{ma^2 U_0}\right). \tag{7.13}$$

In the second approximation, we have used $\xi^2 \approx 2ma^2 U_0/\hbar^2$ as obtained from (7.8) by noting $\xi \gg e^{-2/\xi^2} \sim \eta$ for $0 < \xi \ll 1$. It also follows from (7.5) and (7.6) that $\eta \equiv \kappa a = \sqrt{-2ma^2\epsilon}/\hbar$. Substituting back into (7.13), we obtain the bound-state energy for $U_0 \to 0$ as

$$\epsilon = -\frac{\hbar^2}{2m}\left(\frac{2e^{-\gamma}}{a}\right)^2 \exp\left(-\frac{2\hbar^2}{ma^2 U_0}\right). \tag{7.14}$$

Hence, a bound state is formed for any $U_0 > 0$. We also realize from $\kappa = \sqrt{-2m\epsilon}/\hbar$, (7.11), and $K_0(x) \sim (\pi/2x)^{1/2} e^{-x}$ for $x \to \infty$ [1, 2] that the radius r_0 of the bound state for $\epsilon \to 0$ is quite large as $r_0 \sim \kappa^{-1} = \hbar/\sqrt{-2m\epsilon}$.

7.2 Consideration in Wave Vector Domain

Regarding bound-state formation, we now study the one-particle problem of Sect. 7.1 once more, to trace in the wave vector domain the origin of the qualitative difference between the two- and three-dimensional cases. We shall see that it depends on whether the one-particle density of states is finite at zero energy.

Let us expand the wave function and potential of (7.1) as plane waves,[1]

[1]Factor $V^{-1/2}$ in (7.15) originates from the normalization condition $\langle\phi|\phi\rangle = 1$, whereas V^{-1} in (7.16) is so chosen to ensure that the coefficient $\mathscr{V}_k$ is independent of V in the thermodynamic limit.

$$\phi(\mathbf{r}) = \frac{1}{\sqrt{V}} \sum_{\mathbf{k}} e^{i\mathbf{k}\cdot\mathbf{r}} \phi_{\mathbf{k}}, \tag{7.15}$$

$$\mathscr{V}(r) = \frac{1}{V} \sum_{\mathbf{k}} e^{i\mathbf{k}\cdot\mathbf{r}} \mathscr{V}_k. \tag{7.16}$$

It should be noted that $\phi(\mathbf{r})$ and $\phi_{\mathbf{k}}$ above are different functions distinguished by their arguments, as are $\mathscr{V}(r)$ and $\mathscr{V}_k$. Because $\mathscr{V}(r)$ is isotropic, its Fourier transform $\mathscr{V}_k$ is also isotropic depending only on $k \equiv |\mathbf{k}|$. Let us substitute (7.15) and (7.16) into (7.1), multiply the equation by $V^{-1/2}e^{-i\mathbf{k}'\cdot\mathbf{r}}$, and integrate it over $\mathbf{r}$. Exchanging $\mathbf{k}$ and $\mathbf{k}'$, we obtain the Fourier-transformed Schrödinger equation

$$\varepsilon_k \phi_{\mathbf{k}} + \frac{1}{V} \sum_{\mathbf{k}'} \mathscr{V}_{|\mathbf{k}-\mathbf{k}'|} \phi_{\mathbf{k}'} = \epsilon \phi_{\mathbf{k}}, \tag{7.17}$$

where $\varepsilon_k \equiv \hbar^2 k^2/2m$ is the kinetic energy. Rearranging this equation gives

$$\phi_{\mathbf{k}} = \frac{C_{\mathbf{k}}}{\varepsilon_k - \epsilon}, \qquad C_{\mathbf{k}} \equiv -\frac{1}{V} \sum_{\mathbf{k}'} \mathscr{V}_{|\mathbf{k}-\mathbf{k}'|} \phi_{\mathbf{k}'}.$$

Further, we substitute the first into the second to obtain an integral equation for $C_{\mathbf{k}}$,

$$C_{\mathbf{k}} \equiv -\frac{1}{V} \sum_{\mathbf{k}'} \frac{\mathscr{V}_{|\mathbf{k}-\mathbf{k}'|}}{\varepsilon_{k'} - \epsilon} C_{\mathbf{k}'}. \tag{7.18}$$

Next, we expand $\mathscr{V}_{|\mathbf{k}-\mathbf{k}'|}$ as in (6.51) and retain only the $\ell = 0$ term,

$$\mathscr{V}_{|\mathbf{k}-\mathbf{k}'|} \to \mathscr{V}_0(k, k'), \tag{7.19}$$

which is justified in solely studying the s-wave bound states. Substitution of this equation into (7.18) yields an equation for $C_{\mathbf{k}}$ that depends only on k,

$$C_k = -\int_0^\infty d\varepsilon_{k'} N(\varepsilon_{k'}) \frac{\mathscr{V}_0(k, k')}{\varepsilon_{k'} - \epsilon} C_{k'}, \tag{7.20}$$

where we have transformed the sum over $\mathbf{k}'$ into an integral over $\varepsilon_{k'}$ by using *the density of states per unit volume and spin component* defined by

$$N(\epsilon) \equiv \frac{1}{V} \sum_{\mathbf{k}} \delta(\epsilon - \varepsilon_k). \tag{7.21}$$

Apart from the factor $(2s + 1)V$, this $N(\epsilon)$ is identical to $D(\epsilon)$ in (4.17).

Equation (7.20) is suitable for clarifying the origins for the difference between two- and three-dimensional bound-state formation. In general, the s-wave component $\mathscr{V}_0(k, k')$ of an attractive potential takes a finite negative value for $k, k' \to 0$, whereas it vanishes for $k, k' \to \infty$; see (6.55) with $U_0 < 0$ for example. As a simple model that contains these features and is also tractable analytically in the wave vector domain, we adopt the model potential

$$\mathscr{V}_0(k,k') = -\Gamma_0\,\theta(\varepsilon_c - \varepsilon_k)\theta(\varepsilon_c - \varepsilon_{k'}), \tag{7.22}$$

where $\Gamma_0 > 0$ is a constant, $\varepsilon_c > 0$ denotes a cutoff energy, and $\theta(x)$ is the step function defined in (4.11). Substituting (7.22) into (7.20), we find that C_k can also be written as

$$C_k = C_0\theta(\varepsilon_c - \varepsilon_k). \tag{7.23}$$

Moreover, we have already obtained the density of states in both three and two dimensions as given in (4.17) and (4.54), respectively. Thus, (7.21) per unit volume and spin component vanishes for $\epsilon < 0$ whereas for $\epsilon \geq 0$ it is expressible as

$$N(\epsilon) = \begin{cases} A\epsilon^{1/2} & : \text{three dimensions} \\ N(0) & : \text{two dimensions} \end{cases}, \tag{7.24}$$

where A and $N(0)$ are constants.

First, focusing on the two-dimensional case and substituting (7.22)–(7.24) into (7.20), we obtain an equation that determines the bound-state energy $\epsilon < 0$,

$$\frac{1}{N(0)\Gamma_0} = \int_0^{\varepsilon_c} \frac{1}{\varepsilon_{k'} - \epsilon}\mathrm{d}\varepsilon_{k'} = \ln\frac{\varepsilon_c - \epsilon}{-\epsilon}. \tag{7.25}$$

When the attractive potential is weak, $N(0)\Gamma_0 \ll 1$, the left-hand side of the equation takes a large positive value that diverges as $N(0)\Gamma_0 \to 0$. Meanwhile, the right-hand side as a function of ϵ also diverges logarithmically as $\epsilon \to 0$. Hence, (7.25) always has a solution of $\epsilon < 0$ as long as $N(0)\Gamma_0$ is finite. The eigenenergy in the limit of $N(0)\Gamma_0 \to 0$ may be found analytically by approximating $\varepsilon_c - \epsilon \approx \varepsilon_c$ in (7.25),

$$\epsilon = -\varepsilon_c \exp\left[-\frac{1}{N(0)\Gamma_0}\right]. \tag{7.26}$$

Thus, we confirm that in two dimensions a bound state is formed when there is an infinitesimal attraction.

The same consideration in three dimensions yields an analogous equation that determines the bound-state energy $\epsilon < 0$,

$$\frac{1}{A\Gamma_0} = \int_0^{\varepsilon_c} \frac{\varepsilon_{k'}^{1/2}}{\varepsilon_{k'} - \epsilon}\mathrm{d}\varepsilon_{k'}. \tag{7.27}$$

The left-hand side diverges in the limit $\Gamma_0 \to 0$, whereas the right-hand side as a function of ϵ remains finite in the limit $\epsilon \to 0$ because the density of states $N(\varepsilon_k) \propto \varepsilon_k^{1/2}$ vanishes as $\varepsilon_k \to 0$. Hence, there is no bound-state solution for $\Gamma_0 \to 0$ in three dimensions.

Thus, the present analysis has clarified unambiguously that the bound-state formation in two dimensions by an infinitesimal attractive potential is attributable to a finite density of states at zero energy.

7.3 Cooper's Problem

Besides the repulsive Coulomb force from other electrons, an electron in a metal interacts with quantized lattice vibrations called *phonons*. These phonons may bring a net attraction between each pair of electrons near the Fermi energy, as shown by Fröhrich [6] and Pines and Bardeen [3] before 1956. With this realization, Cooper was able to construct a relatively simple model of two electrons on the Fermi surface attracting each other [4]. Here, we study this *Cooper's problem* to see that they form a bound state by an infinitesimal attraction.

We seek the simplest possibility allowing a pair of electrons on the Fermi surface under a mutual attraction to form an s-wave bound state without a center-of-mass motion. Accordingly, the Schrödinger equation for their orbital motion can be written in terms of their relative coordinates $\mathbf{r}_1 - \mathbf{r}_2$ as

$$\left[\frac{\hat{\mathbf{p}}_1^2}{2m} + \frac{\hat{\mathbf{p}}_2^2}{2m} + \mathscr{V}(\mathbf{r}_1 - \mathbf{r}_2|)\right]\phi(|\mathbf{r}_1 - \mathbf{r}_2|) = (\epsilon + 2\varepsilon_{\mathrm{F}})\phi(|\mathbf{r}_1 - \mathbf{r}_2|). \tag{7.28}$$

where m is the electron mass, $\mathscr{V}$ is the interaction potential, ε_{F} denotes the Fermi energy, and ϵ is the energy of the two electrons measured from their total kinetic energy $2\varepsilon_{\mathrm{F}}$. A solution with $\varepsilon < 0$ corresponds to a bound state. Let us expand the two-particle wave function $\phi(|\mathbf{r}_1 - \mathbf{r}_2|)$ as plane waves,[2]

$$\phi(|\mathbf{r}_1 - \mathbf{r}_2|) = \frac{1}{V}\sum_{\mathbf{k}} \phi_k \, \mathrm{e}^{\mathrm{i}\mathbf{k}\cdot(\mathbf{r}_1 - \mathbf{r}_2)}. \tag{7.29}$$

Substitution of (7.16) and (7.29) into (7.28) gives the Fourier-transformed Schrödinger equation as

$$2(\varepsilon_k - \varepsilon_{\mathrm{F}})\phi_k + \frac{1}{V}\sum_{\mathbf{k}'} \mathscr{V}_{|\mathbf{k}-\mathbf{k}'|}\phi_{k'} = \epsilon\phi_k, \tag{7.30}$$

where $\varepsilon_k = \hbar^2 k^2/2m$ is the kinetic energy of a single electron. Equation (7.30) has the same form as (7.17). Hence, we repeat the procedure from (7.17) through (7.20) for (7.30) to obtain an integral equation for $C_k \equiv [2(\varepsilon_k - \varepsilon_{\mathrm{F}}) - \epsilon]\phi_k$,

$$C_k = -\int_{\varepsilon_{\mathrm{F}}}^{\infty} \mathrm{d}\varepsilon_{k'} N(\varepsilon_{k'}) \frac{\mathscr{V}_0(k, k')}{2(\varepsilon_{k'} - \varepsilon_{\mathrm{F}}) - \epsilon} C_{k'}. \tag{7.31}$$

[2]Factor V^{-1} in (7.29) results from a product of two $V^{-1/2}$ for each of $\mathbf{r}_1$ and $\mathbf{r}_2$.

It differs from (7.20) in only a couple of points, specifically: (i) the denominator is replaced by $2(\varepsilon_{k'} - \varepsilon_{\mathrm{F}}) - \epsilon$ and (ii) the lower limit of the integral is ε_{F}. The latter change reflects the fact that filled one-particle states inside the Fermi sphere cannot be used for forming a bound state because of the Pauli exclusion principle. We also adapt the model potential of (7.22) to the present situation,

$$\mathscr{V}_0(k,k') = -\Gamma_0\,\theta(\varepsilon_{\mathrm{c}} - |\varepsilon_k - \varepsilon_{\mathrm{F}}|)\theta(\varepsilon_{\mathrm{c}} - |\varepsilon_{k'} - \varepsilon_{\mathrm{F}}|), \tag{7.32}$$

with $\varepsilon_{\mathrm{c}} \ll \varepsilon_{\mathrm{F}}$. Substituting (7.32) into (7.31) enables C_k to be written as $C_k = C_0\theta(\varepsilon_{\mathrm{c}} - |\varepsilon_k - \varepsilon_{\mathrm{F}}|)$. Let us put this expression and (7.32) back into (7.31), approximate $N(\varepsilon_{k'}) \approx N(\varepsilon_{\mathrm{F}})$ based on $\varepsilon_{\mathrm{c}} \ll \varepsilon_{\mathrm{F}}$, and make a change of variables as $\xi \equiv \varepsilon_{k'} - \varepsilon_{\mathrm{F}}$. We thereby obtain an equation to determine the bound-state energy as

$$\frac{1}{N(\varepsilon_{\mathrm{F}})\Gamma_0} = \int_0^{\varepsilon_{\mathrm{c}}} \frac{1}{2\xi - \epsilon}\mathrm{d}\xi = \frac{1}{2}\ln\frac{2\varepsilon_{\mathrm{c}} - \epsilon}{-\epsilon}, \tag{7.33}$$

which is essentially equivalent to (7.25) with a replacement of $N(0)$ by $N(\varepsilon_{\mathrm{F}})$. Thus, the presence of the Fermi sphere in Cooper's problem has made: (i) the density of states at the excitation threshold $\epsilon = \varepsilon_{\mathrm{F}}$ finite, and (ii) the formation with an infinitesimal attractive potential of a two-particle bound state on the Fermi surface possible. The bound-state energy for $\Gamma_0 \to 0$ is easily obtained,

$$\epsilon = -2\varepsilon_{\mathrm{c}}\exp\left[-\frac{2}{N(\varepsilon_{\mathrm{F}})\Gamma_0}\right]. \tag{7.34}$$

The formation of this two-particle bound state is now referred to as *Cooper pairing*.

It is worth pointing out that prior to the BCS theory, the idea of pair condensation as the mechanism underlying superconductivity was presented by Schafroth [8] based on the analogy to superfluidity in the charged Bose-gas model [9]. Around 1957, however, it had not established Cooper's finding described above or the BCS wave function (8.6) given below, thereby failing to produce quantitative results testable by experiments.

Problems

7.1. Consider the one-dimensional Schrödinger equation:

$$\left[-\frac{\hbar^2}{2m}\frac{\mathrm{d}^2}{\mathrm{d}x^2} + \mathscr{V}(x)\right]\phi(x) = \epsilon\phi(x)$$

with the attractive potential

$$\mathscr{V}(x) = \begin{cases} -U_0 & : |x| < a \\ 0 & : |x| \geq a \end{cases} \qquad (U_0 > 0).$$

Show that there always exists a bound state of $-U_0 < \epsilon < 0$ for an arbitrary $U_0 > 0$. Obtain an analytic expression of the bound-state energy for $U_0 \to 0$.

References

1. M. Abramowitz, I.A. Stegun (eds.), *Handbook of Mathematical Functions: With Formulas, Graphs, and Mathematical Tables* (Dover, New York, 1965)
2. G.B. Arfken, H.J. Weber, *Mathematical Methods for Physicists* (Academic, New York, 2012)
3. J. Bardeen, D. Pines, Phys. Rev. **99**, 1140 (1955)
4. L.N. Cooper, Phys. Rev. **104**, 1189 (1956)
5. A.R. Edmonds, *Angular Momentum in Quantum Mechanics* (Princeton University Press, Princeton, 1957)
6. H. Fröhlich, Proc. R. Soc. (Lond.) A **215**, 291 (1952)
7. L.D. Landau, E.M. Lifshitz, *Quantum Mechanics: Non-relativistic Theory*, 3rd edn. (Butterworth-Heinemann, Oxford, 1991)
8. M.R. Schafroth, Phys. Rev. **96**, 1442 (1954); see also, M.R. Schafroth, S.T. Butler, J.M. Blatt, Helv. Phys. Acta **30**, 93 (1957)
9. M.R. Schafroth, Phys. Rev. **96**, 1149 (1954); **100**, 463 (1955)

Chapter 8
Mean-Field Equations of Superconductivity

Abstract Cooper's analysis clarified that an ideal gas of identical fermions becomes unstable in the presence of a mutual attractive potential, however small it may be. What is the new ground state, and how is it related to superconductivity? A breakthrough for this fundamental issue was achieved by Schrieffer, then a graduate student of Bardeen. Motivated by Cooper's finding, he finally had the idea of applying Tomonaga's intermediate coupling theory for mesons (Tomonaga, Prog Theor Phys 2:6, 1947) to describe the new ground state (Cooper LN, Feldman D (eds), BCS: 50 years. World Scientific, Hackensack, 2011). In this chapter, we construct this BCS wave function in the coordinate space in such a way that both the pair condensation and phase coherence are manifest. We then derive the *Bogoliubov–Valatin operator* that describes excitations based on the BCS wave function. These two ingredients are subsequently used to obtain the basic mean-field equations of superconductivity, called the *Bogoliubov–de Gennes (BdG) equations*, using the same methods as in Chap. 6 for the Hartree–Fock equations. Besides the Hartree–Fock potential, the BdG equations are characterized by a novel self-consistent potential we call the *pair potential*.

8.1 BCS Wave Function for Cooper-Pair Condensation

Superconductivity may be regarded as a kind of BEC in terms of electrons. However, from the Pauli exclusion principle, it is certainly impossible for electrons with $s = 1/2$ to occupy the same one-particle state macroscopically. Nevertheless, it does not prohibit them from condensing to form identical two-particle bound states. The variational wave function Schrieffer wrote down can be regarded as a mathematical expression of this *Cooper-pair condensation*. First, we present a variational wave function for the macroscopic Cooper-pair condensation, (8.6) below.

The BEC of an ideal Bose gas is characterized by a macroscopic number of particles occupying the one-particle state with the lowest energy, as we have seen in Sect. 4.6. A system of identical fermions may also be able to condense into a two-particle bound state with no fundamental conflicts with the Pauli exclusion principle.

© Springer Japan 2015

T. Kita, *Statistical Mechanics of Superconductivity*, Graduate Texts in Physics,
DOI 10.1007/978-4-431-55405-9_8

Without any consideration, one may write down the wave function immediately as

$$\tilde{\Phi}^{(N)}(\xi_1, \xi_2, \cdots, \xi_N) \propto \phi(\xi_1, \xi_2)\phi(\xi_3, \xi_4)\cdots\phi(\xi_{N-1}, \xi_N), \tag{8.1}$$

where

$$\phi(\xi_1, \xi_2) = -\phi(\xi_2, \xi_1) \tag{8.2}$$

is the Cooper-pair wave function that incorporates the transposition symmetry ($\sigma = -1$) required at the two-particle level. Here we have chosen N ($\sim 10^{23}$) as an even integer for convenience.

However, (8.1) still fails to obey the required permutation symmetry of (3.12) for fermions ($\sigma = -1$). The method to incorporate it has already been given in (3.31). To perform the antisymmetrization concisely for the present case, we introduce the *Cooper-pair creation operator*:

$$\hat{Q}^\dagger \equiv \frac{1}{2}\int d\xi_1 \int d\xi_2 \phi(\xi_1, \xi_2)\hat{\psi}^\dagger(\xi_1)\hat{\psi}^\dagger(\xi_2), \tag{8.3}$$

where $\hat{\psi}^\dagger$ is the field operator for fermions that obeys (3.17) with $\sigma = -1$. Using it, we can easily modify (8.1) to incorporate the required permutation symmetry as [1, 13]

$$|\Phi^{(N)}\rangle \equiv A_N \left(\hat{Q}^\dagger\right)^{N/2} |0\rangle. \tag{8.4}$$

Here A_N is a normalization constant and $|0\rangle$ is defined by (3.18). Indeed, the corresponding wave function is obtained using (3.20) and (3.22),

$$\begin{aligned}\Phi^{(N)}(\xi_1, \xi_2, \cdots, \xi_N) &\equiv \langle \xi_1, \xi_2, \cdots, \xi_N | \Phi^{(N)} \rangle \\ &= \frac{A_N}{2^{N/2}\sqrt{N!}} \sum_{\hat{P}} (-1)^P \phi(\xi_{P1}, \xi_{P2}) \cdots \phi(\xi_{PN-1}, \xi_{PN}),\end{aligned} \tag{8.5}$$

which is appropriately antisymmetrized.

Further, we consider a linear combination of (8.4) in terms of the number of Cooper pairs to make it more convenient for statistical-mechanical calculations in the grand canonical ensemble. The most convenient among various combinations may be the exponential form, called the *BCS wave function* given by [13, 14]

$$|\Phi\rangle \equiv A \sum_{n=0}^{\infty} \frac{\left(\hat{Q}^\dagger\right)^n}{n!} |0\rangle = A \exp\left(\hat{Q}^\dagger\right) |0\rangle, \tag{8.6}$$

where A is the normalization constant and $\hat{Q}^\dagger$ is defined by (8.3).

Equation (8.6) represents an extension of Schrieffer's homogeneous variational wave function with isotropic s-wave symmetry to inhomogeneous cases with arbitrary pairing symmetry (**Problem** 8.1). It has the advantages of (i) being able to describe inhomogeneous superconductors and (ii) the phase coherence of the pair condensation seen manifestly in the wave function. Moreover, (8.6) also includes the *coherent state* (3.71) for BECs and laser beams [10, 15] as the limit where the pair radius is smaller than the mean interparticle spacing (**Problem** 8.2). Thus, (8.6) enables us to study the BEC and Cooper-pair condensation in a unified way.

8.2 Quasiparticle Field for Excitations

Suppose that the ground state is given by (8.6). What kind of excitations may be possible then? Here, we derive the Bogoliubov–Valatin operators that describe excitations and clarify their properties. They are given by (8.13) and (8.14) below and satisfy (i) (8.12) that specifies BCS wave function $|\Phi\rangle$ as the vacuum of the excitations and (ii) fermionic commutation relation (8.18). Fields such as $\hat{\gamma}$ of (8.13), which have properties similar to $\hat{\psi}$ for real particles, are generally called *quasiparticle fields*.

Let us begin with mathematical preliminaries. First, the annihilation operator $\hat{\psi}(\xi)$ satisfies the following commutation relations with (8.3):

$$[\hat{\psi}(\xi), \hat{Q}^\dagger]_+ = \int d\xi_1 \phi(\xi, \xi_1)\hat{\psi}^\dagger(\xi_1) \equiv \phi(\xi, \bar{\xi}_1)\hat{\psi}^\dagger(\bar{\xi}_1), \tag{8.7}$$

$$[[\hat{\psi}(\xi), \hat{Q}^\dagger]_+, \hat{Q}^\dagger]_+ = 0, \tag{8.8}$$

$$[\hat{\psi}(\xi), g(\hat{Q}^\dagger)]_+ = [\hat{\psi}(\xi), \hat{Q}^\dagger]_+ \, g'(\hat{Q}^\dagger), \tag{8.9}$$

where $[\hat{A}, \hat{B}]_\sigma \equiv \hat{A}\hat{B} - \sigma\hat{B}\hat{A}$ ($\sigma = \pm$), and $g(x)$ is a function that is analytic at $x = 0$. To simplify the notation, we shall sometimes express an integration over a variable by an overbar, as in (8.7). Those readers who are familiar with the Einstein summation convention may ignore those bars when reading them. Equations (8.7)–(8.9) can be easily proved as follows. First, (8.7) follows using (3.17) as

$$\begin{aligned}
[\hat{\psi}(\xi), \hat{Q}^\dagger]_+ &= \frac{1}{2}\phi(\bar{\xi}_1, \bar{\xi}_2)[\hat{\psi}(\xi), \hat{\psi}^\dagger(\bar{\xi}_1)\hat{\psi}^\dagger(\bar{\xi}_2)]_+ \\
&= \frac{1}{2}\phi(\bar{\xi}_1, \bar{\xi}_2)\left\{[\hat{\psi}(\xi), \hat{\psi}^\dagger(\bar{\xi}_1)]_- \hat{\psi}^\dagger(\bar{\xi}_2) - \hat{\psi}^\dagger(\bar{\xi}_1)[\hat{\psi}(\xi), \hat{\psi}^\dagger(\bar{\xi}_2)]_-\right\} \\
&= \frac{\phi(\xi, \bar{\xi}_2)\hat{\psi}^\dagger(\bar{\xi}_2) - \phi(\bar{\xi}_1, \xi)\hat{\psi}^\dagger(\bar{\xi}_1)}{2} = \phi(\xi, \bar{\xi}_2)\hat{\psi}^\dagger(\bar{\xi}_2).
\end{aligned}$$

Here the second equality is easily seen to hold by expanding the anticommutator in the curly brackets. The last expression is obtained from the preceding one by changing integration variables as $\xi_1 \to \xi_2$ and using (8.2). Second, (8.8) is seen to hold by noting that the right-hand side of (8.7) is given in terms of $\hat{\psi}^\dagger$ alone and so is $\hat{Q}^\dagger$. Third, we can prove (8.9) using (8.8) by

$$\begin{aligned}[\hat{\psi}(\xi), g(\hat{Q}^\dagger)]_+ &= \sum_{n=0}^{\infty} \frac{g^{(n)}(0)}{n!}[\hat{\psi}(\xi), (\hat{Q}^\dagger)^n]_+ \\ &= \sum_{n=1}^{\infty} \frac{g^{(n)}(0)}{n!} \sum_{\ell=0}^{n-1} (\hat{Q}^\dagger)^\ell [\hat{\psi}(\xi), \hat{Q}^\dagger]_+ (\hat{Q}^\dagger)^{n-1-\ell} \\ &= [\hat{\psi}(\xi), \hat{Q}^\dagger]_+ \sum_{n=1}^{\infty} \frac{g^{(n)}(0)}{n!} n(\hat{Q}^\dagger)^{n-1} = [\hat{\psi}(\xi), \hat{Q}^\dagger]_+ \, g'(\hat{Q}^\dagger),\end{aligned}$$

where the second equality may be confirmed by writing the expression after the equality sign without commutators for $n = 2, 3$.

Next, we consider ξ_1 and ξ_2 in $\phi(\xi_1, \xi_2)$ as row and column indices to introduce a matrix $\underline{\phi} \equiv (\phi(\xi_1, \xi_2))$; its Hermitian conjugate is given by $\underline{\phi}^\dagger \equiv (\phi^*(\xi_2, \xi_1))$. Accordingly, we define the unit matrix by $\underline{1} \equiv (\delta(\xi_1, \xi_2))$. With this notation, (8.2) implies that the matrix $\underline{\phi}$ is antisymmetric as $\underline{\phi}^{\mathrm{T}} = -\underline{\phi}$, where ^{T} denotes the transpose. Using $\underline{\phi}$ and $\underline{1}$, we introduce a couple of new matrices $\underline{u}$ and $\underline{v}$ by [14]

$$\underline{u} \equiv (\underline{1} + \underline{\phi}\,\underline{\phi}^\dagger)^{-1/2}, \qquad \underline{v} \equiv (\underline{1} + \underline{\phi}\,\underline{\phi}^\dagger)^{-1/2}\underline{\phi}, \tag{8.10}$$

which satisfy

$$\underline{u} = \underline{u}^\dagger, \qquad \underline{v} = -\underline{v}^{\mathrm{T}}, \qquad \underline{u}\,\underline{u} + \underline{v}\,\underline{v}^\dagger = \underline{1}, \qquad \underline{u}\,\underline{v} = \underline{v}\,\underline{u}^*. \tag{8.11}$$

The first two identities in (8.11) result from (8.2) and (8.10). The third identity is an extension of the identity $[g(x)]^2 + g(x)x^2 g(x) = 1$ for the scalar function $g(x) \equiv (1 + x^2)^{-1/2}$ to the matrix argument using the Taylor expansion of $g(x)$ at $x = 0$, as also for the fourth identity. Thus, $\underline{u}$ is Hermitian, whereas $\underline{v}$ is antisymmetric. It turns out that $\underline{v}$ may be regarded as an effective condensate wave function, as discussed around (8.47) below.

We are now ready to derive a field operator that describes excitations. Let us operate $\hat{\psi}(\xi)$ on (8.6) and rearrange the resulting expression using (3.18), (8.9), and (8.7) successively as

$$\begin{aligned}\hat{\psi}(\xi)|\Phi\rangle &= A\big[\hat{\psi}(\xi), \mathrm{e}^{\hat{Q}^\dagger}\big]_+|0\rangle + A\mathrm{e}^{\hat{Q}^\dagger}\hat{\psi}(\xi)|0\rangle = \big[\hat{\psi}(\xi), \hat{Q}^\dagger\big]_+ A\mathrm{e}^{\hat{Q}^\dagger}|0\rangle \\ &= \phi(\xi, \bar{\xi}_2)\hat{\psi}^\dagger(\bar{\xi}_2)|\Phi\rangle.\end{aligned}$$

Next, we multiply this equation by $u(\xi_1, \xi)$ from the left, integrate over ξ, and use (8.10) to obtain $u(\xi_1, \bar{\xi})\hat{\psi}(\bar{\xi})|\Phi\rangle - v(\xi_1, \bar{\xi}_2)\hat{\psi}^\dagger(\bar{\xi}_2)|\Phi\rangle = 0$. Thus, we have derived a new field operator that satisfies

$$\hat{\gamma}(\xi_1)|\Phi\rangle = 0, \tag{8.12}$$

in the form

$$\hat{\gamma}(\xi_1) \equiv u(\xi_1, \bar{\xi}_2)\hat{\psi}(\bar{\xi}_2) - v(\xi_1, \bar{\xi}_2)\hat{\psi}^\dagger(\bar{\xi}_2). \tag{8.13}$$

This is a direct extension of the *Bogoliubov–Valatin operator* for homogeneous systems [4, 5, 18] to inhomogeneous situations [14]. According to (8.12), we may characterize the BCS wave function (8.6) as the "vacuum of quasiparticles" in the same terminology as used below (3.59). However, it should be noted that $|\Phi\rangle$ is occupied by particles and should be distinguished clearly from the vacuum state $|0\rangle$ of (3.18) with no particles. The Hermitian conjugate of (8.13) reads

$$\hat{\gamma}^\dagger(\xi_1) = -v^*(\xi_1, \bar{\xi}_2)\hat{\psi}(\bar{\xi}_2) + u^*(\xi_1, \bar{\xi}_2)\hat{\psi}^\dagger(\bar{\xi}_2). \tag{8.14}$$

Further, (8.13) and (8.14) may be expressed concisely in terms of matrices in (8.10),

$$\begin{bmatrix} \hat{\gamma} \\ \hat{\gamma}^\dagger \end{bmatrix} = \begin{bmatrix} \underline{u} & \underline{v} \\ -\underline{v}^* & -\underline{u}^* \end{bmatrix} \begin{bmatrix} \hat{\psi} \\ -\hat{\psi}^\dagger \end{bmatrix}. \tag{8.15}$$

The inverse transformation is

$$\begin{bmatrix} \hat{\psi} \\ -\hat{\psi}^\dagger \end{bmatrix} = \begin{bmatrix} \underline{u} & \underline{v} \\ -\underline{v}^* & -\underline{u}^* \end{bmatrix} \begin{bmatrix} \hat{\gamma} \\ \hat{\gamma}^\dagger \end{bmatrix}, \tag{8.16}$$

which is confirmed by substituting (8.15) into the right-hand side of (8.16) and subsequently using (8.11). Equation (8.16) can be written explicitly as

$$\begin{cases} \hat{\psi}(\xi_1) = u(\xi_1, \bar{\xi}_2)\hat{\gamma}(\bar{\xi}_2) + v(\xi_1, \bar{\xi}_2)\hat{\gamma}^\dagger(\bar{\xi}_2), \\ \hat{\psi}^\dagger(\xi_1) = v^*(\xi_1, \bar{\xi}_2)\hat{\gamma}(\bar{\xi}_2) + u^*(\xi_1, \bar{\xi}_2)\hat{\gamma}^\dagger(\bar{\xi}_2) \end{cases}. \tag{8.17}$$

The quasiparticle field $\hat{\gamma}$ satisfies the commutation relations for fermions:

$$[\hat{\gamma}(\xi), \hat{\gamma}^\dagger(\xi')]_- = \delta(\xi, \xi'), \qquad [\hat{\gamma}(\xi), \hat{\gamma}(\xi')]_- = 0, \tag{8.18}$$

as shown by substituting (8.15) into (8.18) and using (3.17) and (8.11).

8.3 Bogoliubov–de Gennes Equations

For the state (8.6) to be realized, it is necessary for its free energy to be lower than that of the normal state for a given set of independent thermodynamic variables. It has been established for single-element superconductors like Hg and Pb that phonons are responsible for the effective attraction that causes superconductivity. However, thermodynamic properties of "weak-coupling" superconductors do not depend on this source of attraction at all. Hence, we adopt Hamiltonian (6.4), i.e.,

$$\hat{\mathscr{H}} \equiv \int d\xi_1 \hat{\psi}^\dagger(\xi_1)\hat{\mathscr{K}}_1\hat{\psi}(\xi_1)$$
$$+\frac{1}{2}\int d\xi_1 \int d\xi_2 \mathscr{V}(|\mathbf{r}_1-\mathbf{r}_2|)\hat{\psi}^\dagger(\xi_1)\hat{\psi}^\dagger(\xi_2)\hat{\psi}(\xi_2)\hat{\psi}(\xi_1), \tag{8.19}$$

with

$$\hat{\mathscr{K}}_1 \equiv \frac{\hat{\mathbf{p}}_1^2}{2m} + \mathscr{U}(\mathbf{r}_1) - \mu, \tag{8.20}$$

in studying superconductivity. To be more specific, we here assume that the interaction potential $\mathscr{V}(|\mathbf{r}_1-\mathbf{r}_2|)$ has some attractive part to derive the fundamental mean-field equations of superconductivity; that is, the BdG equations. This will be performed by two different methods. The first is based on the variational principle (6.3), whereas the second relies on a self-consistent Wick decomposition technique. Both methods yield an identical set of equations.

The main results are summarized as follows. The BdG equation is given by (8.38), where the operators in it are defined by (8.34)–(8.36) with (8.29) and (8.30). It forms an Hermitian eigenvalue problem, whose eigenfunctions satisfy the orthonormality and completeness given in (8.44) and (8.45). This eigenvalue problem has a particle-hole symmetry given by (8.43). With this fact, we only need to calculate states with $E_q \geq 0$ in (8.38). Equations (8.34), (8.35), and (8.38) form a closed set of self-consistency equations, which for $v_q \to 0$ reduces to the Hartree–Fock equations (6.12) and (6.13).

8.3.1 Derivation Based on Variational Principle

First, we derive the BdG equations based on the variational principle (6.3), i.e.,

$$\Omega[\hat{\rho}] \equiv \mathrm{Tr}\,\hat{\rho}\,(\hat{\mathscr{H}} + \beta^{-1}\ln\hat{\rho}) \geq \Omega_{\mathrm{eq}}, \tag{8.21}$$

where Ω_{eq} is the exact thermodynamic potential. We have already used it in Sect. 6.2.1 to obtain the Hartree–Fock equations by choosing the density matrix as (6.5). The same procedure is followed here, with a key difference being that the quasiparticle field $\hat{\gamma}(\xi)$ now takes the place of $\hat{\psi}(\xi)$ because of (8.12).

To be specific, let us write the variational density matrix in the ideal-gas form in terms of quasiparticles:

$$\hat{\rho} = \exp\left[\beta\left(\Omega_0 - \sum_q E_q\hat{\gamma}_q^\dagger\hat{\gamma}_q\right)\right], \qquad \Omega_0 \equiv -\frac{1}{\beta}\sum_q \ln\left(1+e^{-\beta E_q}\right). \tag{8.22}$$

Here $E_q \geq 0$ is a variational parameter that denotes an excitation energy from the "vacuum" (8.6), and the operator $\hat{\gamma}_q$ is defined through the formal expansion of $\hat{\gamma}(\xi)$ in terms of eigenfunctions $\{\varphi_q(\xi)\}$,

$$\hat{\gamma}(\xi) = \sum_q \hat{\gamma}_q \varphi_q(\xi). \tag{8.23}$$

Using (8.18) and assuming that $\{\varphi_q(\xi)\}$ forms an orthonormal set, one can show that $\hat{\gamma}_q$ obeys the commutation relations:

$$\left[\hat{\gamma}_q, \hat{\gamma}_{q'}^\dagger\right]_- = \delta_{qq'}, \qquad \left[\hat{\gamma}_q, \hat{\gamma}_{q'}\right]_- = 0. \tag{8.24}$$

We use the variational density matrix (8.22) for the system described by Hamiltonian (8.19) to evaluate the grand potential (8.21). A couple of key points for performing this are summarized as follows. First, it follows from (8.22) that the entropy, $S \equiv -k_\mathrm{B}\mathrm{Tr}\,\hat{\rho}\ln\hat{\rho} = -k_\mathrm{B}\langle\ln\hat{\rho}\rangle$, is given in terms of the mean occupation number $\bar{n}_q \equiv \langle\hat{\gamma}_q^\dagger\hat{\gamma}_q\rangle$ as

$$S = -k_\mathrm{B}\beta\left(\Omega_0 - \sum_q E_q\bar{n}_q\right) = k_\mathrm{B}\sum_q\left[-\bar{n}_q\ln\bar{n}_q - (1-\bar{n}_q)\ln(1-\bar{n}_q)\right], \tag{8.25}$$

in exactly the same form as in (6.7) for the Hartree–Fock theory. Second, the Bloch–De Dominicis theorem (5.11) also holds true for the present density matrix (8.22). Hence, the expectation of the four field operators in the interaction can be evaluated by the Wick decomposition procedure into a form that extends (5.18) for the normal state,

$$\begin{aligned}
\rho^{(2)}(\xi_1,\xi_2;\xi_1,\xi_2) &\equiv \langle\hat{\psi}^\dagger(\xi_1)\hat{\psi}^\dagger(\xi_2)\hat{\psi}(\xi_2)\hat{\psi}(\xi_1)\rangle \\
&= \langle\dot{\hat{\psi}}^\dagger(\xi_1)\ddot{\hat{\psi}}^\dagger(\xi_2)\ddot{\hat{\psi}}(\xi_2)\dot{\hat{\psi}}(\xi_1)\rangle + \langle\dot{\hat{\psi}}^\dagger(\xi_1)\ddot{\hat{\psi}}^\dagger(\xi_2)\dot{\hat{\psi}}(\xi_2)\ddot{\hat{\psi}}(\xi_1)\rangle \\
&\quad + \langle\dot{\hat{\psi}}^\dagger(\xi_1)\dot{\hat{\psi}}^\dagger(\xi_2)\ddot{\hat{\psi}}(\xi_2)\ddot{\hat{\psi}}(\xi_1)\rangle \\
&= \langle\hat{\psi}^\dagger(\xi_1)\hat{\psi}(\xi_1)\rangle\langle\hat{\psi}^\dagger(\xi_2)\hat{\psi}(\xi_2)\rangle - \langle\hat{\psi}^\dagger(\xi_1)\hat{\psi}(\xi_2)\rangle\langle\hat{\psi}^\dagger(\xi_2)\hat{\psi}(\xi_1)\rangle \\
&\quad + \langle\hat{\psi}^\dagger(\xi_1)\hat{\psi}^\dagger(\xi_2)\rangle\langle\hat{\psi}(\xi_2)\hat{\psi}(\xi_1)\rangle.
\end{aligned} \tag{8.26}$$

The third term in the final expression is the expectation that characterizes superconductivity. Indeed, each of $\hat{\psi}$ and $\hat{\psi}^\dagger$ in (8.17) is given as a linear combination of $\hat{\gamma}$ and $\hat{\gamma}^\dagger$ so that the density matrix (8.22) yields finite expectations not only for $\langle\hat{\psi}^\dagger(\xi_2)\hat{\psi}(\xi_1)\rangle$ but also for $\langle\hat{\psi}(\xi_2)\hat{\psi}(\xi_1)\rangle$. Specifically, they are expressible in terms of averages

$$\langle\hat{\gamma}_q^\dagger\hat{\gamma}_{q'}\rangle = \frac{\delta_{qq'}}{\mathrm{e}^{\beta E_q}+1} \equiv \delta_{qq'}\bar{n}_q, \qquad \langle\hat{\gamma}_q\hat{\gamma}_{q'}^\dagger\rangle = \delta_{qq'}(1-\bar{n}_q), \tag{8.27}$$

and new functions

$$u_q(\xi_1) \equiv u(\xi_1,\bar{\xi}_2)\varphi_q(\bar{\xi}_2), \qquad v_q(\xi_1) \equiv v^*(\xi_1,\bar{\xi}_2)\varphi_q(\bar{\xi}_2), \tag{8.28}$$

concisely as

$$\rho^{(1)}(\xi_1,\xi_2) \equiv \langle \hat{\psi}^\dagger(\xi_2)\hat{\psi}(\xi_1)\rangle = \rho^{(1)*}(\xi_2,\xi_1)$$
$$= \sum_q [u_q(\xi_1)u_q^*(\xi_2)\bar{n}_q + v_q^*(\xi_1)v_q(\xi_2)(1-\bar{n}_q)], \qquad (8.29)$$

$$\tilde{\rho}^{(1)}(\xi_1,\xi_2) \equiv \langle \hat{\psi}(\xi_1)\hat{\psi}(\xi_2)\rangle = -\tilde{\rho}^{(1)}(\xi_2,\xi_1)$$
$$= \sum_q [u_q(\xi_1)v_q^*(\xi_2) - v_q^*(\xi_1)u_q(\xi_2)]\left(\frac{1}{2}-\bar{n}_q\right). \qquad (8.30)$$

They can be derived by substituting (8.17) into the defining expressions, expanding the quasiparticle field as in (8.23), and using (8.27) and (8.28) to express the quasiparticle expectations in terms of $\bar{n}_q$. For example, (8.29) obtains from

$$\begin{aligned}
&\langle \hat{\psi}^\dagger(\xi_2)\hat{\psi}(\xi_1)\rangle \\
&= \langle [u^*(\xi_2,\bar{\xi}_4)\hat{\gamma}^\dagger(\bar{\xi}_4) + v^*(\xi_2,\bar{\xi}_4)\hat{\gamma}(\bar{\xi}_4)][u(\xi_1,\bar{\xi}_3)\hat{\gamma}(\bar{\xi}_3) + v(\xi_1,\bar{\xi}_3)\hat{\gamma}^\dagger(\bar{\xi}_3)]\rangle \\
&= u^*(\xi_2,\bar{\xi}_4)u(\xi_1,\bar{\xi}_3)\langle\hat{\gamma}^\dagger(\bar{\xi}_4)\hat{\gamma}(\bar{\xi}_3)\rangle + v^*(\xi_2,\bar{\xi}_4)v(\xi_1,\bar{\xi}_3)\langle\hat{\gamma}(\bar{\xi}_4)\hat{\gamma}^\dagger(\bar{\xi}_3)\rangle \\
&= \sum_{qq'} [u^*(\xi_2,\bar{\xi}_4)u(\xi_1,\bar{\xi}_3)\varphi_{q'}^*(\bar{\xi}_4)\varphi_q(\bar{\xi}_3)\langle\hat{\gamma}_{q'}^\dagger\hat{\gamma}_q\rangle \\
&\quad + v^*(\xi_2,\bar{\xi}_4)v(\xi_1,\bar{\xi}_3)\varphi_{q'}(\bar{\xi}_4)\varphi_q^*(\bar{\xi}_3)\langle\hat{\gamma}_{q'}\hat{\gamma}_q^\dagger\rangle] \\
&= \sum_q [u_q(\xi_1)u_q^*(\xi_2)\bar{n}_q + v_q^*(\xi_1)v_q(\xi_2)(1-\bar{n}_q)].
\end{aligned}$$

Equation (8.30) has been derived similarly but involves an additional step that uses the completeness of $\{\varphi_q\}$ and (8.11) to transform $\sum_q u_q(\xi_1)v_q^*(\xi_2)$ as

$$\begin{aligned}
&\sum_q u_q(\xi_1)v_q^*(\xi_2) = u(\xi_1,\bar{\xi}_3)v(\xi_2,\bar{\xi}_4)\sum_q \varphi_q(\bar{\xi}_3)\varphi_q^*(\bar{\xi}_4) \\
&\quad = u(\xi_1,\bar{\xi}_3)v(\xi_2,\bar{\xi}_4)\delta(\bar{\xi}_3,\bar{\xi}_4) = u(\xi_1,\bar{\xi}_3)v(\xi_2,\bar{\xi}_3) = -u(\xi_1,\bar{\xi}_3)v(\bar{\xi}_3,\xi_2) \\
&\quad = -v(\xi_1,\bar{\xi}_3)u^*(\bar{\xi}_3,\xi_2) = -v(\xi_1,\bar{\xi}_3)u(\xi_2,\bar{\xi}_3) \\
&\quad = -v(\xi_1,\bar{\xi}_3)u(\xi_2,\bar{\xi}_4)\delta(\bar{\xi}_3,\bar{\xi}_4) = -v(\xi_1,\bar{\xi}_3)u(\xi_2,\bar{\xi}_4)\sum_q \varphi_q^*(\bar{\xi}_3)\varphi_q(\bar{\xi}_4) \\
&\quad = -\sum_q v_q^*(\xi_1)u_q(\xi_2) = \sum_q \frac{u_q(\xi_1)v_q^*(\xi_2) - v_q^*(\xi_1)u_q(\xi_2)}{2}.
\end{aligned} \qquad (8.31)$$

The resulting expression (8.30) has the advantage that antisymmetry $\tilde{\rho}^{(1)}(\xi_1,\xi_2) = -\tilde{\rho}^{(1)}(\xi_2,\xi_1)$ arising from the commutation relation of $\hat{\psi}(\xi_1)$ and $\hat{\psi}(\xi_2)$ is manifest.

We are now ready to write down the grand potential (8.21) for Hamiltonian (8.19) evaluated by (8.22). Indeed, we can re-express (8.21) using (8.25), (8.26), (8.29), and (8.30),

$$\Omega[\hat{\rho}] = \int d\xi_1 \hat{\mathscr{K}}_1 \rho^{(1)}(\xi_1,\xi_2)\big|_{\xi_2=\xi_1} + \frac{1}{2}\int d\xi_1 \int d\xi_2 \mathscr{V}(|\mathbf{r}_1-\mathbf{r}_2|)\Big[\rho^{(1)}(\xi_1,\xi_1) \\ \times \rho^{(1)}(\xi_2,\xi_2) - \rho^{(1)}(\xi_2,\xi_1)\rho^{(1)}(\xi_1,\xi_2) + \tilde{\rho}^{(1)*}(\xi_2,\xi_1)\tilde{\rho}^{(1)}(\xi_2,\xi_1)\Big] \\ - \frac{1}{\beta}\sum_q [-\bar{n}_q \ln \bar{n}_q - (1-\bar{n}_q)\ln(1-\bar{n}_q)], \tag{8.32}$$

where we have used $\tilde{\rho}^{(1)*}(\xi_2,\xi_1) = \langle \hat{\psi}(\xi_2)\hat{\psi}(\xi_1)\rangle^* = \langle \hat{\psi}^\dagger(\xi_1)\hat{\psi}^\dagger(\xi_2)\rangle$. Equation (8.32) contains an additional contribution $\tilde{\rho}^{(1)}$ compared with (6.8) in the Hartree–Fock approximation. Moreover, the expression for $\rho^{(1)}$ changes from that in (6.9) into (8.29) using u_q and v_q.

Next, we minimize (8.32) with respect to variational parameters $\{E_q\}$. To this end, we note that (8.32) depends on E_q only through $\bar{n}_q$, as seen from (8.27), (8.29), and (8.30). Hence, the minimization with respect to E_q is equivalent to that for $\bar{n}_q$, the necessary condition for which becomes

$$0 = \frac{\delta\Omega[\hat{\rho}]}{\delta\bar{n}_q} \\ = \int d\xi_1 \Big[u_q^*(\xi_1)\hat{\mathscr{K}}_1 u_q(\xi_1) - v_q(\xi_1)\hat{\mathscr{K}}_1 v_q^*(\xi_1)\Big] + \int d\xi_1 \int d\xi_2 \mathscr{V}(|\mathbf{r}_1-\mathbf{r}_2|) \\ \times\Big\{[|u_q(\xi_1)|^2 - |v_q(\xi_1)|^2]\rho^{(1)}(\xi_2,\xi_2) - \Big[u_q^*(\xi_1)u_q(\xi_2) - v_q(\xi_1)v_q^*(\xi_2)\Big] \\ \times \rho^{(1)}(\xi_1,\xi_2) - \frac{1}{2}\Big[u_q(\xi_2)v_q^*(\xi_1) - v_q^*(\xi_2)u_q(\xi_1)\Big]\tilde{\rho}^{(1)*}(\xi_2,\xi_1) \\ -\frac{1}{2}\Big[u_q^*(\xi_2)v_q(\xi_1) - v_q(\xi_2)u_q^*(\xi_1)\Big]\tilde{\rho}^{(1)}(\xi_2,\xi_1)\Big\} - \frac{1}{\beta}\ln\frac{1-\bar{n}_q}{\bar{n}_q} \\ = \int d\xi_1 \Big[u_q^*(\xi_1)\hat{\mathscr{K}}_1 u_q(\xi_1) - v_q^*(\xi_1)\hat{\mathscr{K}}_1^* v_q(\xi_1)\Big] \\ + \int d\xi_1 \int d\xi_2 \Big\{\mathscr{U}_{\rm HF}(\xi_1,\xi_2)\Big[u_q^*(\xi_1)u_q(\xi_2) - v_q(\xi_1)v_q^*(\xi_2)\Big] \\ -\Delta^*(\xi_1,\xi_2)v_q^*(\xi_1)u_q(\xi_2) + \Delta(\xi_1,\xi_2)u_q^*(\xi_1)v_q(\xi_2)\Big\} - E_q. \tag{8.33}$$

In the last equality, we have performed integration by parts twice in terms of the operator (8.20) to change $v_q(\xi_1)\hat{\mathscr{K}}_1 v_q^*(\xi_1) \to [\hat{\mathscr{K}}_1^* v_q(\xi_1)]v_q^*(\xi_1)$, and also simplified the expression using the *Hartree–Fock potential* and *pair potential* defined by

$$\mathscr{U}_{\rm HF}(\xi_1,\xi_2) = \mathscr{U}_{\rm HF}^*(\xi_2,\xi_1)$$
$$\equiv \delta(\xi_1,\xi_2)\int d\xi_3 \mathscr{V}(|\mathbf{r}_1-\mathbf{r}_3|)\rho^{(1)}(\xi_3,\xi_3) - \mathscr{V}(|\mathbf{r}_1-\mathbf{r}_2|)\rho^{(1)}(\xi_1,\xi_2), \quad (8.34)$$

$$\Delta(\xi_1,\xi_2) = -\Delta(\xi_2,\xi_1) \equiv -\mathscr{V}(|\mathbf{r}_1-\mathbf{r}_2|)\tilde{\rho}^{(1)}(\xi_1,\xi_2), \quad (8.35)$$

respectively. In the latter, we have also used a change of variables $\xi_1 \leftrightarrow \xi_2$ and antisymmetry $\Delta(\xi_1,\xi_2) = -\Delta(\xi_2,\xi_1)$ to express the terms with $\Delta(\xi_1,\xi_2)$ and $\Delta^*(\xi_1,\xi_2)$ as

$$\frac{1}{2}\Delta(\bar{\xi}_2,\bar{\xi}_1)\left[u_q^*(\bar{\xi}_2)v_q(\bar{\xi}_1) - v_q(\bar{\xi}_2)u_q^*(\bar{\xi}_1)\right] = \Delta(\bar{\xi}_1,\bar{\xi}_2)u_q^*(\bar{\xi}_1)v_q(\bar{\xi}_2).$$

Introducing the operator

$$\mathscr{K}_{\rm HF}(\xi_1,\xi_2) = \mathscr{K}_{\rm HF}^*(\xi_2,\xi_1) \equiv \hat{\mathscr{K}}_1\delta(\xi_1,\xi_2) + \mathscr{U}_{\rm HF}(\xi_1,\xi_2) \quad (8.36)$$

in terms of (8.20) and (8.34), we can write (8.33) concisely as

$$\int d\xi_1 \int d\xi_2 \left[u_q^*(\xi_1)\ v_q^*(\xi_1)\right]\begin{bmatrix}\mathscr{K}_{\rm HF}(\xi_1,\xi_2) & \Delta(\xi_1,\xi_2)\\ -\Delta^*(\xi_1,\xi_2) & -\mathscr{K}_{\rm HF}^*(\xi_1,\xi_2)\end{bmatrix}\begin{bmatrix}u_q(\xi_2)\\ v_q(\xi_2)\end{bmatrix} = E_q. \quad (8.37)$$

Parameter E_q and functions $u_q(\xi_1)$ and $v_q(\xi_1)$ that obey this equality can be obtained by solving the eigenvalue problem,

$$\int d\xi_2 \begin{bmatrix}\mathscr{K}_{\rm HF}(\xi_1,\xi_2) & \Delta(\xi_1,\xi_2)\\ -\Delta^*(\xi_1,\xi_2) & -\mathscr{K}_{\rm HF}^*(\xi_1,\xi_2)\end{bmatrix}\begin{bmatrix}u_q(\xi_2)\\ v_q(\xi_2)\end{bmatrix} = E_q\begin{bmatrix}u_q(\xi_1)\\ v_q(\xi_1)\end{bmatrix}, \quad (8.38)$$

with normalization,

$$\int d\xi_1 \left[\left|u_q(\xi_1)\right|^2 + \left|v_q(\xi_1)\right|^2\right] = 1. \quad (8.39)$$

That (8.38) with normalization (8.39) satisfies (8.37) is seen by multiplying (8.38) by the row vector $[u_q^*(\xi_1)\ v_q^*(\xi_1)]$ from the left and performing an integration over ξ_1. Equation (8.38) was derived by de Gennes by extending Bogoliubov's quasiparticle method developed for homogeneous systems [4, 5] to inhomogeneous systems [6, 8], and hence are generally referred to as the *Bogoliubov–de Gennes (BdG) equations*.[1] They form the basic mean-field equations of superconductivity.

The properties of (8.38) are summarized as follows. First, the matrix operator on its left-hand side is Hermitian; this follows from the properties of the matrices $\underline{\mathscr{K}}_{\rm HF} \equiv (\mathscr{K}_{\rm HF}(\xi_1,\xi_2)) = \underline{\mathscr{K}}_{\rm HF}^\dagger$ and $\underline{\Delta} \equiv (\Delta(\xi_1,\xi_2)) = -\underline{\Delta}^{\rm T}$,

[1] It should be noted, however, that the same equations had been derived by Andreev [2]. In addition, they have the same content as the Gor'kov equations derived previously in 1959 [11], which will be discussed in Sect. 14.2.

$$\begin{bmatrix} \underline{\mathscr{K}}_{\text{HF}} & \underline{\Delta} \\ -\underline{\Delta}^* & -\underline{\mathscr{K}}^*_{\text{HF}} \end{bmatrix}^\dagger = \begin{bmatrix} \underline{\mathscr{K}}^\dagger_{\text{HF}} & -\underline{\Delta}^{\text{T}} \\ \underline{\Delta}^\dagger & -\underline{\mathscr{K}}^{\text{T}}_{\text{HF}} \end{bmatrix} = \begin{bmatrix} \underline{\mathscr{K}}_{\text{HF}} & \underline{\Delta} \\ -\underline{\Delta}^* & -\underline{\mathscr{K}}^*_{\text{HF}} \end{bmatrix}. \tag{8.40}$$

Hence, its eigenvalues $\{E_q\}$ are real. Second, one can show that the matrix operator also satisfies

$$\underline{\sigma}_x \begin{bmatrix} \underline{\mathscr{K}}_{\text{HF}} & \underline{\Delta} \\ -\underline{\Delta}^* & -\underline{\mathscr{K}}^*_{\text{HF}} \end{bmatrix}^* \underline{\sigma}_x = -\begin{bmatrix} \underline{\mathscr{K}}_{\text{HF}} & \underline{\Delta} \\ -\underline{\Delta}^* & -\underline{\mathscr{K}}^*_{\text{HF}} \end{bmatrix}, \tag{8.41}$$

where $\underline{\sigma}_x$ is the x element of the *Pauli matrices*; for later purposes, we present it here together with the others that form the complete orthonormal set of 2×2 matrices:

$$\underline{\sigma}_0 = \begin{bmatrix} 1 & 0 \\ 0 & 1 \end{bmatrix}, \quad \underline{\sigma}_x = \begin{bmatrix} 0 & 1 \\ 1 & 0 \end{bmatrix}, \quad \underline{\sigma}_y = \begin{bmatrix} 0 & -i \\ i & 0 \end{bmatrix}, \quad \underline{\sigma}_z = \begin{bmatrix} 1 & 0 \\ 0 & -1 \end{bmatrix}. \tag{8.42}$$

Using the symmetry relation (8.41), one can show that the BdG equation has a crucial property called *particle-hole symmetry*. To be specific, let us take the complex conjugate of (8.38), operate $\underline{\sigma}_x$ from the left, and insert the unit matrix $\underline{\sigma}_x^2$ between the matrix operator and eigenvector,

$$\underline{\sigma}_x \begin{bmatrix} \underline{\mathscr{K}}_{\text{HF}} & \underline{\Delta} \\ -\underline{\Delta}^* & -\underline{\mathscr{K}}^*_{\text{HF}} \end{bmatrix}^* \underline{\sigma}_x \underline{\sigma}_x \begin{bmatrix} u_q \\ v_q \end{bmatrix}^* = E_q \underline{\sigma}_x \begin{bmatrix} u_q \\ v_q \end{bmatrix}^*.$$

Simplifying using (8.41) yields

$$\begin{bmatrix} \underline{\mathscr{K}}_{\text{HF}} & \underline{\Delta} \\ -\underline{\Delta}^* & -\underline{\mathscr{K}}^*_{\text{HF}} \end{bmatrix} \begin{bmatrix} v_q^* \\ u_q^* \end{bmatrix} = -E_q \begin{bmatrix} v_q^* \\ u_q^* \end{bmatrix},$$

which may be combined with (8.38) to form a single matrix equation,

$$\begin{bmatrix} \underline{\mathscr{K}}_{\text{HF}} & \underline{\Delta} \\ -\underline{\Delta}^* & -\underline{\mathscr{K}}^*_{\text{HF}} \end{bmatrix} \begin{bmatrix} u_q & v_q^* \\ v_q & u_q^* \end{bmatrix} = \begin{bmatrix} u_q & v_q^* \\ v_q & u_q^* \end{bmatrix} \begin{bmatrix} E_q & 0 \\ 0 & -E_q \end{bmatrix}. \tag{8.43}$$

This equation implies that, once an eigenvalue $E_q \geq 0$ and its eigenfunction $[u_q \; v_q]^{\text{T}}$ are obtained, the rearranged vector $[v_q^* \; u_q^*]^{\text{T}}$ naturally forms an eigenstate that belongs to $-E_q$. Thus, the eigenvalues of the BdG equations are distributed symmetrically with respect to 0, and those satisfying $E_q \geq 0$ represent the excitation energies.

Stated in this matrix form, the orthonormality and completeness of the eigenfunctions of (8.38) are

$$\int d\xi \begin{bmatrix} u_{q'}^*(\xi) & v_{q'}^*(\xi) \\ v_{q'}(\xi) & u_{q'}(\xi) \end{bmatrix} \begin{bmatrix} u_q(\xi) & v_q^*(\xi) \\ v_q(\xi) & u_q^*(\xi) \end{bmatrix} = \begin{bmatrix} \delta_{q'q} & 0 \\ 0 & \delta_{q'q} \end{bmatrix}, \tag{8.44}$$

$$\sum_q \left\{ \begin{bmatrix} u_q(\xi_1) \\ v_q(\xi_1) \end{bmatrix} \begin{bmatrix} u_q^*(\xi_2) & v_q^*(\xi_2) \end{bmatrix} + \begin{bmatrix} v_q^*(\xi_1) \\ u_q^*(\xi_1) \end{bmatrix} \begin{bmatrix} v_q(\xi_2) & u_q(\xi_2) \end{bmatrix} \right\} = \begin{bmatrix} \delta(\xi_1,\xi_2) & 0 \\ 0 & \delta(\xi_1,\xi_2) \end{bmatrix}, \tag{8.45}$$

respectively. The off-diagonal elements in (8.44) result from the orthogonality of the wave functions belonging to different eigenvalues $\pm E_q$. Note also that the first-row elements of (8.45) are equivalent to $\underline{u}\,\underline{u} + \underline{v}\,\underline{v}^\dagger = \underline{1}$ and $\underline{u}\,\underline{v} = \underline{v}\,\underline{u}^*$ in (8.11), respectively. For example, the first statement can be proved by noting (8.28) and assuming completeness for $\{\varphi_q\}$; that is,

$$\begin{aligned}
\delta(\xi_1,\xi_2) &= \sum_q \left[u_q(\xi_1)u_q^*(\xi_2) + v_q^*(\xi_1)v_q(\xi_2) \right] \\
&= u(\xi_1,\bar{\xi}_3)u^*(\xi_2,\bar{\xi}_4)\sum_q \varphi_q(\bar{\xi}_3)\varphi_q^*(\bar{\xi}_4) + v(\xi_1,\bar{\xi}_3)v^*(\xi_2,\bar{\xi}_4)\sum_q \varphi_q^*(\bar{\xi}_3)\varphi_q(\bar{\xi}_4) \\
&= u(\xi_1,\bar{\xi}_3)u^*(\xi_2,\bar{\xi}_3) + v(\xi_1,\bar{\xi}_3)v^*(\xi_2,\bar{\xi}_3) \\
&= u(\xi_1,\bar{\xi}_3)u(\bar{\xi}_3,\xi_2) + v(\xi_1,\bar{\xi}_3)v^\dagger(\bar{\xi}_3,\xi_2).
\end{aligned} \tag{8.46}$$

Thus, the basic properties of the BdG equations have been established. As potentials in (8.34) and (8.35) contain the eigenstates to be obtained as seen in (8.29) and (8.30), (8.38) should be solved self-consistently until convergence is attained.

Although the topic may be rather academic, it is worth pointing out that we can construct matrices $\underline{u} \equiv (u(\xi_1,\xi_2))$ and $\underline{v} \equiv (v(\xi_1,\xi_2))$ from the eigenfunctions of (8.38). First, one establishes equalities:

$$\sum_q u_q(\xi_1)u_q^*(\xi_2) = u(\xi_1,\bar{\xi}_3)u(\bar{\xi}_3,\xi_2), \qquad -\sum_q u_q(\xi_1)v_q^*(\xi_2) = u(\xi_1,\bar{\xi}_3)v(\bar{\xi}_3,\xi_2),$$

in the same way as (8.46). Thus, we obtain $\underline{u}\,\underline{u}$ from the left-hand side of the first equation, which is subsequently used to find $\underline{u} = (\underline{u}\,\underline{u})^{1/2}$. Additionally, the second equality enables us to construct $\underline{v}$ by multiplying $\underline{u}^{-1}$ from the left of $-\sum_q u_q(\xi_1)v_q^*(\xi_2)$. Pair wave function $\underline{\phi}$ can then be constructed based on (8.10) by $\underline{\phi} = \underline{u}^{-1}\underline{v}$. Matrix $\underline{v}$ may be regarded as an effective condensate wave function. Indeed, we can express the one-particle density matrix (8.29) in a manner similar to (8.46),

$$\rho^{(1)}(\xi_1,\xi_2) = \int d\xi_3 v(\xi_1,\xi_3)v^*(\xi_2,\xi_3) + \sum_q \left[u_q(\xi_1)u_q^*(\xi_2) - v_q^*(\xi_1)v_q(\xi_2) \right]\bar{n}_q, \tag{8.47}$$

from which the total particle number N is obtained,

$$N = \int d\xi_1 \rho^{(1)}(\xi_1,\xi_1). \tag{8.48}$$

Thus, we realize that, at $T = 0$ where $\bar{n}_q = 0$ holds, all the particles are in the two-particle bound state described by the effective condensate wave function $v(\xi_1, \xi_2)$.

Finally, we can use (8.38) to simplify the expression for the grand potential (8.32) in equilibrium,

$$\Omega_{\text{BdG}} = -\frac{1}{\beta}\sum_q \ln\left(1 + \mathrm{e}^{-\beta E_q}\right) + \sum_q \int \mathrm{d}\xi_1 \int \mathrm{d}\xi_2 \Big[v_q(\xi_1)\mathscr{K}_{\text{HF}}(\xi_1, \xi_2) v_q^*(\xi_2)$$
$$- \frac{1}{2} u_q^*(\xi_1) v_q(\xi_2) \Delta(\xi_1, \xi_2) - \frac{1}{2} u_q(\xi_1) v_q^*(\xi_2) \Delta^*(\xi_1, \xi_2)$$
$$- \frac{1}{2} \rho^{(1)}(\xi_1, \xi_2) \mathscr{U}_{\text{HF}}(\xi_2, \xi_1) + \frac{1}{2} \tilde{\rho}^{(1)}(\xi_1, \xi_2) \Delta^*(\xi_1, \xi_2) \Big], \tag{8.49}$$

following the procedure used in deriving (6.14).

In summary, we have obtained the basic equation, (8.38), that describes superconductivity together with the corresponding grand potential, (8.49). In the limit $v_q \to 0$, they reduce to (6.12)–(6.14) of the Hartree–Fock theory.

8.3.2 Derivation Based on Wick Decomposition

We now derive the BdG equations in an alternative manner based on the Wick decomposition procedure used in deriving the Hartree–Fock equations in Sect. 6.2.2. Indeed, this is the method de Gennes used in his derivation [8].

First, (6.15) is generalized here to include the contribution of the anomalous pair expectations,

$$\hat{\psi}_1^\dagger \hat{\psi}_2^\dagger \hat{\psi}_2 \hat{\psi}_1 \to \hat{\psi}_1^\dagger \hat{\psi}_1 \langle \hat{\psi}_2^\dagger \hat{\psi}_2 \rangle + \langle \hat{\psi}_1^\dagger \hat{\psi}_1 \rangle \hat{\psi}_2^\dagger \hat{\psi}_2 - \langle \hat{\psi}_1^\dagger \hat{\psi}_1 \rangle \langle \hat{\psi}_2^\dagger \hat{\psi}_2 \rangle$$
$$- \hat{\psi}_1^\dagger \hat{\psi}_2 \langle \hat{\psi}_2^\dagger \hat{\psi}_1 \rangle - \langle \hat{\psi}_1^\dagger \hat{\psi}_2 \rangle \hat{\psi}_2^\dagger \hat{\psi}_1 + \langle \hat{\psi}_1^\dagger \hat{\psi}_2 \rangle \langle \hat{\psi}_2^\dagger \hat{\psi}_1 \rangle$$
$$+ \hat{\psi}_1^\dagger \hat{\psi}_2^\dagger \langle \hat{\psi}_2 \hat{\psi}_1 \rangle + \langle \hat{\psi}_1^\dagger \hat{\psi}_2^\dagger \rangle \hat{\psi}_2 \hat{\psi}_1 - \langle \hat{\psi}_1^\dagger \hat{\psi}_2^\dagger \rangle \langle \hat{\psi}_2 \hat{\psi}_1 \rangle. \tag{8.50}$$

Let us adopt this in (8.19) and use (8.34)–(8.36) to approximate $\hat{\mathscr{H}}$ by the mean-field Hamiltonian:

$$\hat{\mathscr{H}}_{\text{MF}} \equiv \hat{\psi}^\dagger(\bar{\xi}_1) \mathscr{K}_{\text{HF}}(\bar{\xi}_1, \bar{\xi}_2) \hat{\psi}(\bar{\xi}_2) + \frac{1}{2} \hat{\psi}^\dagger(\bar{\xi}_1) \hat{\psi}^\dagger(\bar{\xi}_2) \Delta(\bar{\xi}_1, \bar{\xi}_2)$$
$$- \frac{1}{2} \hat{\psi}(\bar{\xi}_1) \hat{\psi}(\bar{\xi}_2) \Delta^*(\bar{\xi}_1, \bar{\xi}_2) - \frac{1}{2} \mathscr{U}_{\text{HF}}(\bar{\xi}_1, \bar{\xi}_2) \rho^{(1)}(\bar{\xi}_2, \bar{\xi}_1)$$
$$+ \frac{1}{2} \tilde{\rho}^{(1)}(\bar{\xi}_1, \bar{\xi}_2) \Delta^*(\bar{\xi}_1, \bar{\xi}_2). \tag{8.51}$$

We then use the commutation relation for $\hat{\psi}$ and symmetry in (8.36) to convert the term with $\mathscr{K}_{\rm HF}$ to

$$\begin{aligned}
&\hat{\psi}^\dagger(\bar{\xi}_1)\mathscr{K}_{\rm HF}(\bar{\xi}_1,\bar{\xi}_2)\hat{\psi}(\bar{\xi}_2) = \mathscr{K}_{\rm HF}(\bar{\xi}_1,\bar{\xi}_2)\hat{\psi}^\dagger(\bar{\xi}_1)\hat{\psi}(\bar{\xi}_2)\\
&= \mathscr{K}^*_{\rm HF}(\bar{\xi}_2,\bar{\xi}_1)\left[\delta(\bar{\xi}_1,\bar{\xi}_2) - \hat{\psi}(\bar{\xi}_2)\hat{\psi}^\dagger(\bar{\xi}_1)\right]\\
&= \mathscr{K}_{\rm HF}(\bar{\xi}_1,\bar{\xi}_2)\delta(\bar{\xi}_1,\bar{\xi}_2) - \hat{\psi}(\bar{\xi}_2)\mathscr{K}^*_{\rm HF}(\bar{\xi}_2,\bar{\xi}_1)\hat{\psi}^\dagger(\bar{\xi}_1)\\
&= \frac{1}{2}\Big[\hat{\psi}^\dagger(\bar{\xi}_1)\mathscr{K}_{\rm HF}(\bar{\xi}_1,\bar{\xi}_2)\hat{\psi}(\bar{\xi}_2) + \mathscr{K}_{\rm HF}(\bar{\xi}_1,\bar{\xi}_2)\delta(\bar{\xi}_2,\bar{\xi}_1)\\
&\quad - \hat{\psi}(\bar{\xi}_1)\mathscr{K}^*_{\rm HF}(\bar{\xi}_1,\bar{\xi}_2)\hat{\psi}^\dagger(\bar{\xi}_2)\Big].
\end{aligned}$$

Equation (8.51) is then written

$$\begin{aligned}
\hat{\mathscr{H}}_{\rm MF} &= \frac{1}{2}\left[\hat{\psi}^\dagger(\bar{\xi}_1)\ \hat{\psi}(\bar{\xi}_1)\right]\begin{bmatrix}\mathscr{K}_{\rm HF}(\bar{\xi}_1,\bar{\xi}_2) & \Delta(\bar{\xi}_1,\bar{\xi}_2)\\ -\Delta^*(\bar{\xi}_1,\bar{\xi}_2) & -\mathscr{K}^*_{\rm HF}(\bar{\xi}_1,\bar{\xi}_2)\end{bmatrix}\begin{bmatrix}\hat{\psi}(\bar{\xi}_2)\\ \hat{\psi}^\dagger(\bar{\xi}_2)\end{bmatrix}\\
&\quad + \frac{1}{2}\mathscr{K}_{\rm HF}(\bar{\xi}_1,\bar{\xi}_2)\delta(\bar{\xi}_2,\bar{\xi}_1) - \frac{1}{2}\mathscr{U}_{\rm HF}(\bar{\xi}_1,\bar{\xi}_2)\rho^{(1)}(\bar{\xi}_2,\bar{\xi}_1)\\
&\quad + \frac{1}{2}\tilde{\rho}^{(1)}(\bar{\xi}_1,\bar{\xi}_2)\Delta^*(\bar{\xi}_1,\bar{\xi}_2).
\end{aligned} \tag{8.52}$$

The last three terms on the right-hand side are constants. In contrast, the first term has a bilinear form with respect to $\hat{\psi}^\dagger$ and $\hat{\psi}$; its 2×2 matrix is Hermitian as shown in (8.40). Hence, we can diagonalize it given the expansions

$$\begin{bmatrix}\hat{\psi}(\xi)\\ \hat{\psi}^\dagger(\xi)\end{bmatrix} = \sum_q \begin{bmatrix}u_q(\xi) & v_q^*(\xi)\\ v_q(\xi) & u_q^*(\xi)\end{bmatrix}\begin{bmatrix}\hat{\gamma}_q\\ \hat{\gamma}_q^\dagger\end{bmatrix}, \tag{8.53}$$

$$\left[\hat{\psi}^\dagger(\xi)\ \hat{\psi}(\xi)\right] = \sum_q \left[\hat{\gamma}_q^\dagger\ \hat{\gamma}_q\right]\begin{bmatrix}u_q^*(\xi) & v_q^*(\xi)\\ v_q(\xi) & u_q(\xi)\end{bmatrix}, \tag{8.54}$$

and using (8.43) and (8.44),

$$\begin{aligned}
&\frac{1}{2}\left[\hat{\psi}^\dagger(\bar{\xi}_1)\ \hat{\psi}(\bar{\xi}_1)\right]\begin{bmatrix}\mathscr{K}_{\rm HF}(\bar{\xi}_1,\bar{\xi}_2) & \Delta(\bar{\xi}_1,\bar{\xi}_2)\\ -\Delta^*(\bar{\xi}_1,\bar{\xi}_2) & -\mathscr{K}^*_{\rm HF}(\bar{\xi}_1,\bar{\xi}_2)\end{bmatrix}\begin{bmatrix}\hat{\psi}(\bar{\xi}_2)\\ \hat{\psi}^\dagger(\bar{\xi}_2)\end{bmatrix}\\
&= \frac{1}{2}\sum_q\left[\hat{\psi}^\dagger(\bar{\xi}_1)\ \hat{\psi}(\bar{\xi}_1)\right]\begin{bmatrix}\mathscr{K}_{\rm HF}(\bar{\xi}_1,\bar{\xi}_2) & \Delta(\bar{\xi}_1,\bar{\xi}_2)\\ -\Delta^*(\bar{\xi}_1,\bar{\xi}_2) & -\mathscr{K}^*_{\rm HF}(\bar{\xi}_1,\bar{\xi}_2)\end{bmatrix}\begin{bmatrix}u_q(\bar{\xi}_2) & v_q^*(\bar{\xi}_2)\\ v_q(\bar{\xi}_2) & u_q^*(\bar{\xi}_2)\end{bmatrix}\begin{bmatrix}\hat{\gamma}_q\\ \hat{\gamma}_q^\dagger\end{bmatrix}\\
&= \frac{1}{2}\sum_{q'q}\left[\hat{\gamma}_{q'}^\dagger\ \hat{\gamma}_{q'}\right]\begin{bmatrix}u_{q'}^*(\bar{\xi}_1) & v_{q'}^*(\bar{\xi}_1)\\ v_{q'}(\bar{\xi}_1) & u_{q'}(\bar{\xi}_1)\end{bmatrix}\begin{bmatrix}u_q(\bar{\xi}_1) & v_q^*(\bar{\xi}_1)\\ v_q(\bar{\xi}_1) & u_q^*(\bar{\xi}_1)\end{bmatrix}\begin{bmatrix}E_q & 0\\ 0 & -E_q\end{bmatrix}\begin{bmatrix}\hat{\gamma}_q\\ \hat{\gamma}_q^\dagger\end{bmatrix}
\end{aligned}$$

$$
\begin{aligned}
&= \frac{1}{2}\sum_q \begin{bmatrix} \hat{\gamma}_q^\dagger & \hat{\gamma}_q \end{bmatrix} \begin{bmatrix} E_q & 0 \\ 0 & -E_q \end{bmatrix} \begin{bmatrix} \hat{\gamma}_q \\ \hat{\gamma}_q^\dagger \end{bmatrix} = \frac{1}{2}\sum_q \left(E_q \hat{\gamma}_q^\dagger \hat{\gamma}_q - E_q \hat{\gamma}_q \hat{\gamma}_q^\dagger \right) \\
&= \sum_q E_q \hat{\gamma}_q^\dagger \hat{\gamma}_q - \frac{1}{2}\sum_q E_q .
\end{aligned} \tag{8.55}
$$

The second term in the final expression is divergent. Combining it with the second term in (8.52) and using (8.38), (8.44), and (8.45), we obtain

$$
\begin{aligned}
&\frac{1}{2}\mathscr{K}_{\mathrm{HF}}(\bar{\xi}_1,\bar{\xi}_2)\delta(\bar{\xi}_2,\bar{\xi}_1) - \frac{1}{2}\sum_q E_q \\
&\quad = \frac{1}{2}\mathscr{K}_{\mathrm{HF}}(\bar{\xi}_1,\bar{\xi}_2)\sum_q \left[u_q(\bar{\xi}_2)u_q^*(\bar{\xi}_1) + v_q^*(\bar{\xi}_2)v_q(\bar{\xi}_1) \right] \\
&\quad\quad - \frac{1}{2}\sum_q \begin{bmatrix} u_q^*(\bar{\xi}_1) & v_q^*(\bar{\xi}_1) \end{bmatrix} \begin{bmatrix} \mathscr{K}_{\mathrm{HF}}(\bar{\xi}_1,\bar{\xi}_2) & \Delta(\bar{\xi}_1,\bar{\xi}_2) \\ -\Delta^*(\bar{\xi}_1,\bar{\xi}_2) & -\mathscr{K}_{\mathrm{HF}}^*(\bar{\xi}_1,\bar{\xi}_2) \end{bmatrix} \begin{bmatrix} u_q(\bar{\xi}_2) \\ v_q(\bar{\xi}_2) \end{bmatrix} \\
&\quad = \sum_q \Big[v_q(\bar{\xi}_1)\mathscr{K}_{\mathrm{HF}}(\bar{\xi}_1,\bar{\xi}_2)v_q^*(\bar{\xi}_2) - \frac{1}{2}u_q^*(\bar{\xi}_1)\Delta(\bar{\xi}_1,\bar{\xi}_2)v_q(\bar{\xi}_2) \\
&\quad\quad - \frac{1}{2}u_q(\bar{\xi}_1)\Delta^*(\bar{\xi}_1,\bar{\xi}_2)v_q^*(\bar{\xi}_2) \Big],
\end{aligned} \tag{8.56}
$$

where we have used $\mathscr{K}_{\mathrm{HF}}(\bar{\xi}_1,\bar{\xi}_2) = \mathscr{K}_{\mathrm{HF}}^*(\bar{\xi}_2,\bar{\xi}_1)$ and $\Delta^*(\bar{\xi}_1,\bar{\xi}_2) = -\Delta^*(\bar{\xi}_2,\bar{\xi}_1)$ and also performed a change of variables for the last equality. Using (8.55) and (8.56), we can expand (8.52) as

$$
\begin{aligned}
\hat{\mathscr{H}}_{\mathrm{MF}} &= \sum_q \Big[E_q \hat{\gamma}_q^\dagger \hat{\gamma}_q + v_q(\bar{\xi}_1)\mathscr{K}_{\mathrm{HF}}(\bar{\xi}_1,\bar{\xi}_2)v_q^*(\bar{\xi}_2) - \frac{1}{2}u_q^*(\bar{\xi}_1)\Delta(\bar{\xi}_1,\bar{\xi}_2)v_q(\bar{\xi}_2) \\
&\quad - \frac{1}{2}u_q(\bar{\xi}_1)\Delta^*(\bar{\xi}_1,\bar{\xi}_2)v_q^*(\bar{\xi}_2) \Big] - \frac{1}{2}\mathscr{U}_{\mathrm{HF}}(\bar{\xi}_1,\bar{\xi}_2)\rho^{(1)}(\bar{\xi}_2,\bar{\xi}_1) \\
&\quad + \frac{1}{2}\tilde{\rho}^{(1)}(\bar{\xi}_1,\bar{\xi}_2)\Delta^*(\bar{\xi}_1,\bar{\xi}_2).
\end{aligned} \tag{8.57}
$$

This expression and (8.25) are used to estimate $\langle \hat{\mathscr{H}}_{\mathrm{MF}} \rangle - TS$. We thereby reproduce (8.49) for the grand potential. Thus, the BdG formalism has also been derived by the Wick decomposition procedure.

8.3.3 Matrix Representation of Spin Variables

It is convenient for practical purposes to express every spin variable in (8.38) separately as a matrix.

Specifically, let us introduce the pair of vectors

$$\mathbf{u}_q(\mathbf{r}) \equiv \begin{bmatrix} u_q(\mathbf{r}\uparrow) \\ u_q(\mathbf{r}\downarrow) \end{bmatrix}, \qquad \mathbf{v}_q(\mathbf{r}) \equiv \begin{bmatrix} v_q(\mathbf{r}\uparrow) \\ v_q(\mathbf{r}\downarrow) \end{bmatrix}, \tag{8.58}$$

where $\uparrow$ and $\downarrow$ represent $\alpha = 1/2$ and $\alpha = -1/2$, respectively. Accordingly, the one-particle density matrices (8.29) and (8.30) are transformed into the 2×2 matrices:

$$\underline{\rho}^{(1)}(\mathbf{r}_1, \mathbf{r}_2) = \left[\underline{\rho}^{(1)}(\mathbf{r}_2, \mathbf{r}_1)\right]^{\dagger} \equiv \sum_q \left[\mathbf{u}_q(\mathbf{r}_1)\mathbf{u}_q^{\dagger}(\mathbf{r}_2)\bar{n}_q + \mathbf{v}_q^{*}(\mathbf{r}_1)\mathbf{v}_q^{\mathrm{T}}(\mathbf{r}_2)(1-\bar{n}_q)\right], \tag{8.59}$$

$$\underline{\tilde{\rho}}^{(1)}(\mathbf{r}_1, \mathbf{r}_2) = -\left[\underline{\tilde{\rho}}^{(1)}(\mathbf{r}_2, \mathbf{r}_1)\right]^{\mathrm{T}} \equiv \sum_q \left[\mathbf{u}_q(\mathbf{r}_1)\mathbf{v}_q^{\dagger}(\mathbf{r}_2) - \mathbf{v}_q^{*}(\mathbf{r}_1)\mathbf{u}_q^{\mathrm{T}}(\mathbf{r}_2)\right]\left(\frac{1}{2} - \bar{n}_q\right). \tag{8.60}$$

Indeed, each term on the right are all composed of *dyadics*, i.e., direct products of 2×1 and 1×2 vectors forming a 2×2 matrix. The self-consistent potentials in (8.34) and (8.35) are also expressible as 2×2 matrices,

$$\begin{aligned} \underline{\mathscr{U}}_{\mathrm{HF}}(\mathbf{r}_1, \mathbf{r}_2) = \underline{\mathscr{U}}_{\mathrm{HF}}^{\dagger}(\mathbf{r}_2, \mathbf{r}_1) &\equiv \underline{\sigma}_0 \delta(\mathbf{r}_1, \mathbf{r}_2) \int d^3 r_3 \mathscr{V}(|\mathbf{r}_1 - \mathbf{r}_3|) \mathrm{Tr}\, \underline{\rho}^{(1)}(\mathbf{r}_3, \mathbf{r}_3) \\ &\quad - \mathscr{V}(|\mathbf{r}_1 - \mathbf{r}_2|)\, \underline{\rho}^{(1)}(\mathbf{r}_1, \mathbf{r}_2), \end{aligned} \tag{8.61}$$

$$\underline{\Delta}(\mathbf{r}_1, \mathbf{r}_2) = -\underline{\Delta}^{\mathrm{T}}(\mathbf{r}_2, \mathbf{r}_1) \equiv -\mathscr{V}(|\mathbf{r}_1 - \mathbf{r}_2|)\, \underline{\tilde{\rho}}^{(1)}(\mathbf{r}_1, \mathbf{r}_2), \tag{8.62}$$

where $\underline{\sigma}_0$ is the 2×2 unit matrix, given in (8.42), and Tr denotes the trace. Similarly, the Hartree–Fock operator (8.36) becomes

$$\underline{\mathscr{K}}_{\mathrm{HF}}(\mathbf{r}_1, \mathbf{r}_2) \equiv \left[\frac{\hat{\mathbf{p}}_1^2}{2m} + \mathscr{U}(\mathbf{r}_1) - \mu\right] \delta(\mathbf{r}_1, \mathbf{r}_2)\underline{\sigma}_0 + \underline{\mathscr{U}}_{\mathrm{HF}}(\mathbf{r}_1, \mathbf{r}_2). \tag{8.63}$$

With this notation, (8.38) reads

$$\int d^3 r_2 \begin{bmatrix} \underline{\mathscr{K}}_{\mathrm{HF}}(\mathbf{r}_1, \mathbf{r}_2) & \underline{\Delta}(\mathbf{r}_1, \mathbf{r}_2) \\ -\underline{\Delta}^{*}(\mathbf{r}_1, \mathbf{r}_2) & -\underline{\mathscr{K}}_{\mathrm{HF}}^{*}(\mathbf{r}_1, \mathbf{r}_2) \end{bmatrix} \begin{bmatrix} \mathbf{u}_q(\mathbf{r}_2) \\ \mathbf{v}_q(\mathbf{r}_2) \end{bmatrix} = E_q \begin{bmatrix} \mathbf{u}_q(\mathbf{r}_1) \\ \mathbf{v}_q(\mathbf{r}_1) \end{bmatrix}, \tag{8.64}$$

with

$$\int \left[|\mathbf{u}_q(\mathbf{r})|^2 + |\mathbf{v}_q(\mathbf{r})|^2\right] d^3 r = 1. \tag{8.65}$$

We only need to consider states with $E_q \geq 0$ in consequence of the particle-hole symmetry (8.43). Indeed, the eigenstate with $-E_q$ is given by $\left[\mathbf{v}_q^\dagger(\mathbf{r})\ \mathbf{u}_q^\dagger(\mathbf{r})\right]^{\mathrm{T}}$ in terms of $\mathbf{u}_q$ and $\mathbf{v}_q$ above with E_q.

With (8.58)–(8.63), the grand potential (8.49) in equilibrium now becomes

$$\Omega_{\mathrm{BdG}} = -\frac{1}{\beta}\sum_q \ln\left(1+\mathrm{e}^{-\beta E_q}\right) + \sum_q \int \mathrm{d}^3 r_1 \int \mathrm{d}^3 r_2 \Big[\mathbf{v}_q^{\mathrm{T}}(\mathbf{r}_1)\underline{\mathscr{K}}_{\mathrm{HF}}(\mathbf{r}_1,\mathbf{r}_2)\mathbf{v}_q^*(\mathbf{r}_2)$$
$$+\frac{1}{2}\mathrm{Tr}\,\mathbf{u}_q^*(\mathbf{r}_1)\mathbf{v}_q^{\mathrm{T}}(\mathbf{r}_2)\underline{\Delta}(\mathbf{r}_2,\mathbf{r}_1) + \frac{1}{2}\mathrm{Tr}\,\mathbf{u}_q(\mathbf{r}_1)\mathbf{v}_q^\dagger(\mathbf{r}_2)\underline{\Delta}^*(\mathbf{r}_2,\mathbf{r}_1)$$
$$-\frac{1}{2}\mathrm{Tr}\,\underline{\rho}^{(1)}(\mathbf{r}_1,\mathbf{r}_2)\underline{\mathscr{U}}_{\mathrm{HF}}(\mathbf{r}_2,\mathbf{r}_1) - \frac{1}{2}\mathrm{Tr}\,\underline{\tilde{\rho}}^{(1)}(\mathbf{r}_1,\mathbf{r}_2)\underline{\Delta}^*(\mathbf{r}_2,\mathbf{r}_1)\Big]. \tag{8.66}$$

Finally, the particle number (8.48) is rewritten given the one-particle density matrix of (8.59),

$$N = \int \mathrm{d}^3 r\ \mathrm{Tr}\,\underline{\rho}^{(1)}(\mathbf{r},\mathbf{r}). \tag{8.67}$$

8.3.4 BdG Equations for Homogeneous Cases

When the system is homogeneous with no external potential ($\mathscr{U} = 0$), we can simplify the BdG equations considerably.

Adopting periodic boundary conditions, a quasiparticle eigenstate q can be specified with a wave vector $\mathbf{k}$ and spin index $\tilde{\alpha} = 1, 2$ as $q \equiv \mathbf{k}\tilde{\alpha}$, where $\tilde{\alpha}$ is generally some linear combination of $\alpha = \uparrow, \downarrow$. The corresponding eigenfunction can be expressed as a plane wave,

$$\begin{bmatrix}\mathbf{u}_{\mathbf{k}\tilde{\alpha}}(\mathbf{r})\\ \mathbf{v}_{\mathbf{k}\tilde{\alpha}}(\mathbf{r})\end{bmatrix} = \frac{1}{\sqrt{V}}\mathrm{e}^{\mathrm{i}\mathbf{k}\cdot\mathbf{r}}\begin{bmatrix}\mathbf{u}_{\tilde{\alpha}}(\mathbf{k})\\ \mathbf{v}_{\tilde{\alpha}}(\mathbf{k})\end{bmatrix}. \tag{8.68}$$

Substituting this expression into (8.59) and (8.60), we obtain expansions of the one-particle density matrices

$$\underline{\rho}^{(1)}(\mathbf{r}_1,\mathbf{r}_2) = \frac{1}{V}\sum_{\mathbf{k}} \underline{\rho}^{(1)}(\mathbf{k})\,\mathrm{e}^{\mathrm{i}\mathbf{k}\cdot(\mathbf{r}_1-\mathbf{r}_2)}, \tag{8.69}$$

$$\underline{\tilde{\rho}}^{(1)}(\mathbf{r}_1,\mathbf{r}_2) = \frac{1}{V}\sum_{\mathbf{k}} \underline{\tilde{\rho}}^{(1)}(\mathbf{k})\,\mathrm{e}^{\mathrm{i}\mathbf{k}\cdot(\mathbf{r}_1-\mathbf{r}_2)}, \tag{8.70}$$

with

$$\underline{\rho}^{(1)}(\mathbf{k}) = \left[\underline{\rho}^{(1)}(\mathbf{k})\right]^\dagger = \sum_{\tilde{\alpha}}\left[\mathbf{u}_{\tilde{\alpha}}(\mathbf{k})\mathbf{u}_{\tilde{\alpha}}^\dagger(\mathbf{k})\bar{n}_{\mathbf{k}\tilde{\alpha}} + \mathbf{v}_{\tilde{\alpha}}^*(-\mathbf{k})\mathbf{v}_{\tilde{\alpha}}^{\mathrm{T}}(-\mathbf{k})(1-\bar{n}_{-\mathbf{k}\tilde{\alpha}})\right], \tag{8.71}$$

$$\underline{\tilde{\rho}}^{(1)}(\mathbf{k}) = -\left[\underline{\tilde{\rho}}^{(1)}(-\mathbf{k})\right]^{\mathrm{T}}$$
$$= \sum_{\tilde{\alpha}} \left[\mathbf{u}_{\tilde{\alpha}}(\mathbf{k})\mathbf{v}_{\tilde{\alpha}}^{\dagger}(\mathbf{k})\left(\frac{1}{2} - \bar{n}_{\mathbf{k}\tilde{\alpha}}\right) - \mathbf{v}_{\tilde{\alpha}}^{*}(-\mathbf{k})\mathbf{u}_{\tilde{\alpha}}^{\mathrm{T}}(-\mathbf{k})\left(\frac{1}{2} - \bar{n}_{-\mathbf{k}\tilde{\alpha}}\right)\right]. \quad (8.72)$$

Next, we substitute (6.18), (8.69), and (8.70) into (8.61) and (8.62) to expand the self-consistent potentials as

$$\underline{\mathscr{U}}_{\mathrm{HF}}(\mathbf{r}_1, \mathbf{r}_2) = \frac{1}{V}\sum_{\mathbf{k}} \underline{\mathscr{U}}_{\mathrm{HF}}(\mathbf{k})\, \mathrm{e}^{\mathrm{i}\mathbf{k}\cdot(\mathbf{r}_1-\mathbf{r}_2)}, \quad (8.73)$$

$$\underline{\Delta}(\mathbf{r}_1, \mathbf{r}_2) = \frac{1}{V}\sum_{\mathbf{k}} \underline{\Delta}(\mathbf{k})\, \mathrm{e}^{\mathrm{i}\mathbf{k}\cdot(\mathbf{r}_1-\mathbf{r}_2)}, \quad (8.74)$$

with

$$\underline{\mathscr{U}}_{\mathrm{HF}}(\mathbf{k}) = \underline{\mathscr{U}}_{\mathrm{HF}}^{\dagger}(\mathbf{k}) = \frac{1}{V}\sum_{\mathbf{k}'} \left[\underline{\sigma}_0 \mathscr{V}_0 \mathrm{Tr}\, \underline{\rho}^{(1)}(\mathbf{k}') - \mathscr{V}_{|\mathbf{k}-\mathbf{k}'|}\, \underline{\rho}^{(1)}(\mathbf{k}')\right], \quad (8.75)$$

$$\underline{\Delta}(\mathbf{k}) = -\underline{\Delta}^{\mathrm{T}}(-\mathbf{k}) = -\frac{1}{V}\sum_{\mathbf{k}'} \mathscr{V}_{|\mathbf{k}-\mathbf{k}'|}\, \underline{\tilde{\rho}}^{(1)}(\mathbf{k}'). \quad (8.76)$$

It also follows from (6.18) and (8.73) that the Hartree–Fock operator (8.63) with $\mathscr{U} = 0$ is expressible as

$$\underline{\mathscr{H}}_{\mathrm{HF}}(\mathbf{r}_1, \mathbf{r}_2) = \frac{1}{V}\sum_{\mathbf{k}} \underline{\mathscr{H}}_{\mathrm{HF}}(\mathbf{k})\, \mathrm{e}^{\mathrm{i}\mathbf{k}\cdot(\mathbf{r}_1-\mathbf{r}_2)}, \quad (8.77)$$

with

$$\underline{\mathscr{H}}_{\mathrm{HF}}(\mathbf{k}) \equiv \left(\frac{\hbar^2 k^2}{2m} - \mu\right)\underline{\sigma}_0 + \underline{\mathscr{U}}_{\mathrm{HF}}(\mathbf{k}). \quad (8.78)$$

Substituting (8.68), (8.74), and (8.77) into (8.64) and performing the integration, we thereby obtain the BdG equation for homogeneous systems,

$$\begin{bmatrix} \underline{\mathscr{H}}_{\mathrm{HF}}(\mathbf{k}) & \underline{\Delta}(\mathbf{k}) \\ -\underline{\Delta}^{*}(-\mathbf{k}) & -\underline{\mathscr{H}}_{\mathrm{HF}}^{*}(-\mathbf{k}) \end{bmatrix} \begin{bmatrix} \mathbf{u}_{\tilde{\alpha}}(\mathbf{k}) \\ \mathbf{v}_{\tilde{\alpha}}(\mathbf{k}) \end{bmatrix} = E_{\tilde{\alpha}}(\mathbf{k}) \begin{bmatrix} \mathbf{u}_{\tilde{\alpha}}(\mathbf{k}) \\ \mathbf{v}_{\tilde{\alpha}}(\mathbf{k}) \end{bmatrix}, \quad (8.79)$$

with

$$\left|\mathbf{u}_{\tilde{\alpha}}(\mathbf{k})\right|^2 + \left|\mathbf{v}_{\tilde{\alpha}}(\mathbf{k})\right|^2 = 1. \quad (8.80)$$

Equation (8.79) constitutes an eigenvalue problem of a 4×4 matrix, which may be solved without much difficulty. However, the potentials in the 4×4 space are given by (8.75) and (8.76) in terms of the one-particle density matrices (8.71) and (8.72), which include the eigenvalues $E_{\tilde{\alpha}}(\mathbf{k}) \geq 0$ and eigenfunctions $\left[\mathbf{u}_{\tilde{\alpha}}^{\mathrm{T}}(\mathbf{k})\, \mathbf{v}_{\tilde{\alpha}}^{\mathrm{T}}(\mathbf{k})\right]^{\mathrm{T}}$ to be obtained. Thus, (8.75), (8.76), and (8.79) have to be solved self-consistently until convergence is reached. Finally, from (8.43), the eigenvector of (8.79) belonging to $-E_{\tilde{\alpha}}(\mathbf{k})$ is given by $\left[\mathbf{v}_{\tilde{\alpha}}^{\dagger}(-\mathbf{k})\ \mathbf{u}_{\tilde{\alpha}}^{\dagger}(-\mathbf{k})\right]^{\mathrm{T}}$.

The grand potential (8.66) is also simplified using (8.68)–(8.70), (8.73), (8.74), and (8.77), and then performing the integration,

$$\begin{aligned}\Omega_{\mathrm{BdG}} = \sum_{\mathbf{k}\tilde{\alpha}} \Big\{ & -\frac{1}{\beta}\ln\left[1+\mathrm{e}^{-\beta E_{\tilde{\alpha}}(\mathbf{k})}\right] + \mathbf{v}_{\tilde{\alpha}}^{\mathrm{T}}(-\mathbf{k})\underline{\mathscr{K}}_{\mathrm{HF}}(\mathbf{k})\mathbf{v}_{\tilde{\alpha}}^{*}(-\mathbf{k}) \\ & + \frac{1}{2}\mathrm{Tr}\,\mathbf{u}_{\tilde{\alpha}}^{*}(\mathbf{k})\mathbf{v}_{\tilde{\alpha}}^{\mathrm{T}}(\mathbf{k})\underline{\Delta}(-\mathbf{k}) + \frac{1}{2}\mathrm{Tr}\,\mathbf{u}_{\tilde{\alpha}}(\mathbf{k})\mathbf{v}_{\tilde{\alpha}}^{\dagger}(\mathbf{k})\underline{\Delta}^{*}(-\mathbf{k}) \\ & - \frac{1}{2}\mathrm{Tr}\,\underline{\rho}^{(1)}(\mathbf{k})\underline{\mathscr{U}}_{\mathrm{HF}}(\mathbf{k}) - \frac{1}{2}\mathrm{Tr}\,\underline{\tilde{\rho}}^{(1)}(\mathbf{k})\underline{\Delta}^{*}(-\mathbf{k}) \Big\}. \end{aligned} \tag{8.81}$$

It follows from (8.67) and (8.69) that the particle number N is expressible in terms of (8.71) as

$$N = \sum_{\mathbf{k}} \mathrm{Tr}\,\underline{\rho}^{(1)}(\mathbf{k}). \tag{8.82}$$

8.4 Expansion of Pairing Interaction

8.4.1 Isotropic Cases

Equation (8.76) for homogeneous systems is called the *gap equation*, which for isotropic cases can be simplified further.

First, the Fourier component $\mathscr{V}_{|\mathbf{k}-\mathbf{k}'|}$ is expanded as a harmonic series

$$\mathscr{V}_{|\mathbf{k}-\mathbf{k}'|} = \sum_{\ell=0}^{\infty} \mathscr{V}_{\ell}(k,k') \sum_{m=-\ell}^{\ell} 4\pi Y_{\ell m}(\hat{\mathbf{k}})Y_{\ell m}^{*}(\hat{\mathbf{k}}'), \tag{8.83}$$

where $\hat{\mathbf{k}} \equiv \mathbf{k}/k$ is a unit vector, and $Y_{\ell m}(\hat{\mathbf{k}})$ are the spherical harmonic functions [3, 9]; adopting polar coordinates $\hat{\mathbf{k}} = (\sin\theta_{\mathbf{k}}\cos\varphi_{\mathbf{k}}, \sin\theta_{\mathbf{k}}\sin\varphi_{\mathbf{k}}, \cos\theta_{\mathbf{k}})$, they are defined as

$$Y_{\ell m}(\hat{\mathbf{k}}) = \frac{(-1)^{\ell}}{2^{\ell}\ell!}\sqrt{\frac{(2\ell+1)(\ell+m)!}{4\pi(\ell-m)!}}\,\frac{1}{(1-z^2)^{m/2}}\frac{\mathrm{d}^{\ell-m}}{\mathrm{d}z^{\ell-m}}(1-z^2)^{\ell}\bigg|_{z=\cos\theta_{\mathbf{k}}} \mathrm{e}^{\mathrm{i}m\varphi_{\mathbf{k}}}. \tag{8.84}$$

They obey orthonormality conditions given with respect to an integration over solid angle $\mathrm{d}\Omega_{\mathbf{k}}$

$$\int \mathrm{d}\Omega_{\mathbf{k}} \equiv \int_0^{\pi} \mathrm{d}\theta_{\mathbf{k}}\sin\theta_{\mathbf{k}} \int_0^{2\pi} \mathrm{d}\varphi_{\mathbf{k}}, \tag{8.85}$$

as

$$\int \mathrm{d}\Omega_{\mathbf{k}} Y^*_{\ell' m'}(\hat{\mathbf{k}}) Y_{\ell m}(\hat{\mathbf{k}}) = \delta_{\ell' \ell}\delta_{m' m}. \tag{8.86}$$

Moreover, $Y_{\ell m}$ obeys a sum rule in connection with the Legendre polynomial $P_\ell(x)$ defined in (6.35) as [3, 9]

$$P_\ell(\hat{\mathbf{k}} \cdot \hat{\mathbf{k}}') = \frac{4\pi}{2\ell + 1} \sum_{m=-\ell}^{\ell} Y_{\ell m}(\hat{\mathbf{k}}) Y^*_{\ell m}(\hat{\mathbf{k}}').$$

Indeed, (8.83) with this sum rule produces (6.51).

Substitution of (8.83) into (8.76) yields an expansion for $\underline{\Delta}(\mathbf{k})$

$$\underline{\Delta}(\mathbf{k}) = \sum_{\ell=0}^{\infty} \sum_{m=-\ell}^{\ell} \underline{\Delta}_{\ell m}(k) \sqrt{4\pi}\, Y_{\ell m}(\hat{\mathbf{k}}), \tag{8.87}$$

with

$$\underline{\Delta}_{\ell m}(k) = -\frac{1}{V} \sum_{\mathbf{k}'} \mathscr{V}_\ell(k, k') \sqrt{4\pi}\, Y^*_{\ell m}(\hat{\mathbf{k}}') \underline{\tilde{\rho}}^{(1)}(\mathbf{k}'). \tag{8.88}$$

In most cases, only a single ℓ is known to yield finite $\{\underline{\Delta}_{\ell m}(k)\}$. Cases $\ell = 0, 1, 2$ are called s-wave, p-wave, and d-wave pairing, respectively. Making a change of integration variables $\mathbf{k}' \to -\mathbf{k}'$ in (8.88) and using the symmetry (8.72) and $Y_{\ell m}(-\hat{\mathbf{k}}') = (-1)^\ell Y_{\ell m}(\hat{\mathbf{k}}')$, we realize that $\underline{\Delta}_{\ell m}(k)$ satisfies

$$\underline{\Delta}_{\ell m}(k) = (-1)^{\ell+1} \underline{\Delta}^{\mathrm{T}}_{\ell m}(k). \tag{8.89}$$

The sum over $\mathbf{k}$ in (8.88) can be replaced by an integral as

$$\frac{1}{V}\sum_{\mathbf{k}} = \int \frac{\mathrm{d}^3 k}{(2\pi)^3} = \int_{-\infty}^{\infty} \mathrm{d}\varepsilon_k N(\varepsilon_k) \int \frac{\mathrm{d}\Omega_{\mathbf{k}}}{4\pi}, \tag{8.90}$$

where $\mathrm{d}\Omega_{\mathbf{k}}$ is given by (8.85) and $N(\varepsilon)$ is the density of states per unit volume and spin component defined by (7.21), which is further related to (4.17) by $N(\epsilon) = D(\epsilon)/(2s+1)V$.

8.4.2 Anisotropic Cases

We discuss how the expansion (8.83) for isotropic systems may be generalized for anisotropic systems; this part can be omitted completely for proceeding to later chapters.

Conduction electrons in a metal move in the lattice potential created by the protons in the nucleus and the core electrons. Accordingly, Hamiltonian $\hat{\mathscr{H}}$ for conduction electrons now commute only with those discrete symmetry operations $\{\hat{R}\}$ that keep the crystal structure invariant. They form the group G that maintain $\hat{\mathscr{H}}$ invariant as $\hat{R}\hat{\mathscr{H}}\hat{R}^{-1} = \hat{\mathscr{H}}$. The symmetry operations relevant here are the coset G/T of G by the translation subgroup T, which coincides for simple crystal structures with the point group [12, 16]. In many cases, the time-reversal operator can also be added to these symmetry operations.

The one-particle energy $\varepsilon^b_{\mathbf{k},\alpha}$ of the conduction electrons in zero magnetic field may be specified by the wave vector $\mathbf{k}$ in the first Brillouin zone, "spin" index α, and band index b. It satisfies $\hat{R}\varepsilon^b_{\mathbf{k}\alpha} = \varepsilon^b_{\mathbf{k}\alpha}$ in the wave vector domain reflecting the symmetry of $\hat{\mathscr{H}}$.

Similarly, the interaction potential relevant to the pairing may also be written in terms of state labels $(\mathbf{k}, \alpha, b)$ as $\mathscr{V}_{\mathbf{k}\mathbf{k}'}$ for a single-band model with no spin dependence or $\mathscr{V}_{\mathbf{k}\alpha b,\mathbf{k}'\alpha' b'}$ for the most general situations. The first-principle calculations for $\mathscr{V}_{\mathbf{k}\mathbf{k}'}$ are generally difficult to perform, but we may assume that it is Hermitian $\mathscr{V}^*_{\mathbf{k}\mathbf{k}'} = \mathscr{V}_{\mathbf{k}'\mathbf{k}}$ and satisfies symmetry relations $\hat{R}\mathscr{V}_{\mathbf{k}\mathbf{k}'}\hat{R}^{-1} = \mathscr{V}_{\mathbf{k}\mathbf{k}'}$. With these properties, we can expand it as

$$\mathscr{V}_{\mathbf{k}\mathbf{k}'} = \sum_{\Gamma j \gamma} \mathscr{V}_{\Gamma j}\, \phi_{\Gamma j \gamma}(\mathbf{k})\, \phi^*_{\Gamma j \gamma}(\mathbf{k}'), \tag{8.91}$$

where $\mathscr{V}_{\Gamma j}$ denotes the jth eigenvalue associated with the irreducible representation Γ for G/T [12, 16] and $\phi_{\Gamma j \gamma}(\mathbf{k})$ is its γth eigenfunction, so that the transformed basis $\hat{R}\phi_{\Gamma j \gamma}(\mathbf{k})$ for any $R \in G/T$ can be expressed as a linear combination of the bases $\{\phi_{\Gamma j \gamma'}(\mathbf{k})\}_{\gamma'}$ within the same (Γ, j). We remark that (8.83) is included as a special case of (8.91) under the mapping $\Gamma \to \ell$, $\gamma \to m$, $\phi_{\Gamma j \gamma} \to \sqrt{4\pi} Y_{\ell m}$; index j can be removed for isotropic systems because there is only a single eigenvalue for each ℓ. For general cases, (8.91) should be modified with the replacement $\mathbf{k} \to \mathbf{k}\alpha b$.

Similarly, (8.90) for the isotropic model is now replaced by

$$\frac{1}{N_a}\sum_{\mathbf{k}} = \int \frac{\mathrm{d}^3 k}{(2\pi)^3} = \int_{-\infty}^{\infty} \mathrm{d}\varepsilon_{\mathbf{k}} N(\varepsilon_{\mathbf{k}}) \int \mathrm{d}S_{\mathbf{k}}, \tag{8.92}$$

where N_a is the number of unit cells in the system,

$$N(\epsilon) \equiv \frac{1}{N_a}\sum_{\mathbf{k}} \delta(\epsilon - \varepsilon_{\mathbf{k}}) \tag{8.93}$$

denotes *the density of states per unit cell and spin component*, and $\mathrm{d}S_{\mathbf{k}}$ is the 'volume' element on the equal-energy surface $\varepsilon_{\mathbf{k}} = \epsilon$ with normalization condition $\int \mathrm{d}S_{\mathbf{k}} = 1$.

Problems

8.1. The pair wave function ϕ in (8.3) for the homogeneous spin-singlet pairing can be expanded in terms of the relative coordinates $\mathbf{r}_1 - \mathbf{r}_2$ as

$$\phi(\mathbf{r}_1\alpha_1, \mathbf{r}_2\alpha_2) = \frac{1}{V}\sum_{\mathbf{k}} \phi_{\mathbf{k}}\, e^{i\mathbf{k}\cdot(\mathbf{r}_1-\mathbf{r}_2)}\left(\delta_{\alpha_1\uparrow}\delta_{\alpha_2\downarrow} - \delta_{\alpha_1\downarrow}\delta_{\alpha_2\uparrow}\right). \tag{8.94}$$

where $\uparrow$ and $\downarrow$ denote $\alpha = 1/2$ and $-1/2$, respectively. It also follows from (8.2) that $\phi_{\mathbf{k}} = \phi_{-\mathbf{k}}$ holds.

(a) Show that the Cooper-pair creation operator (8.3) can be written as

$$\hat{Q}^\dagger = \sum_{\mathbf{k}} \phi_{\mathbf{k}} \hat{c}^\dagger_{\mathbf{k}\uparrow}\hat{c}^\dagger_{-\mathbf{k}\downarrow}, \qquad \hat{c}^\dagger_{\mathbf{k}\alpha} \equiv \int \hat{\psi}^\dagger(\mathbf{r}\alpha)\frac{1}{\sqrt{V}} e^{i\mathbf{k}\cdot\mathbf{r}} d^3r.$$

(b) Show that the condensate wave function (8.6) is expressible as

$$|\Phi\rangle = \prod_{\mathbf{k}} \left(u_{\mathbf{k}} + v_{\mathbf{k}} \hat{c}^\dagger_{\mathbf{k}\uparrow}\hat{c}^\dagger_{-\mathbf{k}\downarrow}\right)|0\rangle, \tag{8.95}$$

with $u_{\mathbf{k}} \equiv 1/\sqrt{1+|\phi_{\mathbf{k}}|^2}$ and $v_{\mathbf{k}} \equiv \phi_{\mathbf{k}}/\sqrt{1+|\phi_{\mathbf{k}}|^2}$. This is exactly the wave function Schrieffer wrote down.

8.2. In the context of Cooper-pair condensation, answer the following questions.

(a) Show that the Cooper-pair creation operator (8.3) satisfies

$$[\hat{Q}, \hat{Q}^\dagger]_+ = \frac{1}{2}|\phi(\bar{\xi}_1, \bar{\xi}_2)|^2 + \hat{\psi}^\dagger(\bar{\xi}_1)\phi(\bar{\xi}_1, \bar{\xi}_2)\phi^*(\bar{\xi}_2, \bar{\xi}_1')\hat{\psi}(\bar{\xi}_1'). \tag{8.96}$$

(b) The second term in (8.96) is proportional to the overlap integral

$$\phi(\bar{\xi}_1, \bar{\xi}_2)\phi^*(\bar{\xi}_2, \bar{\xi}_1'),$$

so that it may be negligible when the radius of the bound state is smaller than the mean interparticle spacing. Show that (8.6) for this case reduces to the coherent state (3.71), where Θ is now a complex number with magnitude $\left[|\phi(\bar{\xi}_1, \bar{\xi}_2)|^2/2\right]^{1/2}$, and operator $\hat{c}^\dagger \equiv \hat{Q}^\dagger/\Theta$ satisfies $[\hat{c}, \hat{c}^\dagger]_+ = 1$.

References

1. V. Ambegaokar, in *Superconductivity*, vol. I, ed. R.D. Parks (Marcel Dekker, New York, 1969), Chap. 5
2. A.F. Andreev, J. Exp. Theor. Phys. **46**, 1823 (1964). (Sov. Phys. JETP **19**, 1228 (1964))

3. G.B. Arfken, H.J. Weber, *Mathematical Methods for Physicists* (Academic, New York, 2012)
4. N.N. Bogoliubov, J. Exp. Theor. Phys. **34**, 58 (1958). (Sov. Phys. JETP **7**, 41 (1958)); Nuovo Cimento **7**, 794 (1958)
5. N.N. Bogoliubov, V.V. Tolmachev, D.V. Shirkov, *A New Method in the Theory of Superconductivity* (Consultants Bureau, New York, 1959)
6. C. Caroli, P.G. de Gennes, J. Matricon, Phys. Lett. **9**, 307 (1964)
7. L.N. Cooper, D. Feldman (eds.), *BCS: 50 Years* (World Scientific, Hackensack, 2011)
8. P.G. de Gennes, *Superconductivity of Metals and Alloys* (W.A. Benjamin, New York, 1966)
9. A.R. Edmonds, *Angular Momentum in Quantum Mechanics* (Princeton University Press, Princeton, 1957)
10. R.J. Glauber, Phys. Rev. **131**, 2766 (1963)
11. L.P. Gor'kov, J. Exp. Theor. Phys. **36**, 1918 (1959). (Sov. Phys. JETP **9**, 1364 (1959)); J. Exp. Theor. Phys. **37**, 1407 (1959). (Sov. Phys. JETP **10**, 998 (1960))
12. T. Inui, Y. Tanabe, Y. Onodera, *Group Theory and Its Applications in Physics* (Springer, Berlin, 1990)
13. M. Ishikawa, Prog. Theor. Phys. **57**, 1836 (1977)
14. T. Kita, J. Phys. Soc. Jpn. **65**, 1355 (1996)
15. E.C.G. Sudarshan, Phys. Rev. Lett. **10**, 277 (1963)
16. M. Tinkham, *Group Theory and Quantum Mechanics* (McGraw-Hill, New York, 1964)
17. S. Tomonaga, Prog. Theor. Phys. **2**, 6 (1947)
18. J.G. Valatin, Nuovo Cimento **7**, 843 (1958)

Chapter 9
BCS Theory

Abstract Immediately noting the correctness of Schrieffer's variational wave function, Bardeen, with his deep knowledge of the phenomenon (Cooper LN, Feldman D (eds) BCS: 50 years. World Scientific, Hackensack, 2011), teamed up with Cooper and Schrieffer to construct a microscopic theory of superconductivity. The BCS theory was thereby developed quite rapidly to bring remarkable agreement between theory and various experiments on single-element superconductors. This is because in these superconductors the relevant attraction in Cooper-pair condensation is weak, making the mean-field description appropriate. In this chapter, we derive the main thermodynamic results of the BCS theory for homogeneous s-wave superconductors based on the formalism developed in the previous chapter.

9.1 Self-Consistency Equations

We shall derive the quasiparticle eigenenergies and eigenstates of the homogenous s-wave pairing as (9.5) and (9.6), and the self-consistency equations for the Hartree-Fock and pair potentials as (9.12) and (9.13).

The BCS theory considers the possibility of homogeneous s-wave pairing [2]. It follows from (8.87), (8.89), and $Y_{00}(\hat{\mathbf{k}}) = (4\pi)^{-1/2}$ [1, 4] that the gap matrix of this s-wave pairing is isotropic as $\underline{\Delta}(\mathbf{k}) = \underline{\Delta}_{00}(k)$ with symmetry $\underline{\Delta}_{00}(k) = -\underline{\Delta}_{00}^{\mathrm{T}}(k)$. Hence, we can express $\underline{\Delta}(\mathbf{k})$ as

$$\underline{\Delta}(\mathbf{k}) = \begin{bmatrix} 0 & \Delta_k \\ -\Delta_k & 0 \end{bmatrix} = i\underline{\sigma}_y \Delta_k, \tag{9.1}$$

where $\underline{\sigma}_y$ is the y component of the Pauli matrices in (8.42). Thus, $\Delta_{\uparrow\downarrow}(\mathbf{k}) = -\Delta_{\downarrow\uparrow}(\mathbf{k}) = \Delta_k$ whereas $\Delta_{\uparrow\uparrow}(\mathbf{k}) = \Delta_{\downarrow\downarrow}(\mathbf{k}) = 0$, implying that the s-wave superconductivity is caused by Cooper pairs composed of a pair of $\uparrow$-spin and $\downarrow$-spin electrons in the spin-singlet state; see also **Problem** 8.1 on this point. In the absence of magnetic fields, the self-consistent Hartree–Fock potential is also expected to be diagonal and isotropic; that is, $\underline{\mathscr{U}}_{\mathrm{HF}}(\mathbf{k}) = \underline{\sigma}_0 \mathscr{U}_k^{\mathrm{HF}}$. Accordingly, we can express (8.78) as

© Springer Japan 2015

T. Kita, *Statistical Mechanics of Superconductivity*, Graduate Texts in Physics,
DOI 10.1007/978-4-431-55405-9_9

$$\underline{\mathscr{K}}_{\mathrm{HF}}(\mathbf{k}) = \underline{\sigma}_0\,\xi_k, \qquad \xi_k \equiv \frac{\hbar^2 k^2}{2m} + \mathscr{U}_k^{\mathrm{HF}} - \mu. \tag{9.2}$$

Substituting (9.1) and (9.2) into (8.79) yields the eigenvalue problem:

$$\begin{bmatrix} \xi_k & 0 & 0 & \Delta_k \\ 0 & \xi_k & -\Delta_k & 0 \\ 0 & -\Delta_k^* & -\xi_k & 0 \\ \Delta_k^* & 0 & 0 & -\xi_k \end{bmatrix} \begin{bmatrix} u_{\tilde{\alpha}}(k\uparrow) \\ u_{\tilde{\alpha}}(k\downarrow) \\ v_{\tilde{\alpha}}(k\uparrow) \\ v_{\tilde{\alpha}}(k\downarrow) \end{bmatrix} = E_{\tilde{\alpha}}(k) \begin{bmatrix} u_{\tilde{\alpha}}(k\uparrow) \\ u_{\tilde{\alpha}}(k\downarrow) \\ v_{\tilde{\alpha}}(k\uparrow) \\ v_{\tilde{\alpha}}(k\downarrow) \end{bmatrix}. \tag{9.3}$$

Diagonalization of this 4×4 matrix is equivalent to that for the (1,4)- and (2,3)-submatrices. The above equation separates into two equations of the form

$$\begin{bmatrix} \xi_k & \pm\Delta_k \\ \pm\Delta_k^* & -\xi_k \end{bmatrix} \begin{bmatrix} u_k \\ \pm v_k \end{bmatrix} = E_k \begin{bmatrix} u_k \\ \pm v_k \end{bmatrix}, \tag{9.4}$$

where the upper and lower signs correspond to the (1,4)- and (2,3)-submatrices, respectively. Here we have adopted a simplified notation for the eigenvector. The eigenvalue equation is identical for both, $(\xi_k - E_k)(-\xi_k - E_k) - |\Delta_k|^2 = 0$, which yields a positive eigenvalue

$$E_k = \sqrt{\xi_k^2 + |\Delta_k|^2}. \tag{9.5}$$

Its eigenvector can be determined based on the second row of (9.4) and normalization (8.80), which read

$$\Delta_k^* u_k - (E_k + \xi_k) v_k = 0, \qquad |u_k|^2 + |v_k|^2 = 1,$$

respectively. They yield

$$u_k = \frac{E_k + \xi_k}{\sqrt{(E_k + \xi_k)^2 + |\Delta_k|^2}} = \sqrt{\frac{E_k + \xi_k}{2E_k}}, \qquad v_k = \frac{\Delta_k^*}{\sqrt{2E_k(E_k + \xi_k)}}, \tag{9.6}$$

where we have chosen u_k as real. The original eigenvalue problem (9.3) can now be diagonalized in terms of the unitary matrix:

$$\hat{U}_k \equiv \begin{bmatrix} u_k & 0 & 0 & -v_k^* \\ 0 & u_k & v_k^* & 0 \\ 0 & -v_k & u_k & 0 \\ v_k & 0 & 0 & u_k \end{bmatrix} = \begin{bmatrix} \underline{\sigma}_0 u_k & -i\underline{\sigma}_y v_k^* \\ -i\underline{\sigma}_y v_k & \underline{\sigma}_0 u_k \end{bmatrix} \tag{9.7}$$

as

$$\begin{bmatrix} \underline{\sigma}_0 \xi_k & i\underline{\sigma}_y \Delta_k \\ -i\underline{\sigma}_y \Delta_k^* & -\underline{\sigma}_0 \xi_k \end{bmatrix} \hat{U}_k = \hat{U}_k \begin{bmatrix} \underline{\sigma}_0 E_k & \underline{0} \\ \underline{0} & -\underline{\sigma}_0 E_k \end{bmatrix}, \tag{9.8}$$

with $\underline{0}$ denoting the 2×2 zero matrix. The first two columns of $\hat{U}_k$ represent the eigenvectors of $\tilde{\alpha} = 1, 2$ with an identical positive eigenvalue E_k. The latter two columns for $-E_k$ have been obtained based on the procedure described below (8.80).

Comparing (9.3) and (9.8), we can express the eigenvectors of $\tilde{\alpha} = 1, 2$ in (9.3) as

$$\mathbf{u}_1(\mathbf{k}) = \begin{bmatrix} u_k \\ 0 \end{bmatrix}, \quad \mathbf{v}_1(\mathbf{k}) = \begin{bmatrix} 0 \\ v_k \end{bmatrix}, \quad \mathbf{u}_2(\mathbf{k}) = \begin{bmatrix} 0 \\ u_k \end{bmatrix}, \quad \mathbf{v}_2(\mathbf{k}) = \begin{bmatrix} -v_k \\ 0 \end{bmatrix}. \tag{9.9}$$

Substituting them into (8.71) and (8.72) yields the one-particle density matrices

$$\underline{\rho}^{(1)}(\mathbf{k}) = \underline{\sigma}_0 \left[u_k^2 \bar{n}_k + |v_k|^2 (1 - \bar{n}_k) \right], \tag{9.10}$$

$$\underline{\tilde{\rho}}^{(1)}(\mathbf{k}) = \begin{bmatrix} 0 & u_k v_k^* (1 - 2\bar{n}_k) \\ -u_k v_k^* (1 - 2\bar{n}_k) & 0 \end{bmatrix} = i\underline{\sigma}_y \frac{\Delta_k}{2E_k} \tanh \frac{\beta E_k}{2}, \tag{9.11}$$

where we have used $1 - 2\bar{n}_k = \tanh(\beta E_k/2)$ and $u_k v_k^* = \Delta_k / 2E_k$ as obtained using (9.6).

We next substitute (9.10) into (8.75). We then confirm that the Hartree–Fock potential is indeed diagonal and isotropic as $\underline{\mathscr{U}}_{\text{HF}}(\mathbf{k}) = \underline{\sigma}_0 \mathscr{U}_k^{\text{HF}}$ with

$$\mathscr{U}_k^{\text{HF}} = \mathscr{V}_0 \frac{N}{V} - \frac{1}{V} \sum_{\mathbf{k}'} \mathscr{V}_{|\mathbf{k}-\mathbf{k}'|} \left[u_{k'}^2 \bar{n}_{k'} + |v_{k'}|^2 (1 - \bar{n}_{k'}) \right], \tag{9.12}$$

where N denotes the particle number of (8.82). Also using (9.1) and (9.11) in (8.88) for $\ell = m = 0$ and noting $Y_{00}(\hat{\mathbf{k}}) = (4\pi)^{-1/2}$, we obtain the *gap equation* to determine Δ_k as

$$\Delta_k = -\int \frac{\mathrm{d}^3 k'}{(2\pi)^3} \mathscr{V}_0(k, k') \frac{\Delta_{k'}}{2E_{k'}} \tanh \frac{\beta E_{k'}}{2}. \tag{9.13}$$

Equations (9.12) and (9.13) form a set of nonlinear equations for $(\mathscr{U}_k^{\text{HF}}, \Delta_k)$. Indeed, the quantities (u_k, v_k, E_k) used there are expressible in terms of $(\mathscr{U}_k^{\text{HF}}, \Delta_k)$ as (9.2), (9.5), and (9.6).

By substituting (9.1), (9.2), (9.9), (9.10), and (9.11) into (8.81) and subsequently using (9.5) and (9.6), the corresponding grand potential in equilibrium is obtained,

$$\begin{aligned}\Omega_{\text{BdG}} &= \sum_{\mathbf{k}}\Big\{-\frac{2}{\beta}\ln\left(1+e^{-\beta E_k}\right)+2|v_k|^2\xi_k - u_k v_k \Delta_k - u_k v_k^* \Delta_k^* \\ &\quad - \mathscr{U}_k^{\text{HF}}\left[u_k^2\bar{n}_k + |v_k|^2(1-\bar{n}_k)\right] + u_k v_k^*(1-2\bar{n}_k)\Delta_k^*\Big\} \\ &= \sum_{\mathbf{k}}\Big[-\frac{2}{\beta}\ln\left(1+e^{-\beta E_k}\right)+\xi_k - E_k + \frac{|\Delta_k|^2}{2E_k}(1-2\bar{n}_k) \\ &\quad - \mathscr{U}_k^{\text{HF}}\frac{E_k - \xi_k + 2\xi_k\bar{n}_k}{2E_k}\Big].\end{aligned} \tag{9.14}$$

The particle number N is obtained from (8.82) and (9.10) yielding

$$N = 2\sum_{\mathbf{k}}\left[u_k^2\bar{n}_k + |v_k|^2(1-\bar{n}_k)\right]. \tag{9.15}$$

9.2 Effective Pairing Interaction

Focusing on the *weak-coupling* regime where inequality

$$|\mathscr{V}_k|\frac{N}{V} \ll \frac{\hbar^2 k_F^2}{2m} \equiv \varepsilon_F \tag{9.16}$$

holds, we simplify (9.13) further to obtain (9.24) below given in terms of the effective pairing interaction (9.26) near the Fermi surface. Those who are interested more in physics than in mathematical consistency may skip to the next section.

When (9.16) is satisfied, we can neglect the Hartree–Fock potential because $\mathscr{U}_k^{\text{HF}} \to 0$ to an excellent approximation. Hence, we replace ξ_k in (9.2) by the ideal-gas form

$$\xi_k = \frac{\hbar^2 k^2}{2m} - \mu, \tag{9.17}$$

and consider only (9.13). The corresponding superconducting transition temperature (or *critical temperature*) T_c is expected to be much smaller than the Fermi temperature $T_F \equiv \varepsilon_F/k_B$. For single-element superconductors, for example, $T_F \approx 10^4 \sim 10^5$ K (see Cu in Table 4.1 with $T_F \sim T_Q$), whereas $T_c \approx 1 \sim 10$ K (see Table 9.3 below). On the basis of this observation, we introduce a *cutoff energy* ε_c in such a way as to satisfy

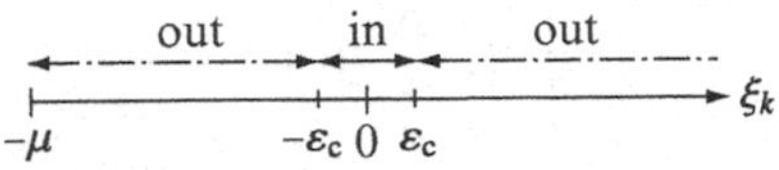

Fig. 9.1 Decomposition of the energy domain, where "in" and "out" denote the domains with $|\xi_k| \le \varepsilon_c$ near the Fermi surface and $|\xi_k| > \varepsilon_c$, respectively. Inequality $\varepsilon_c \ll \mu \sim \varepsilon_F$ holds

Table 9.1 Debye temperatures of single-element metals and semiconductors [5]

Solid	Al	Cd	Cr	Cu	Fe	Ni	Si
T_D (K)	428	209	630	343	467	450	640

$$|\Delta_k| \sim k_B T_c \ll \varepsilon_c \ll \varepsilon_F, \tag{9.18}$$

and divide the energy domain into "in" for $|\xi_k| \le \varepsilon_c$ and "out" for $|\xi_k| > \varepsilon_c$ as depicted in Fig. 9.1. The quantity ε_c/k_B for single-element superconductors, where phonons are responsible for the attraction between electrons, is estimated near the Debye temperature of order $T_D \approx 200 \sim 600$ K, as listed in Table 9.1. Thus, we can choose ε_c as in (9.18) also in real systems.

At low temperatures where $T \lesssim T_c$ holds, we introduce two approximations into (9.13): (i) $\tanh x \approx 1$ for $x \gg 1$ and (ii) $E_k = (\xi_k^2 + |\Delta_k|^2)^{1/2} \approx |\xi_k|$ for $|\xi_k| \ge \varepsilon_c$ based on (9.18) to transform the key factor in its integrand to

$$\frac{1}{E_{k'}} \tanh \frac{\beta E_{k'}}{2} \approx \frac{\theta(\varepsilon_c - |\xi_{k'}|)}{E_{k'}} \tanh \frac{\beta E_{k'}}{2} + \frac{\theta(|\xi_{k'}| - \varepsilon_c)}{|\xi_{k'}|},$$

where $\theta(x)$ denotes the step function (4.11). We substitute this into (9.13) and perform an integration over the solid angle, as in (4.17). Furthermore, we approximate the k' integral by a discrete sum over k' with interval dk' to obtain

$$\Delta_k = -\sum_{k'} dk' \frac{\mathcal{V}_0(k,k')k'^2}{4\pi^2} \left[\theta(\varepsilon_c - |\xi_{k'}|) \frac{\Delta_{k'}}{E_{k'}} \tanh \frac{\beta E_{k'}}{2} + \theta(|\xi_{k'}| - \varepsilon_c) \frac{\Delta_{k'}}{|\xi_{k'}|} \right]. \tag{9.19}$$

Next, we introduce a matrix $\underline{M} \equiv (M_{kk'})$ with elements

$$M_{kk'} \equiv -dk' \frac{\mathcal{V}_0(k,k')k'^2}{4\pi^2 |\xi_{k'}|}, \tag{9.20}$$

and recast (9.19) as

$$\begin{bmatrix} \mathbf{\Delta}^{\text{in}} \\ \mathbf{\Delta}^{\text{out}} \end{bmatrix} = \begin{bmatrix} \underline{M}^{\text{in,in}} & \underline{M}^{\text{in,out}} \\ \underline{M}^{\text{out,in}} & \underline{M}^{\text{out,out}} \end{bmatrix} \begin{bmatrix} \boldsymbol{\Lambda}^{\text{in}} \\ \mathbf{\Delta}^{\text{out}} \end{bmatrix}, \tag{9.21}$$

where $\mathbf{\Delta}^{\text{in}}$ and $\mathbf{\Delta}^{\text{out}}$ are vectors composed of elements Δ_k in the domain $|\xi_k| \le \varepsilon_c$ and $|\xi_k| > \varepsilon_c$, respectively, and $\boldsymbol{\Lambda}^{\text{in}}$ is a vector with elements

$$f_k \equiv |\xi_k| \frac{\Delta_k}{E_k} \tanh \frac{\beta E_k}{2} \tag{9.22}$$

for $|\xi_k| \le \varepsilon_c$.

Next, we eliminate $\mathbf{\Delta}^{\text{out}}$ from (9.21) to derive a closed equation for $\mathbf{\Delta}^{\text{in}}$ as follows. The lower element of (9.21) reads

$$\mathbf{\Delta}^{\text{out}} = \underline{M}^{\text{out,in}} \mathbf{\Lambda}^{\text{in}} + \underline{M}^{\text{out,out}} \mathbf{\Delta}^{\text{out}},$$

which can be solved in terms of $\mathbf{\Delta}^{\text{out}}$ using the unit matrix $\underline{1}^{\text{out,out}} \equiv (\delta_{kk'})$ for $|\xi_k| > \varepsilon_c$,

$$\mathbf{\Delta}^{\text{out}} = \left(\underline{1}^{\text{out,out}} - \underline{M}^{\text{out,out}}\right)^{-1} \underline{M}^{\text{out,in}} \mathbf{\Lambda}^{\text{in}}. \tag{9.23}$$

Substituting (9.23) into the upper element of (9.21), we obtain

$$\mathbf{\Delta}^{\text{in}} = \left[\underline{M}^{\text{in,in}} + \underline{M}^{\text{in,out}} \left(\underline{1}^{\text{out,out}} - \underline{M}^{\text{out,out}}\right)^{-1} \underline{M}^{\text{out,in}}\right] \mathbf{\Lambda}^{\text{in}}.$$

This equation is further modified using (9.20) and (9.22) to a form similar to (9.13),

$$\Delta_k^{\text{in}} = -\int \frac{\mathrm{d}^3 k'}{(2\pi)^3} \mathscr{V}_0^{(\text{eff})}(k,k') \frac{\Delta_{k'}^{\text{in}}}{2E_{k'}} \tanh \frac{\beta E_{k'}}{2}, \tag{9.24}$$

where $\mathscr{V}_0^{(\text{eff})}(k,k')$ is an *effective pairing interaction* near the Fermi surface defined by

$$\mathscr{V}_0^{(\text{eff})}(k,k') \equiv \mathscr{V}_0(k,k') - {\sum_{k_1 k_2}}' \frac{\mathscr{V}_0(k,k_1) k_1^2}{4\pi^2 |\xi_{k_1}|} \mathrm{d}k_1 \left(\underline{1}^{\text{out,out}} - \underline{M}^{\text{out,out}}\right)^{-1}_{k_1 k_2} \mathscr{V}_0(k_2,k'). \tag{9.25}$$

Here the primed sum is over $|\xi_{k_1}|,\ |\xi_{k_2}| > \varepsilon_c$ with the limits $\mathrm{d}k_1,\ \mathrm{d}k_2 \to 0$ implied. In practical calculations, one may reduce the value of $\mathrm{d}k$ successively to confirm convergence.

Arguments k and k' in (9.25) lie in a thin shell near the Fermi surface satisfying $|\xi_k|,\ |\xi_{k'}| \le \varepsilon_c \ll \varepsilon_F$. Hence, to an excellent approximation, we can set $\mathscr{V}_0^{(\text{eff})}(k,k') \approx \mathscr{V}_0^{(\text{eff})}(k_F,k_F)$. Alternatively, $\mathscr{V}_0^{(\text{eff})}(k,k')$ may be expressed in terms of constant $\mathscr{V}_0^{(\text{eff})} \equiv \mathscr{V}_0^{(\text{eff})}(k_F,k_F)$ and cutoff energy ε_c,

$$\mathscr{V}_0^{(\text{eff})}(k,k') = \mathscr{V}_0^{(\text{eff})} \theta(\varepsilon_c - |\xi_k|) \theta(\varepsilon_c - |\xi_{k'}|), \tag{9.26}$$

where $\theta(x)$ is the step function (4.11), and ε_c denotes a cutoff energy chosen subject to (9.18). We shall see that the pair condensation does occur when $\mathscr{V}_0^{(\text{eff})} < 0$ is satisfied.

Table 9.2 Effective s-wave pairing interaction $V_0^{(\mathrm{eff})}$ and transition temperature T_c in units of $k_F = \varepsilon_F = k_B = 1$ calculated for the interaction potential (9.27)

a_1	r_1	a_2	r_2	$V_0(k_F, k_F)$	ε_c	$V_0^{(\mathrm{eff})}$	T_c
−0.12	0.1	0.0	–	−2.90	0.01	−8.61	1.16×10^{-4}
−0.13	0.1	0.0	–	−3.14	0.01	−11.22	3.36×10^{-4}
−0.14	0.1	0.0	–	−3.38	0.01	−15.17	8.40×10^{-4}
−0.05	0.016	0.1	0.001	1.26	0.01	−6.14	1.83×10^{-5}
−0.05	0.015	0.1	0.001	1.26	0.01	−8.06	8.48×10^{-5}
−0.05	0.014	0.1	0.001	1.26	0.01	−12.45	4.76×10^{-4}
−0.05	0.014	0.1	0.001	1.26	0.005	−15.94	4.76×10^{-4}

To see how $\mathscr{V}_0^{(\mathrm{eff})} < 0$ may be realized, we consider the model interaction potential described by a linear combination of two exponential functions given by

$$\mathscr{V}(r) = \sum_{j=1}^{2} \frac{\hbar^2 a_j}{2mr_j^3} \mathrm{e}^{-r/r_j}, \tag{9.27}$$

where $r_j > 0$ and a_j $(j = 1, 2)$ are parameters with units of length that specify the range and strength of the potential, respectively. Expanding its Fourier coefficient $\mathscr{V}_k$ as (8.83), we obtain the s-wave component $\mathscr{V}_0(k, k')$ as (see (6.55))

$$\mathscr{V}_0(k, k') = \frac{4\pi\hbar^2}{m} \sum_{j=1}^{2} \frac{a_j}{(1 + r_j^2 k^2 + r_j^2 k'^2)^2 - 4r_j^4 k^2 k'^2}. \tag{9.28}$$

Using (6.25), we can transform the weak-coupling condition of (9.16) into $k_F|a_1 + a_2| \ll 1$. Table 9.2 gives $\mathscr{V}_0^{(\mathrm{eff})}$ calculated numerically for several sets of parameters (a_1, r_1, a_2, r_2) based on (9.25). The first three rows are values for pure attractive potentials with $a_1 < 0$ and $a_2 = 0$, whereas the last four rows are for potentials with an additional short-range repulsion $(a_1 < 0, a_2 > 0)$. For reference, we have also given values of T_c obtained from (9.33) below. Thus, the potential (9.27) does produce pair condensation. We also notice that the renormalized potential $V_0^{(\mathrm{eff})}$ can be negative even when the original $V_0(k_F, k_F)$ on the Fermi surface is repulsive. The values of T_c obtained for the same values of parameters (a_1, r_1, a_2, r_2) in the last two rows are seen to be identical irrespective of the cutoff energy ε_c, as is expected. One may also check that T_c thereby obtained coincides with the value from the original equation (9.13) in the weak-coupling regime.

As will be seen shortly, introducing the effective pairing interaction enables us to perform analytic calculations of thermodynamic quantities in the weak-coupling regime using only the states of $|\xi_k| \le \varepsilon_c$. Nevertheless, we should keep in mind that the pair correlation extends also to $|\xi_k| > \varepsilon_c$, as seen from (9.23).

9.3 Gap Equation and Its Solution

The gap equation (9.24) written in terms of the effective pairing interaction (9.26) is equivalent to the original (9.13). Moreover, because only states of $|\xi_k| \le \varepsilon_c$ are relevant, (9.24) can be treated analytically. Indeed, this effective-interaction method can describe not only single-element superconductors like Hg and Pb quantitatively, but also the p-wave superfluidity in liquid ^{3}He with large spin fluctuations to the first approximation, as we shall see in Chap. 13. Here, we solve (9.24) and clarify basic thermodynamic properties of s-wave pair-condensed states.

Equation (9.24) with (9.26) implies that the energy gap $\Delta_k^{\rm in}$ near the Fermi surface also depends solely on constant Δ,

$$\Delta_k^{\rm in} = \Delta\,\theta(\varepsilon_c - |\xi_k|). \tag{9.29}$$

We substitute it together with (9.26) back into (9.24), writing the integral using the density of states $N(\varepsilon)$ defined by (7.21), and then perform a change of variable as (9.17). Equation (9.24) thus reduces to

$$\begin{aligned}\Delta\,\theta(\varepsilon_c - |\xi_k|) &= -\int_0^\infty d\varepsilon_{k'} N(\varepsilon_{k'}) \mathscr{V}_0^{(\rm eff)}(k,k') \frac{\Delta\,\theta(\varepsilon_c - |\xi_{k'}|)}{2E_{k'}} \tanh\frac{\beta E_{k'}}{2} \\ &= -\mathscr{V}_0^{(\rm eff)}\theta(\varepsilon_c - |\xi_k|)\int_{-\varepsilon_c}^{\varepsilon_c} d\xi_{k'} N(\xi_{k'}+\mu)\frac{\Delta}{2E_{k'}}\tanh\frac{\beta E_{k'}}{2}.\end{aligned}$$

As cutoff energy ε_c satisfies (9.18), we can replace $N(\xi_{k'}+\mu) \approx N(\mu) \approx N(\varepsilon_F)$ to an excellent approximation. Dividing the resulting equation by Δ and choosing k to satisfy $|\xi_k| \le \varepsilon_c$, we obtain a simplified gap equation as

$$\frac{1}{g_0} = \int_0^{\varepsilon_c} \frac{1}{E}\tanh\frac{\beta E}{2} d\xi, \tag{9.30}$$

where $E = \sqrt{\xi^2 + |\Delta|^2}$, and g_0 is a dimensionless coupling constant defined by

$$g_0 \equiv -N(\varepsilon_F)\mathscr{V}_0^{(\rm eff)}. \tag{9.31}$$

As will be seen shortly, (9.30) has a non-trivial solution when $g_0 > 0$ (i.e., $\mathscr{V}_0^{(\rm eff)} < 0$) is satisfied. Alternatively stated, pair condensation is realized if there is a net attraction near the Fermi level.

In the following, we choose the phase of Δ equal to zero so that $\Delta \ge 0$. To determine the transition temperature, we set $T = T_c$ and $\Delta = 0$ in (9.30). The resulting equation for T_c can be transformed as

$$\frac{1}{g_0} = \int_0^{\varepsilon_c} \frac{1}{\xi}\tanh\frac{\xi}{2k_B T_c} d\xi = \int_0^{\varepsilon_c/2k_B T_c} \frac{\tanh x}{x} dx$$

$$= \tanh x \ln x \Big|_0^{\varepsilon_c/2k_B T_c} - \int_0^{\varepsilon_c/2k_B T_c} \frac{\ln x}{\cosh^2 x} dx$$
$$\approx \ln \frac{\varepsilon_c}{2k_B T_c} - \int_0^{\infty} \frac{\ln x}{\cosh^2 x} dx = \ln \frac{\varepsilon_c}{2k_B T_c} + \ln \frac{4e^{\gamma}}{\pi}$$
$$= \ln \frac{2e^{\gamma}\varepsilon_c}{\pi k_B T_c}, \tag{9.32}$$

where $\gamma = 0.57721\cdots$ is Euler's constant, and we have set $\varepsilon_c/2k_B T_c \to \infty$ in the upper limit of the integral by noting (9.18) and $\ln x/\cosh^2 x \approx 0$ for $x \gg 1$. We thereby obtain an expression for $k_B T_c$,

$$k_B T_c = \frac{2e^{\gamma}}{\pi}\varepsilon_c e^{-1/g_0} \approx 1.13\varepsilon_c e^{-1/g_0}. \tag{9.33}$$

Similarly, setting $T = 0$ in (9.30) gives an equation for the zero-temperature energy gap $\Delta_0 \equiv \Delta(T = 0)$ as

$$\frac{1}{g_0} = \int_0^{\varepsilon_c} \frac{d\xi}{\sqrt{\xi^2 + \Delta_0^2}} = \ln\left(\xi + \sqrt{\xi^2 + \Delta_0^2}\right)\Big|_0^{\varepsilon_c} \approx \ln \frac{2\varepsilon_c}{\Delta_0}, \tag{9.34}$$

which yields

$$\Delta_0 = 2\varepsilon_c e^{-1/g_0}. \tag{9.35}$$

Dividing (9.35) by one-half of (9.33) gives

$$\frac{2\Delta_0}{k_B T_c} = 2\pi e^{-\gamma} \approx 3.53. \tag{9.36}$$

This relation without (g_0, ε_c) is an important prediction of the BCS theory that can be directly tested in experiments (see Table 9.3 below).

To obtain the energy gap $\Delta \equiv \Delta(T)$ for $0 \le T \le T_c$, we subtract (9.32) from (9.30) to obtain

$$0 = \int_0^{\varepsilon_c} \left(\frac{1}{E}\tanh\frac{E}{2k_B T} - \frac{1}{\xi}\tanh\frac{\xi}{2k_B T}\right) d\xi + \int_0^{\varepsilon_c} \frac{1}{\xi}\left(\tanh\frac{\xi}{2k_B T} - \tanh\frac{\xi}{2k_B T_c}\right) d\xi$$
$$\approx \int_0^{\infty} \left(\frac{1}{E}\tanh\frac{E}{2k_B T} - \frac{1}{\xi}\tanh\frac{\xi}{2k_B T}\right) d\xi + \ln\frac{T_c}{T},$$

where we have used (9.32) and $\varepsilon_c/2k_B T \gg 1$. This equation can be expressed as

$$\ln\frac{T_c}{T} = \int_0^{\infty} \left(\frac{1}{\xi}\tanh\frac{\xi}{2k_B T} - \frac{1}{E}\tanh\frac{E}{2k_B T}\right) d\xi. \tag{9.37}$$

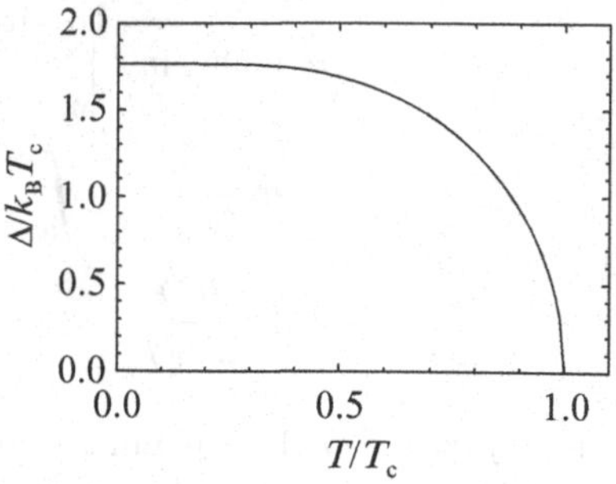

Fig. 9.2 Energy gap as a function of reduced temperature

Similarly, subtracting (9.34) from (9.30) gives an alternative expression

$$\ln\frac{\Delta_0}{\Delta} = 2\int_0^\infty \frac{1}{E}\frac{1}{e^{\beta E}+1}d\xi. \tag{9.38}$$

With a change of variable $\xi \to x \equiv \xi/k_B T_c$, one sees that these integrals depend only on the ratios T/T_c and $\Delta/k_B T_c$. That is, each of (9.37) and (9.38) determines the dimensionless energy gap $\Delta/k_B T_c$ as a function of reduced temperature T/T_c. The two equations are equivalent; (9.37) is useful for $T \lesssim T_c$, whereas (9.38) may be more convenient for $T \gtrsim 0$.

Figure 9.2 plots energy gap Δ as a function of reduced temperature T/T_c. We observe that Δ grows rapidly for $T \lesssim T_c$, which is typical of *second-order phase transitions* described by mean-field theories. We now solve (9.37) analytically for $T \lesssim T_c$ to study this behavior in detail. To this end, we use the series

$$\frac{1}{x}\tanh\frac{x}{2} = \sum_{n=0}^\infty \frac{4}{x^2+(2n+1)^2\pi^2}, \tag{9.39}$$

which may be seen to hold by noting that both sides share poles with the same residues [1]. Using it and introducing $\varepsilon_n \equiv (2n+1)\pi k_B T$, we expand the second term in the integrand of (9.37) in terms of $\Delta/k_B T \ll 1$ to obtain

$$\begin{aligned}\frac{1}{E}\tanh\frac{E}{2k_B T} &= \sum_{n=0}^\infty \frac{4k_B T}{\xi^2+\Delta^2+\varepsilon_n^2} = \sum_{n=0}^\infty \frac{4k_B T}{\xi^2+\varepsilon_n^2}\left(1-\frac{\Delta^2}{\xi^2+\varepsilon_n^2}+\cdots\right)\\ &= \frac{1}{\xi}\tanh\frac{\xi}{2k_B T} - 4k_B T\sum_{n=0}^\infty \frac{\Delta^2}{\left(\xi^2+\varepsilon_n^2\right)^2}+\cdots,\end{aligned} \tag{9.40}$$

which we substitute into (9.37). Retaining only the leading term, and performing the integration over ξ by applying the residue theorem [1],

$$\int_0^\infty \frac{d\xi}{\left(\xi^2+\varepsilon_n^2\right)^2} = \frac{1}{2}\int_{-\infty}^\infty \frac{d\xi}{\left(\xi^2+\varepsilon_n^2\right)^2} = \frac{2\pi i}{2}\lim_{\xi\to i\varepsilon_n}\frac{d}{d\xi}\frac{1}{(\xi+i\varepsilon_n)^2} = \frac{\pi}{4\varepsilon_n^3},$$

we thereby simplify (9.37) for $T \lesssim T_c$ to derive

$$\ln\frac{T_c}{T} \approx \frac{\Delta^2}{(\pi k_B T)^2}\sum_{n=0}^{\infty}\frac{1}{(2n+1)^3} = \frac{\Delta^2}{(\pi k_B T)^2}\left(1-\frac{1}{2^3}\right)\sum_{n=1}^{\infty}\frac{1}{n^3} = \frac{7\zeta(3)\Delta^2}{8(\pi k_B T)^2}, \tag{9.41}$$

where $\zeta(3) = 1.202\cdots$ is the Riemann zeta function (4.40). The left-hand side of this equation can be approximated as $\ln(T_c/T) = -\ln[1-(T_c-T)/T_c] \approx (T_c-T)/T_c$ to leading order, whereas we may set $k_B T \approx k_B T_c$ in the rightmost expression. Hence, for the energy gap of $T \lesssim T_c$, we obtain the analytic expression

$$\Delta(T \lesssim T_c) \approx \pi k_B T_c\left[\frac{8}{7\zeta(3)}\right]^{1/2}\left(\frac{T_c-T}{T_c}\right)^{1/2}, \tag{9.42}$$

which rapidly grows proportional to $(T_c-T)^{1/2}$. This temperature dependence of Δ just below the transition temperature is characteristic of the mean-field second-order phase transition; see Sect. 9.5 below on this point.

9.4 Thermodynamic Properties

Having obtained the energy gap Δ, we proceed to clarify the temperature dependences of the heat capacity, chemical potential, and free energy.

9.4.1 Heat Capacity

First, we focus on heat capacity. Entropy in the mean-field description of superconductivity is given by (8.25), which is formally identical to (4.7) for ideal Fermi gases with $\sigma = -1$. The difference lies in the quasiparticle energy E_q, as seen from (8.27). Hence, the heat capacity $C = T(\partial S/\partial T)$ is obtained from the first expression of (4.8) for ideal gases by replacement $\varepsilon_q - \mu \to E_q$. Noting $q = \mathbf{k}\tilde{\alpha}$ ($\tilde{\alpha} = 1, 2$) for homogeneous superconductors and using the density of states (4.17) with $s = 1/2$, we can transform the resulting expression as

$$\begin{aligned} C &= \sum_{\mathbf{k}\tilde{\alpha}} E_k\frac{\partial \bar{n}_k}{\partial T} = \int_{-\infty}^{\infty} d\varepsilon_k\, D(\varepsilon_k) E_k\frac{\partial \bar{n}_k}{\partial T} = \int_{-\infty}^{\infty} d\xi_k\, D(\xi_k+\mu) E_k\frac{\partial \bar{n}_k}{\partial T} \\ &\approx k_B D(\varepsilon_F)\int_{-\infty}^{\infty} d\xi \left(x^2 - \frac{1}{2k_B^2 T}\frac{d\Delta^2}{dT}\right)\frac{e^x}{(e^x+1)^2}\bigg|_{x=\beta E}, \end{aligned} \tag{9.43}$$

where we have used (9.5) and also approximated $D(\xi+\mu) \approx D(\varepsilon_F)$.

Next, we focus on the temperature just below T_c where $E = |\xi|$ holds. Making the change of variable $\xi \to x \equiv \beta\xi$ in (9.43), and using (9.42) to calculate $d\Delta^2/dT$, we obtain the heat capacity just below T_c,

$$C(T_c) = D(\varepsilon_F)k_B^2 T_c \int_{-\infty}^{\infty} \left(x^2 - \frac{1}{2k_B^2 T} \frac{d\Delta^2}{dT}\bigg|_{T=T_c} \right) \frac{e^x}{(e^x+1)^2} dx$$
$$= C_n(T_c) + D(\varepsilon_F)k_B^2 T_c \frac{4\pi^2}{7\zeta(3)}, \tag{9.44}$$

where $C_n(T_c)$ denotes the normal heat capacity given by (6.31). Thus, we obtain an estimate of the magnitude of the discontinuity $\Delta C \equiv C(T_c) - C_n(T_c)$ relative to the normal heat capacity $C_n(T_c)$ as

$$\frac{\Delta C}{C_n(T_c)} = \frac{12}{7\zeta(3)} = 1.43. \tag{9.45}$$

This is another important prediction of the BCS theory directly testable by experiments. Table 9.3 summarizes T_c, $2\Delta_0/k_B T_c$, and $\Delta C/C_n$ at $T = T_c$ for single-element superconductors. We observe good agreement between experiments and theoretical predictions of (9.36) and (9.45).

Figure 9.3 presents the temperature dependence of the superconducting heat capacity (9.43) normalized by the normal heat capacity $C_n \propto T$ given explicitly by (6.31). A jump at $T = T_c$ is followed by an exponential decrease of the heat capacity as $T \to 0$. This is because there remain no excitations of order $k_B T$ at low temperatures in consequence of the widening of the energy gap $\Delta \gg k_B T$. This

Table 9.3 Properties of single-element superconductors [8]

	Al	Hg	In	Nb	Pb	V
T_c (K)	1.2	4.16	0.4	8.8	7.22	4.9
$2\Delta_0/k_B T_c$	3.53	3.95	3.65	3.65	3.95	3.50
$\Delta C/C_n$	1.29–1.59	2.37	1.73	1.87	2.71	1.49

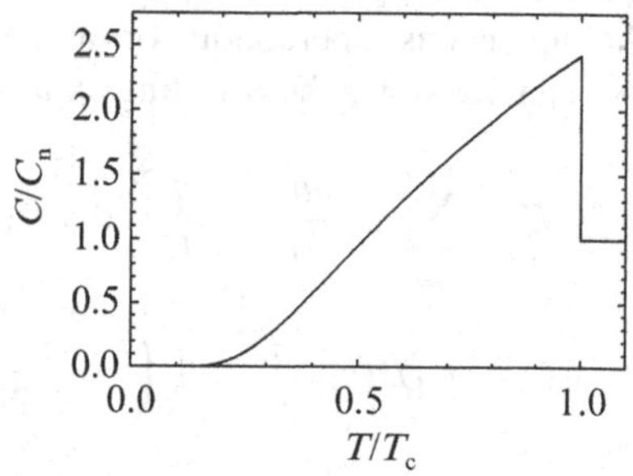

Fig. 9.3 Superconducting heat capacity relative to normal heat capacity $C_n \propto T$

low-temperature behavior can be understood intuitively in terms of the quasiparticle density of states defined generally by[1]

$$D_{\rm s}(E) \equiv \sum_{\mathbf{k}\tilde{\alpha}} [\delta(E - E_{\mathbf{k}\tilde{\alpha}}) + \delta(E + E_{\mathbf{k}\tilde{\alpha}})], \tag{9.46}$$

where we have extended the domain of E to negative energies for later purposes. Using this and noting $E_{\mathbf{k}\tilde{\alpha}} \to E_k > 0$ in the present case, we can express the heat capacity (9.43) as

$$C = \int_0^\infty {\rm d}E D_{\rm s}(E) E \frac{\partial \bar{n}(E)}{\partial T}, \tag{9.47}$$

with $\bar{n}(E) \equiv ({\rm e}^{\beta E} + 1)^{-1}$. Calculations show that (9.46) for the s-wave excitation spectrum (9.5) becomes (**Problem** 9.1)

$$D_{\rm s}(E) = \frac{|E|}{(E^2 - \Delta^2)^{1/2}} \theta(|E| - \Delta) D(\varepsilon_{\rm F}), \tag{9.48}$$

where $D(\varepsilon_{\rm F})$ is the normal density of states at the Fermi energy. As plotted in Fig. 9.4 for $E \geq 0$, there are no states for $0 \leq E < \Delta$. Combining this fact with $\bar{n}(E) \approx {\rm e}^{-\beta\Delta}$ for $E \gtrsim \Delta$ and $T \to 0$, we conclude that the low-temperature heat capacity is proportional to ${\rm e}^{-\beta\Delta}$. A more detailed calculation yields (**Problem** 9.2)

$$C(T \to 0) \approx k_{\rm B} D(\varepsilon_{\rm F}) \sqrt{\frac{\pi \Delta^5}{2(k_{\rm B}T)^3}}\, {\rm e}^{-\Delta/k_{\rm B}T}. \tag{9.49}$$

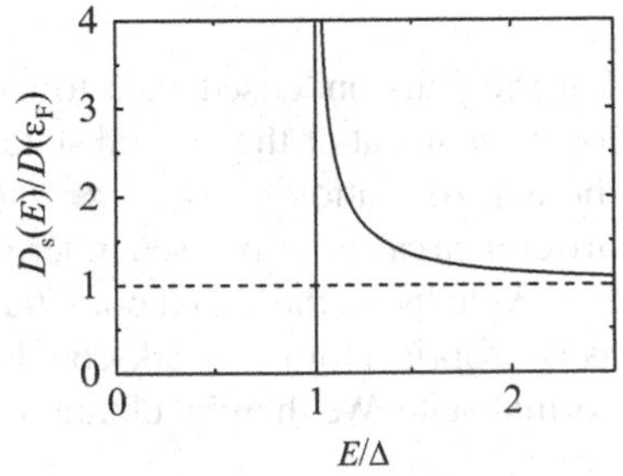

Fig. 9.4 Quasiparticle density of states for s-wave superconductors as a function of the excitation energy $E \geq 0$

[1]Variable E should be distinguished from eigenvalues $E_{\mathbf{k}\tilde{\alpha}}$.

9.4.2 Chemical Potential

We now show that, within the approximation of $D(\varepsilon) \approx D(\varepsilon_F)$, chemical potential μ in the superconducting phase is the same as μ_n for the normal state.

Expressing (9.15) in terms of the density of states (4.17) for $s = 1/2$, then subtracting (4.12) for the normal state, we use (9.6) to transform the resulting equation as

$$\begin{aligned}
0 &= \int_0^\infty D(\varepsilon_k)\left[|v_k|^2 + \frac{u_k^2 - |v_k|^2}{e^{\beta E_k}+1} - \frac{1}{e^{\beta(\varepsilon_k-\mu_n)}+1}\right] d\varepsilon_k \\
&\approx D(\varepsilon_F)\int_0^\infty \left[\frac{1}{2}\left(1-\frac{\xi}{E}\right) + \frac{\xi}{E}\frac{1}{e^{\beta E}+1} - \frac{1}{e^{\beta(\varepsilon-\mu_n)}+1}\right] d\varepsilon \\
&\approx D(\varepsilon_F)\int_{-\infty}^\infty \frac{1}{2}\left[-\frac{\xi}{E}\tanh\frac{\beta E}{2} + \tanh\frac{\beta(\xi+\mu-\mu_n)}{2}\right] d\xi,
\end{aligned}$$

where we have made a change of variable, $\varepsilon \to \xi = \varepsilon - \mu$, and also replaced the lower limit of integration by $-\infty$ as the main contribution to the integral stems from region $|\xi|/\mu \ll 1$. The first term in the final integrand is odd in ξ so that it gives a null contribution. We thereby conclude that the above equality holds when

$$\mu = \mu_n \tag{9.50}$$

is satisfied.

9.4.3 Free Energy

For the pair-condensed state to be stable, its free energy $F = \Omega + \mu N$ must be lower than that of the normal state, $F_n = \Omega_n + \mu_n N$. We shall confirm this within the approximation of $D(\varepsilon) \approx D(\varepsilon_F)$, where (9.50) holds so that the free-energy difference can be expressed in terms of the grand potential Ω, $F_{sn} \equiv F - F_n = \Omega - \Omega_n$. We express the grand potential (9.14) using the density of states, set $\mathscr{U}_k^{HF} = 0$ as appropriate for the weak-coupling regime, and subtract $\Omega_n \equiv \Omega_{BdG}\big|_{\Delta=0}$ for the normal state. We thereby obtain the free-energy difference as

$$F_{sn}(T) = \frac{D(\varepsilon_F)}{2}\int_{-\infty}^\infty \left(-\frac{2}{\beta}\ln\frac{1+e^{-\beta E}}{1+e^{-\beta|\xi|}} + |\xi| - E + \frac{\Delta^2}{2E}\tanh\frac{\beta E}{2}\right) d\xi. \tag{9.51}$$

This definite integral can be calculated analytically at $T = 0$,

$$F_{sn}(0) = D(\varepsilon_F)\int_0^\infty \left(\xi - E + \frac{\Delta_0^2}{2E}\right) d\xi = -\frac{1}{4}D(\varepsilon_F)\Delta_0^2. \tag{9.52}$$

Additionally, one can show that $F_{\rm sn}(T \lesssim T_{\rm c})$ decreases continuously from $F_{\rm sn}(T_{\rm c}) = 0$ as (**Problem** 9.3)

$$F_{\rm sn}(T \lesssim T_{\rm c}) \approx -\frac{2D(\varepsilon_{\rm F})(\pi k_{\rm B})^2}{7\zeta(3)}(T - T_{\rm c})^2. \tag{9.53}$$

Thus, the pair-condensed state with a finite Δ has been confirmed to have a lower free energy than the normal state.

9.5 Landau Theory of Second-Order Phase Transition

Superconductivity presents a prototype of a second-order phase transition where the symmetry changes spontaneously. It is also distinctive in that the Landau theory of second-order phase transitions [6] can be applied even quantitatively. Hence, a brief outline of the Landau theory for the superconducting phase transition is worth presenting. Nevertheless, this section can be skipped without loss of continuity.

The Landau theory is relevant in describing continuous phase transitions with spontaneous symmetry breaking without latent heat. For an isotropic ferromagnet, for example, rotational symmetry is broken spontaneously because of the emergence of magnetization $\mathbf{M}$, i.e., the magnetic moment per unit volume, which is realized by a cooperative alignment of electron spins. The corresponding free energy may be written as a function of $\mathbf{M}$ as $F = F[\mathbf{M}]$. Landau assumed that it is expandable near the transition temperature $T_{\rm c}$ in terms of the lowest-order scalar $M^2 = \mathbf{M} \cdot \mathbf{M}$ with $\mathbf{M}$; hence,

$$\frac{F}{V} = \frac{F_{\rm n}}{V} + a_2 M^2 + \frac{a_4}{2} M^4 + \cdots. \tag{9.54}$$

Here $F_{\rm n}$ is the normal free energy, a_2 changes its sign at the transition temperature $T_{\rm c}$ as $a_2 = \alpha(T - T_{\rm c})$ with $\alpha > 0$, and a_4 is a positive constant. This free energy as a functional of M is plotted schematically in Fig. 9.5 by setting $F_{\rm n} = 0$ and extending the domain of M to the negative region. Thus, the state $M \neq 0$ may be stabilized

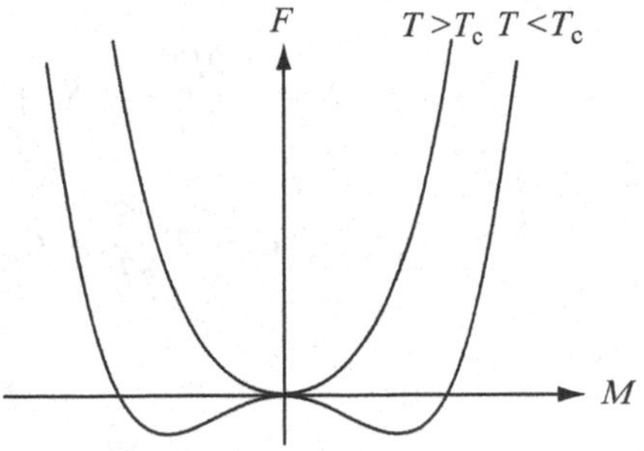

Fig. 9.5 Landau free energy for a ferromagnet as a functional of magnetization M

below T_c, but the resulting $F[\mathbf{M}]$ still has a large degeneracy with respect to the direction of **M**. This degeneracy needs to be broken spontaneously to lower the free energy by the emergence of **M**. This phenomenon is called *spontaneous symmetry breaking*, with **M** called the *order parameter* that distinguishes the ferromagnetic state from the normal state. Unfortunately, the basic assumption that "$F[\mathbf{M}]$ can be expanded analytically in terms of M^2" is not correct in the strict sense because of large spatial and temporal fluctuations near T_c. Alternatively stated, $M = 0$ is a singular point of the free energy. However, this *fluctuation range* near T_c where (9.54) does not apply can be very narrow for some systems [6] to be unobservable in practice.

Classic superconductors form a typical example for which an expansion of the type (9.54) holds true even quantitatively with negligible fluctuation range. The order parameter for this case can be identified as the energy gap Δ, which is generally a complex number. The scalar of the lowest order with respect to Δ is $|\Delta|^2$. Hence, the quantity that is broken spontaneously is the degeneracy with respect to the phase of Δ, which is sometimes called *spontaneously broken gauge symmetry* [7]. Now, we expand the free energy per unit volume F/V near T_c as

$$\frac{F}{V} = \frac{F_n}{V} + a_2|\Delta|^2 + \frac{a_4}{2}|\Delta|^4 + \cdots, \tag{9.55}$$

choosing a_2 and a_4 in the form

$$a_2 = \frac{D(\varepsilon_F)}{2V}\frac{T - T_c}{T_c}, \qquad a_4 = \frac{D(\varepsilon_F)}{2V}\frac{7\zeta(3)}{8(\pi k_B T_c)^2}. \tag{9.56}$$

Then, the extremal condition, $0 = \partial F/\partial|\Delta|^2 = a_2 + a_4|\Delta|^2$, yields (9.42) for the equilibrium energy gap near T_c. In addition, substitution of (9.42) into (9.55) gives

$$F^{eq} = F_n - \frac{2D(\varepsilon_F)(\pi k_B)^2}{7\zeta(3)}(T - T_c)^2, \tag{9.57}$$

as the equilibrium free energy near T_c, which agrees with (9.53) for the condensation energy. The first and second derivatives of F^{eq} with respect to T yield the entropy and heat capacity,

$$S = -\frac{\partial F^{eq}}{\partial T} = S_n + \frac{4D(\varepsilon_F)(\pi k_B)^2}{7\zeta(3)}(T - T_c), \tag{9.58}$$

$$C = T\frac{\partial S}{\partial T} \approx C_n + T_c\frac{4D(\varepsilon_F)(\pi k_B)^2}{7\zeta(3)}. \tag{9.59}$$

Equation (9.59) also coincides with (9.44) from the BCS theory. Thus, the superconducting phase transition within the BCS theory is describable in terms of the Landau theory of second-order phase transitions.

It follows from (9.57) and (9.58) that the free energy F and its first-order derivative $-S$ are both continuous, implying no latent heat upon the transition. In contrast, heat capacity C of (9.59), determined from the *second-order* derivative of F, is discontinuous. These are general features of the *second-order phase transition* predicted by the Landau theory. We shall see an inhomogeneous extension of the Landau theory in (14.95) below.

Problems

9.1. Substitute (9.5) into $E_{\mathbf{k}\tilde{\alpha}}$ in (9.46) to obtain the s-wave quasiparticle density of states (9.48).

9.2. Show (9.49).

9.3. By multiplying (9.39) by x, using $\tanh(x/2) = 1-2/(\mathrm{e}^x+1)$, and integrating the resulting expression over $x_1 \le x \le x_2$, we thereby obtain

$$x_2 - x_1 + 2\ln\frac{1+\mathrm{e}^{-x_2}}{1+\mathrm{e}^{-x_1}} = 2\sum_{n=0}^{\infty}\ln\frac{x_2^2+(2n+1)^2\pi^2}{x_1^2+(2n+1)^2\pi^2}.$$

Use this equality together with (9.39) and (9.42) to show that the condensation energy (9.51) can be approximated for $T \lesssim T_c$ as in (9.53).

References

1. G.B. Arfken, H.J. Weber, *Mathematical Methods for Physicists* (Academic, New York, 2012)
2. J. Bardeen, L.N. Cooper, J.R. Schrieffer, Phys. Rev. **108**, 1175 (1957)
3. L.N. Cooper, D. Feldman (eds.), *BCS: 50 Years* (World Scientific, Hackensack, 2011)
4. A.R. Edmonds, *Angular Momentum in Quantum Mechanics* (Princeton University Press, Princeton, 1957)
5. D.E. Gray (ed.), *American Institute of Physics Handbook*, 3rd edn. (McGraw-Hill, New York, 1972)
6. L.D. Landau, in *Collected Papers of L.D. Landau*, ed. by D. Ter Haar (Pergamon, Oxford, 1965), p. 193; L.D. Landau, E.M. Lifshitz, *Statistical Physics Part 1* (Pergamon, New York, 1980), p. 446
7. A.J. Leggett, F. Sols, Found. Phys. **21**, 353 (1991)
8. G. Rickayzen, in *Superconductivity*, ed. R.D. Parks, chap. 3, vol. I (Marcel Dekker, New York, 1969)

Chapter 10
Superfluidity, Meissner Effect, and Flux Quantization

Abstract One of the most outstanding features of superconductivity is undoubtedly the persistence of a current without dissipation. However, in obeying Ampère's law, the flow of charged particles necessarily produces magnetic fields, thereby complicating the phenomenon. With this observation, we first consider neutral systems to reveal the origin of superfluidity, i.e., the persistence of flow without dissipation, caused by the phase coherence of the Cooper-pair condensate. Subsequently, we discuss the Meissner effect concerning the flow of charged systems that expels weak magnetic fields from the bulk of superconductors. Finally, we study flux quantization arising from the single-valuedness of the macroscopic wave function.

10.1 Superfluid Density and Spin Susceptibility

We consider a neutral Cooper-pair condensate to clarify the origin of superfluidity. We shall also study spin paramagnetism to obtain expressions for s-wave pairing of the spin susceptibility and superfluid density, (10.18) and (10.22), respectively, in terms of the Yosida function (10.16). It is also shown that molecular-field effects modify these expressions; (10.28) and (10.29).

Let us express the pair wave function as a 2×2 matrix using the spin degrees of freedom, $\underline{\phi}(\mathbf{r}_1, \mathbf{r}_2) \equiv \big(\phi(\mathbf{r}_1\alpha_1, \mathbf{r}_2\alpha_2)\big)$. Then the state in which the pair moves with center-of-mass momentum $\hbar\mathbf{q}$ is expressible by incorporating its contribution to the expansion of (8.94) in the form

$$\underline{\phi}(\mathbf{r}_1, \mathbf{r}_2) = \frac{1}{V}\sum_{\mathbf{k}} \underline{\phi}(\mathbf{k})\, e^{i\mathbf{k}\cdot(\mathbf{r}_1-\mathbf{r}_2)} e^{i\mathbf{q}\cdot(\mathbf{r}_1+\mathbf{r}_2)/2} = \frac{1}{V}\sum_{\mathbf{k}} \underline{\phi}(\mathbf{k})\, e^{i\mathbf{k}_+\cdot\mathbf{r}_1 - i\mathbf{k}_-\cdot\mathbf{r}_2}, \tag{10.1}$$

where $\mathbf{k}_\pm$ is defined by

$$\mathbf{k}_\pm \equiv \mathbf{k} \pm \frac{\mathbf{q}}{2}. \tag{10.2}$$

© Springer Japan 2015
T. Kita, *Statistical Mechanics of Superconductivity*, Graduate Texts in Physics,
DOI 10.1007/978-4-431-55405-9_10

Its Hermitian conjugate $\underline{\phi}^\dagger(\mathbf{r}_1,\mathbf{r}_2) \equiv \left(\phi^*(\mathbf{r}_2\alpha_2,\mathbf{r}_1\alpha_1)\right)$ is given by

$$\underline{\phi}^\dagger(\mathbf{r}_1,\mathbf{r}_2) = \frac{1}{V}\sum_{\mathbf{k}}\underline{\phi}^\dagger(\mathbf{k})\,\mathrm{e}^{-\mathrm{i}\mathbf{k}\cdot(\mathbf{r}_2-\mathbf{r}_1)-\mathrm{i}\mathbf{q}\cdot(\mathbf{r}_1+\mathbf{r}_2)/2} = \frac{1}{V}\sum_{\mathbf{k}}\underline{\phi}^\dagger(\mathbf{k})\,\mathrm{e}^{\mathrm{i}\mathbf{k}_-\cdot\mathbf{r}_1-\mathrm{i}\mathbf{k}_+\cdot\mathbf{r}_2}.$$

Let us substitute the two expansions together with that of the delta function in (6.18) into (8.10). We then find that the 2×2 matrices $\underline{u}(\mathbf{r}_1,\mathbf{r}_2)$ and $\underline{v}(\mathbf{r}_1,\mathbf{r}_2)$ are also expressible as

$$\underline{u}(\mathbf{r}_1,\mathbf{r}_2) = \frac{1}{V}\sum_{\mathbf{k}}\underline{u}(\mathbf{k}_+)\,\mathrm{e}^{\mathrm{i}\mathbf{k}_+\cdot(\mathbf{r}_1-\mathbf{r}_2)}, \qquad \underline{v}(\mathbf{r}_1,\mathbf{r}_2) = \frac{1}{V}\sum_{\mathbf{k}}\underline{v}(\mathbf{k})\,\mathrm{e}^{\mathrm{i}\mathbf{k}_+\cdot\mathbf{r}_1-\mathrm{i}\mathbf{k}_-\cdot\mathbf{r}_2}, \tag{10.3}$$

with $\underline{u}(\mathbf{k}_+) \equiv \left[\underline{\sigma}_0 + \underline{\phi}(\mathbf{k})\underline{\phi}^\dagger(\mathbf{k})\right]^{-1/2}$ and $\underline{v}(\mathbf{k}) \equiv \underline{u}(\mathbf{k}_+)\underline{\phi}(\mathbf{k})$. The rationale for using the arguments $\mathbf{k}_+$ and $\mathbf{k}$ for $\underline{u}$ and $\underline{v}$, respectively, will become clear shortly.

Noting that $\underline{u}$ and $\underline{v}$ in (10.3) are the first-row elements of the transformation matrix in (8.15), one may also expect that the first-row elements of the BdG matrix (8.64) are expressible as

$$\underline{\hat{\mathscr{H}}}_{\mathrm{HF}}(\mathbf{r}_1,\mathbf{r}_2) = \frac{1}{V}\sum_{\mathbf{k}}\underline{\hat{\mathscr{H}}}_{\mathrm{HF}}(\mathbf{k}_+)\,\mathrm{e}^{\mathrm{i}\mathbf{k}_+\cdot(\mathbf{r}_1-\mathbf{r}_2)}, \tag{10.4}$$

$$\underline{\Delta}(\mathbf{r}_1,\mathbf{r}_2) = \frac{1}{V}\sum_{\mathbf{k}}\underline{\Delta}(\mathbf{k})\,\mathrm{e}^{\mathrm{i}\mathbf{k}_+\cdot\mathbf{r}_1-\mathrm{i}\mathbf{k}_-\cdot\mathbf{r}_2}. \tag{10.5}$$

For example, the pair potential of (8.35) is given at $T=0$ by

$$\underline{\Delta}(\mathbf{r}_1,\mathbf{r}_2) = \mathscr{V}(|\mathbf{r}_1-\mathbf{r}_2|)\underline{u}(\mathbf{r}_1,\bar{\mathbf{r}}_3)\underline{v}(\bar{\mathbf{r}}_3,\mathbf{r}_2),$$

as can be shown using (8.30) and (8.31), which is expressible as (10.5). The expansion (10.4) may be confirmed similarly based on (8.36). Additionally, one can show that the complex conjugates of (10.4) and (10.5) can be written with a change $\mathbf{k}\to-\mathbf{k}$ of summation variables as

$$\underline{\hat{\mathscr{H}}}^*_{\mathrm{HF}}(\mathbf{r}_1,\mathbf{r}_2) = \frac{1}{V}\sum_{\mathbf{k}}\underline{\hat{\mathscr{H}}}^*_{\mathrm{HF}}(-\mathbf{k}_-)\,\mathrm{e}^{\mathrm{i}\mathbf{k}_-\cdot(\mathbf{r}_1-\mathbf{r}_2)}, \tag{10.6}$$

$$\underline{\Delta}^*(\mathbf{r}_1,\mathbf{r}_2) = \frac{1}{V}\sum_{\mathbf{k}}\underline{\Delta}^*(-\mathbf{k})\,\mathrm{e}^{\mathrm{i}\mathbf{k}_-\cdot\mathbf{r}_1-\mathrm{i}\mathbf{k}_+\cdot\mathbf{r}_2}. \tag{10.7}$$

Equations (10.4)–(10.7) can now be combined in matrix form as

$$\begin{bmatrix} \underline{\hat{\mathscr{H}}}_{\mathrm{HF}}(\mathbf{r}_1,\mathbf{r}_2) & \underline{\Delta}(\mathbf{r}_1,\mathbf{r}_2) \\ -\underline{\Delta}^*(\mathbf{r}_1,\mathbf{r}_2) & -\underline{\hat{\mathscr{H}}}^*_{\mathrm{HF}}(\mathbf{r}_1,\mathbf{r}_2) \end{bmatrix}$$

$$= \frac{1}{V}\sum_{\mathbf{k}}\begin{bmatrix} \underline{\sigma}_0\, e^{i\mathbf{k}_+\cdot\mathbf{r}_1} & \underline{0} \\ \underline{0} & \underline{\sigma}_0\, e^{i\mathbf{k}_-\cdot\mathbf{r}_1}\end{bmatrix}\begin{bmatrix} \hat{\underline{\mathscr{K}}}_{\mathrm{HF}}(\mathbf{k}_+) & \underline{\Delta}(\mathbf{k}) \\ -\underline{\Delta}^*(-\mathbf{k}) & \hat{\underline{\mathscr{K}}}^*_{\mathrm{HF}}(-\mathbf{k}_-)\end{bmatrix}\begin{bmatrix} \underline{\sigma}_0\, e^{-i\mathbf{k}_+\cdot\mathbf{r}_2} & \underline{0} \\ \underline{0} & \underline{\sigma}_0\, e^{-i\mathbf{k}_-\cdot\mathbf{r}_2}\end{bmatrix}, \tag{10.8}$$

where $\underline{\sigma}_0$ and $\underline{0}$ are the 2×2 unit and zero matrices, respectively.

Substituting (10.8) into (8.64), we can choose $q \equiv \mathbf{k}\tilde{\alpha}$ ($\tilde{\alpha} = 1, 2$) and express the eigenfunction as

$$\begin{bmatrix}\mathbf{u}_{\mathbf{k}\tilde{\alpha}}(\mathbf{r}) \\ \mathbf{v}_{\mathbf{k}\tilde{\alpha}}(\mathbf{r})\end{bmatrix} = \frac{1}{\sqrt{V}}\begin{bmatrix} \underline{\sigma}_0\, e^{i\mathbf{k}_+\cdot\mathbf{r}} & \underline{0} \\ \underline{0} & \underline{\sigma}_0\, e^{i\mathbf{k}_-\cdot\mathbf{r}}\end{bmatrix}\begin{bmatrix}\mathbf{u}_{\tilde{\alpha}}(\mathbf{k}) \\ \mathbf{v}_{\tilde{\alpha}}(\mathbf{k})\end{bmatrix}. \tag{10.9}$$

Vectors $\mathbf{u}_{\tilde{\alpha}}(\mathbf{k})$ and $\mathbf{v}_{\tilde{\alpha}}(\mathbf{k})$ now obey the 4×4 eigenvalue problem:

$$\begin{bmatrix} \underline{\mathscr{K}}_{\mathrm{HF}}(\mathbf{k}_+) & \underline{\Delta}(\mathbf{k}) \\ -\underline{\Delta}^*(-\mathbf{k}) & -\underline{\mathscr{K}}^*_{\mathrm{HF}}(-\mathbf{k}_-)\end{bmatrix}\begin{bmatrix}\mathbf{u}_{\tilde{\alpha}}(\mathbf{k}) \\ \mathbf{v}_{\tilde{\alpha}}(\mathbf{k})\end{bmatrix} = E_{\tilde{\alpha}}(\mathbf{k})\begin{bmatrix}\mathbf{u}_{\tilde{\alpha}}(\mathbf{k}) \\ \mathbf{v}_{\tilde{\alpha}}(\mathbf{k})\end{bmatrix} \tag{10.10}$$

and normalization condition (8.80). Equation (10.10) extends (8.79) for homogeneous systems to describe a uniform flow with momentum $\hbar\mathbf{q}$.

From now on we focus on s-wave Cooper pairing and consider a situation where a weak homogeneous magnetic field of flux density B is also present along the z direction. The resulting one-particle energy acquires an additional contribution $\delta\varepsilon_{\mathbf{k}\alpha} = -\mu^0_{\mathrm{m}}\alpha B$ proportional to both B and $\alpha = \pm 1/2$ as in (6.38). Incorporating this effect into (9.2) and setting $\mathscr{U}^{\mathrm{HF}}_k \to 0$ as before, we can express $\underline{\mathscr{K}}_{\mathrm{HF}}(\pm\mathbf{k}_\pm)$ as

$$\begin{aligned}\underline{\mathscr{K}}_{\mathrm{HF}}(\pm\mathbf{k}_\pm) &= \left[\frac{\hbar^2(\pm\mathbf{k}+\mathbf{q}/2)^2}{2m} - \mu\right]\underline{\sigma}_0 - \frac{\mu^0_{\mathrm{m}}}{2}B\underline{\sigma}_z \\ &\approx \left[\xi_k \pm \frac{\hbar^2\mathbf{k}\cdot\mathbf{q}}{2m}\right]\underline{\sigma}_0 - \frac{\mu^0_{\mathrm{m}}}{2}B\underline{\sigma}_z,\end{aligned} \tag{10.11}$$

where $\underline{\sigma}_z$ is given in (8.42), ξ_k is defined by (9.17), and we have neglected terms of order q^2 based on the assumption $q \ll k \sim k_{\mathrm{F}}$. Let us substitute (10.11) together with (9.1) into (10.10). We then find that the matrix to be diagonalized is of the form

$$\begin{bmatrix} \xi_k + C_{\mathbf{q}} - C_B & 0 & 0 & \Delta_k \\ 0 & \xi_k + C_{\mathbf{q}} + C_B & -\Delta_k & 0 \\ 0 & -\Delta^*_k & -\xi_k + C_{\mathbf{q}} + C_B & 0 \\ \Delta^*_k & 0 & 0 & -\xi_k + C_{\mathbf{q}} - C_B\end{bmatrix}, \tag{10.12}$$

with $C_{\mathbf{q}} \equiv \hbar^2\mathbf{k}\cdot\mathbf{q}/2m$ and $C_B \equiv \mu^0_{\mathrm{m}}B/2$. The corresponding 4×4 eigenvalue problem is reduced to those of the (1,4) and (2,3) submatrices given by

$$\begin{bmatrix} \xi_k + C_{\mathbf{q}} \mp C_B & \pm\Delta_k \\ \pm\Delta_k^* & -\xi_k + C_{\mathbf{q}} \mp C_B \end{bmatrix} \begin{bmatrix} u_k \\ \pm v_k \end{bmatrix} = E_{\mathbf{k}\tilde{\alpha}} \begin{bmatrix} u_k \\ \pm v_k \end{bmatrix}.$$

Compared with (9.4), this equation contains an extra term $C_{\mathbf{q}} \mp C_B$, which lies equally on the diagonal. Hence, we only need to add $C_{\mathbf{q}} \mp C_B$ to (9.5) to obtain its eigenvalues,

$$E_{\mathbf{k}\tilde{\alpha}} = E_k + \frac{\hbar^2 \mathbf{k}\cdot\mathbf{q}}{2m} + \frac{(-1)^{\tilde{\alpha}}}{2}\mu_{\mathrm{m}}^0 B \qquad (\tilde{\alpha} = 1, 2). \tag{10.13}$$

Moreover, the eigenvectors are still given by (9.9) as they remain invariant in adding contribution proportional to the unit matrix. The corresponding one-particle density matrix is obtained by substituting (10.9) and (10.13) into (8.59) with $q \to \mathbf{k}\tilde{\alpha}$,

$$\underline{\rho}^{(1)}(\mathbf{r}_1, \mathbf{r}_2) = \frac{1}{V}\sum_{\mathbf{k}} \begin{bmatrix} u_k^2 \bar{n}_{\mathbf{k}1} + |v_k|^2(1 - \bar{n}_{-\mathbf{k}2}) & 0 \\ 0 & u_k^2 \bar{n}_{\mathbf{k}2} + |v_k|^2(1 - \bar{n}_{-\mathbf{k}1}) \end{bmatrix} \mathrm{e}^{\mathrm{i}\mathbf{k}_+\cdot(\mathbf{r}_1 - \mathbf{r}_2)}, \tag{10.14}$$

where u_k and v_k are given by (9.6), and we have made a change of summation variables $\mathbf{k} \to -\mathbf{k}$ for the $\mathbf{v}_{\tilde{\alpha}}(\mathbf{k})$ terms.

10.1.1 Spin Susceptibility

First, we set $\mathbf{q} = \mathbf{0}$ and consider the limit $B \to 0$ to obtain the spin susceptibility. The spin magnetic moment operator along the z axis is given by $(\mu_{\mathrm{m}}^0/2)\underline{\sigma}_z$. Operating with it on the one-particle density matrix (10.14), we then take the trace, and use the relation $n_{-\mathbf{k}\tilde{\alpha}} = n_{\mathbf{k}\tilde{\alpha}}$ valid for $\mathbf{q} = \mathbf{0}$ and $u_k^2 + |v_k|^2 = 1$. We thereby obtain the spin magnetic moment as

$$M = \int \mathrm{d}^3 r \,\mathrm{Tr}\, \frac{\mu_{\mathrm{m}}^0}{2}\underline{\sigma}_z \underline{\rho}^{(1)}(\mathbf{r}, \mathbf{r}) = \frac{\mu_{\mathrm{m}}^0}{2}\sum_{\mathbf{k}}(\bar{n}_{\mathbf{k}1} - \bar{n}_{\mathbf{k}2}).$$

Recalling (10.13), we then approximate the mean occupation number,

$$\bar{n}_{\mathbf{k}\tilde{\alpha}} \approx \bar{n}_k + \frac{\partial \bar{n}_k}{\partial E_k}\frac{(-1)^{\tilde{\alpha}}}{2}\mu_{\mathrm{m}}^0 B \qquad (\tilde{\alpha} = 1, 2),$$

with which we obtain the moment as

$$M \approx \frac{(\mu_{\mathrm{m}}^0)^2}{4} B \sum_{\mathbf{k}} 2\left(-\frac{\partial \bar{n}_k}{\partial E_k}\right) \approx \frac{(\mu_{\mathrm{m}}^0)^2}{4} BD(\varepsilon_{\mathrm{F}}) \int_{-\infty}^{\infty}\left(-\frac{\partial \bar{n}_k}{\partial E_k}\right)\mathrm{d}\xi_k = \chi_{\mathrm{n}} Y(T) B. \tag{10.15}$$

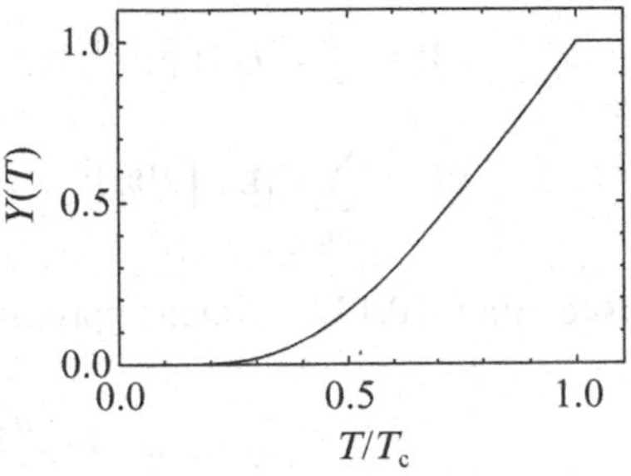

Fig. 10.1 Yosida function as a function of temperature

Here, $\chi_{\rm n} \equiv \left(\mu_{\rm m}^0/2\right)^2 D(\varepsilon_{\rm F})$ is the normal susceptibility derivable from (6.44) by $F_0^{\rm a} \to 0$ as appropriate for $\mathscr{U}_k^{\rm HF} \to 0$, and $Y(T)$ is the *Yosida function* [10] defined by

$$Y(T) \equiv \int_{-\infty}^{\infty}\left(-\frac{\partial \bar{n}_k}{\partial E_k}\right) d\xi_k = \int_{-\infty}^{\infty} \frac{1}{4k_{\rm B}T}\operatorname{sech}^2\frac{\sqrt{\xi^2+[\Delta(T)]^2}}{2k_{\rm B}T}d\xi. \tag{10.16}$$

Various response functions of the pair-condensed state are expressible in terms of the Yosida function, which describes quasiparticle excitations. Figure 10.1 plots $Y(T)$ as a function of temperature. This satisfies $Y(T_{\rm c}) = 1$ and $Y(0) = 0$, and can be written alternatively as (**Problem** 10.1)

$$Y(T) = 1 - 2\pi k_{\rm B}T\sum_{n=0}^{\infty}\frac{|\Delta(T)|^2}{\left[\varepsilon_n^2+|\Delta(T)|^2\right]^{3/2}}, \tag{10.17}$$

with $\varepsilon_n \equiv (2n+1)\pi k_{\rm B}T$.

The spin susceptibility $\chi \equiv M/B$ is obtained from (10.15),

$$\chi(T) = \chi_{\rm n} Y(T). \tag{10.18}$$

The reduction of χ as $T \to 0$ can be understood as a condensing of developing pairs into the $(\uparrow, \downarrow)$ singlet bound state as described below (9.1), which does not contribute to the susceptibility.

10.1.2 Superfluid Density

Next, we set $B = 0$ and consider the limit $\mathbf{q} \to \mathbf{0}$ to calculate the total momentum $\mathbf{P}$. We operate with momentum operator $\hat{\mathbf{p}}_1 \equiv -i\hbar\nabla_1$ on the one-particle density matrix (10.14) and then take the trace. We thereby obtain the total momentum as

$$\mathbf{P} = \int \mathrm{d}^3 r_1 \mathrm{Tr}\, \hat{\mathbf{p}}_1 \underline{\rho}^{(1)}(\mathbf{r}_1, \mathbf{r}_2)\bigg|_{\mathbf{r}_2=\mathbf{r}_1}$$

$$= \sum_{\mathbf{k}} \hbar \mathbf{k}_+ \left[2|v_k|^2 + u_k^2(\bar{n}_{\mathbf{k}1} + \bar{n}_{\mathbf{k}2}) - |v_k|^2(\bar{n}_{-\mathbf{k}1} + \bar{n}_{-\mathbf{k}2})\right]. \quad (10.19)$$

Recalling (10.13), we then approximate the mean occupation number by

$$\bar{n}_{\mathbf{k}\tilde{\alpha}} \approx \bar{n}_k + \frac{\partial \bar{n}_k}{\partial E_k}\frac{\hbar^2 \mathbf{k}\cdot\mathbf{q}}{2m} \qquad (\tilde{\alpha} = 1, 2),$$

with which we obtain the total momentum

$$\mathbf{P} \approx \sum_{\mathbf{k}} \hbar\left(\mathbf{k} + \frac{\mathbf{q}}{2}\right)\left[2|v_k|^2 + 2(u_k^2 - |v_k|^2)\bar{n}_k + 2(u_k^2 + |v_k|^2)\frac{\partial \bar{n}_k}{\partial E_k}\frac{\hbar^2 \mathbf{k}\cdot\mathbf{q}}{2m}\right]$$

$$= \frac{\hbar \mathbf{q}}{2}\sum_{\mathbf{k}}\left[2|v_k|^2 + 2(u_k^2 - |v_k|^2)\bar{n}_k\right] + 2\sum_{\mathbf{k}}\frac{\partial \bar{n}_k}{\partial E_k}\hbar\mathbf{k}\frac{\hbar^2\mathbf{k}\cdot\mathbf{q}}{2m}.$$

$$= \frac{\hbar \mathbf{q}}{2}N + 2\sum_{\mathbf{k}}\frac{\partial \bar{n}_k}{\partial E_k}\hbar\mathbf{k}\frac{\hbar^2\mathbf{k}\cdot\mathbf{q}}{2m}. \quad (10.20)$$

The second equality results from the fact that terms odd in $\mathbf{k}$ yield a null contribution after the integration over the solid angle, and we have used (9.15) for the total particle number N. Subsequently, we convert the second term into a three-dimensional integral as (8.90) and perform its angular integration using

$$\int \frac{\mathrm{d}\Omega_{\mathbf{k}}}{4\pi} k_\eta k_{\eta'} = \delta_{\eta\eta'}\frac{k^2}{3} \qquad (\eta, \eta' = x, y, z).$$

The factor $\partial \bar{n}_k / \partial E_k$ in (10.20) implies that only the region $\xi \approx 0$ contributes to the integral, so that we can approximate the density of states as $D(\varepsilon) \approx D(\varepsilon_\mathrm{F})$, where $D(\varepsilon) = 2VN(\varepsilon)$ for $s = 1/2$. The η element of the second term in (10.20) thereby reduces to

$$2\sum_{\mathbf{k}}\frac{\partial \bar{n}_k}{\partial E_k}\hbar k_\eta \frac{\hbar^2\mathbf{k}\cdot\mathbf{q}}{2m} = \int_{-\infty}^{\infty} \mathrm{d}\varepsilon_k D(\varepsilon_k)\frac{\partial \bar{n}_k}{\partial E_k}\sum_{\eta'=x,y,z}\int\frac{\mathrm{d}\Omega_{\mathbf{k}}}{4\pi}k_\eta k_{\eta'}\frac{\hbar^3 q_{\eta'}}{2m}$$

$$= \int_{-\infty}^{\infty} \mathrm{d}\xi_k D(\xi_k + \mu)\frac{\partial \bar{n}_k}{\partial E_k}\sum_{\eta'=x,y,z}\delta_{\eta\eta'}\frac{k^2}{3}\frac{\hbar^3 q_{\eta'}}{2m}$$

$$\approx \frac{\hbar q_\eta}{2}D(\varepsilon_\mathrm{F})\frac{\hbar^2 k_\mathrm{F}^2}{3m}\int_{-\infty}^{\infty}\mathrm{d}\xi_k \frac{\partial \bar{n}_k}{\partial E_k}$$

$$= -\frac{\hbar q_\eta}{2}N\int_{-\infty}^{\infty}\left(-\frac{\partial \bar{n}_k}{\partial E_k}\right)\mathrm{d}\xi_k.$$

In the last equality, we have substituted (6.29) for $s = 1/2$ and $m^* = m$ together with (4.35) for the Fermi wave number to express $D(\varepsilon_F)\hbar^2 k_F^2/3m = N$. Substituting the result into (10.20), we obtain the total momentum in terms of the Yosida function (10.16) as

$$\mathbf{P} = mN\,[1 - Y(T)]\,\mathbf{v}_s, \tag{10.21}$$

where $\mathbf{v}_s \equiv \hbar\mathbf{q}/2m$ is the superfluid velocity. Dividing the coefficient of $\mathbf{v}_s$ in (10.21) by V and m defines the *superfluid density*[1]:

$$n_s \equiv \frac{N}{V}\,[1 - Y(T)]\,. \tag{10.22}$$

Expression (10.21) may be understood as follows. At $T = 0$ where $Y(0) = 0$ holds, all the particles condense into the pair bound state of (10.1), and they can move coherently without dissipation. At finite temperatures, the presence of quasiparticle excitations causes a reduction in the superfluid density, which eventually vanishes at $T = T_c$ where $Y(T_c) = 1$. Thus, the present formalism starting from (8.6) enables us to understand the superfluidity naturally. The origin of the persistence of flow without dissipation may be attributed to the phase coherence over quite a large number of particles ($N \sim 10^{23}$) originating from condensation into an identical two-particle bound state, which blocks any small perturbation from affecting the motion.

10.1.3 Leggett's Theory of Superfluid Fermi Liquids

We now incorporate the Hartree–Fock potential $\underline{\mathscr{U}}_{\rm HF}$ into our consideration to see how it alters the results for the spin susceptibility and superfluid density obtained above. This issue was considered by Larkin and Migdal at $T = 0$ [5], and studied thoroughly by Leggett [6] at all temperatures below T_c. The content may be regarded as an extension of Landau's Fermi-liquid theory considered in Sect. 6.3 to superfluid phases.

With the presence of superflow and external magnetic field, the quasiparticle energy $E_{\mathbf{k}\tilde{\alpha}}$ for $\underline{\mathscr{U}}_{\rm HF} \to \underline{0}$ is given by (10.13). There is a shift in the eigenenergy, which to first order in $\hbar\mathbf{q}$ and B is given by

$$\delta E^0_{\mathbf{k}\tilde{\alpha}} \equiv \frac{\hbar^2\mathbf{k}\cdot\mathbf{q}}{2m} + \frac{(-1)^{\tilde{\alpha}}}{2}\mu^0_{\rm m} B \qquad (\tilde{\alpha} = 1, 2). \tag{10.23}$$

Similarly, we expect that the first-order variation $\delta E_{\mathbf{k}\tilde{\alpha}}$ for $\underline{\mathscr{U}}_{\rm HF} \neq \underline{0}$ is also expressible as

[1]Quantity $\rho_s \equiv m n_s$ is also called *superfluid density* in the literature.

$$\delta E_{\mathbf{k}\tilde{\alpha}} \equiv \frac{\hbar^2 \mathbf{k}\cdot\mathbf{q}}{2m_{\rm s}^*} + \frac{(-1)^{\tilde{\alpha}}}{2}\mu_{\rm m}^{\rm s} B \qquad (\tilde{\alpha} = 1, 2), \tag{10.24}$$

where $m_{\rm s}^*$ and $\mu_{\rm m}^{\rm s}$ are parameters that must be determined self-consistently. We shall see that a solution of this form does exist and obtain expressions for $m_{\rm s}^*$ and $\mu_{\rm m}^{\rm s}$.

The one-particle density matrix with the first-order variation (10.24) is still given by (10.14). Substituting it together with (6.18) into (8.61), we find that $\underline{\mathscr{U}}_{\rm HF}(\mathbf{r}_1, \mathbf{r}_2)$ is expressible in the same way as in (10.4) with "coefficient"

$$\underline{\mathscr{U}}_{\rm HF}(\mathbf{k}_+) = \frac{1}{V}\sum_{\mathbf{k}'}\Bigg\{\underline{\sigma}_0 \mathscr{V}_0[2|v_{k'}|^2 + u_{k'}^2(\bar{n}_{\mathbf{k}'1} + \bar{n}_{\mathbf{k}'2}) - |v_{k'}|^2(\bar{n}_{-\mathbf{k}'1} + \bar{n}_{-\mathbf{k}'2})]$$

$$-\mathscr{V}_{|\mathbf{k}-\mathbf{k}'|}\begin{bmatrix} u_{k'}^2\bar{n}_{\mathbf{k}'1} + |v_{k'}|^2(1-\bar{n}_{-\mathbf{k}'2}) & 0 \\ 0 & u_{k'}^2\bar{n}_{\mathbf{k}'2} + |v_{k'}|^2(1-\bar{n}_{-\mathbf{k}'1})\end{bmatrix}\Bigg\}.$$

Next, we expand the mean occupation number,

$$\bar{n}_{\mathbf{k}\tilde{\alpha}} = \bar{n}_k + \frac{\partial \bar{n}_k}{\partial E_k}\delta E_{\mathbf{k}\tilde{\alpha}} \qquad (\tilde{\alpha} = 1, 2),$$

and use the symmetry $\delta E_{-\mathbf{k}\tilde{\alpha}} = -\delta E_{\mathbf{k},3-\tilde{\alpha}}$ of (10.24) and equality $u_k^2 + |v_k|^2 = 1$ to obtain a first-order expression for the Hartree–Fock potential,

$$\delta\underline{\mathscr{U}}_{\rm HF}(\mathbf{k}_+) = \frac{1}{V}\sum_{\mathbf{k}'}\frac{\partial \bar{n}_{k'}}{\partial E_{k'}}\left\{\underline{\sigma}_0\mathscr{V}_0(\delta E_{\mathbf{k}'1} + \delta E_{\mathbf{k}'2}) - \mathscr{V}_{|\mathbf{k}-\mathbf{k}'|}\begin{bmatrix}\delta E_{\mathbf{k}'1} & 0 \\ 0 & \delta E_{\mathbf{k}'2}\end{bmatrix}\right\}.$$

It is diagonal and expressible in terms of the Landau f function (6.23) as

$$\delta\mathscr{U}_{\mathbf{k}\tilde{\alpha}}^{\rm HF} = \frac{1}{V}\sum_{\mathbf{k}'\tilde{\alpha}'} f_{\tilde{\alpha}\tilde{\alpha}'}(\mathbf{k},\mathbf{k}')\frac{\partial \bar{n}_{k'}}{\partial E_{k'}}\delta E_{\mathbf{k}'\tilde{\alpha}'}. \tag{10.25}$$

As first-order variation (10.24) is equal to the sum of (10.23) and (10.25) by definition, we then have

$$\delta E_{\mathbf{k}\tilde{\alpha}} = \delta E_{\mathbf{k}\tilde{\alpha}}^0 + \frac{1}{V}\sum_{\mathbf{k}'\tilde{\alpha}'} f_{\tilde{\alpha}\tilde{\alpha}'}(\mathbf{k},\mathbf{k}')\frac{\partial \bar{n}_{k'}}{\partial E_{k'}}\delta E_{\mathbf{k}'\tilde{\alpha}'}. \tag{10.26}$$

This is the self-consistency equation for $\delta E_{\mathbf{k}\tilde{\alpha}}$ with the same form as (6.22) in the normal state. Hence, to determine the unknown parameters $m_{\rm s}^*$ and $\mu_{\rm m}^{\rm s}$, we only need to repeat the arguments of Sect. 6.3.3 for $B = 0$ and Sect. 6.3.4 for $\mathbf{q} = \mathbf{0}$ using (10.26) as the first-order variation of the eigenenergy. In this respect, we notice that the change from $\partial \bar{n}_{k'}/\partial \varepsilon_{k'} = -\delta(\varepsilon_{k'} - \varepsilon_{\rm F})$ to $\partial \bar{n}_{k'}/\partial E_{k'}$ in the superfluid phase yields an extra factor $Y(T)$, i.e., the Yosida function given by (10.16). This observation implies that (6.37) and (6.44) for the normal state are now replaced by

$$\frac{m_s^*}{m} = 1 + \frac{1}{3}F_1^s Y(T), \tag{10.27}$$

$$\chi = \left(\mu_m^0/2\right)^2 D(\varepsilon_F)\frac{Y(T)}{1 + F_0^a Y(T)}, \tag{10.28}$$

respectively. Thus, we have derived the sought-after expression for the spin susceptibility.

To derive the superfluid density n_s, we start once again from (10.20) by replacing $m \to m_s^*$, then converting its sum over $\mathbf{k}$ into a three-dimensional integral as in (8.90), and expressing the density of states in (6.29) as $D(\varepsilon_F) = (m^*/m)D^0(\varepsilon_F) = (1+F_1^s/3)D^0(\varepsilon_F)$ in terms of $D^0(\varepsilon_F)$ for the ideal gas. Repeating the considerations for (10.21), we then obtain the superfluid density

$$n_s = \frac{N}{V}\left[1 - \frac{\left(1 + \frac{1}{3}F_1^s\right)Y(T)}{1 + \frac{1}{3}F_1^s Y(T)}\right]. \tag{10.29}$$

This coincides with N/V at $T = 0$ as $Y(0) = 0$, indicating that a coherent flow with all particles is realized. In contrast, n_s vanishes at $T = T_c$ because $Y(T_c) = 1$ in accordance with our expectation. Interaction effects on n_s, which are embodied in F_1^s, may become substantial at finite temperatures between 0 and T_c.

10.2 Meissner Effect and Flux Quantization

Superconductivity is caused by condensing Cooper pairs of electrons each of which carries charge $e < 0$. Their superflow necessarily obeys Maxwell's equations for electromagnetism. Among them is Ampère's law for steady currents given by

$$\nabla \times \mathbf{B}(\mathbf{r}) = \mu_0 \mathbf{j}(\mathbf{r}), \tag{10.30}$$

where $\mathbf{B}$ denotes the microscopic magnetic flux density, $\mu_0 = 4\pi \times 10^{-7}$ N·A^{-2} is the *vacuum permeability*, and $\mathbf{j}(\mathbf{r})$ is the current density. This coupling between the supercurrent and magnetic field yields a unique phenomenon called the *Meissner effect*, whereby weak magnetic fields are excluded from the bulk of superconductors. Discovered by Meissner and Ochsenfeld in 1933 [9], we discuss this effect together with the flux quantization, using the *London equation* to be developed from (10.30).

10.2.1 Ampère's Law

To begin, we derive Ampère's law (10.30) itself based on a variational principle to find a microscopic expression of the current density $\mathbf{j}(\mathbf{r})$.

Let us incorporate the effects of the magnetic field into the grand potential (8.32). First, operator (8.20) needs to embrace the vector potential $\mathbf{A}_1 \equiv \mathbf{A}(\mathbf{r}_1)$ as

$$\hat{\mathscr{K}}_1 \equiv \frac{(\hat{\mathbf{p}}_1 - e\mathbf{A}_1)^2}{2m} + \mathscr{U}(\mathbf{r}_1) - \mu. \tag{10.31}$$

Second, the energy of the magnetic field

$$\mathscr{H}_{\text{mag}} = \frac{1}{2\mu_0} \int d^3r_1 \, (\nabla_1 \times \mathbf{A}_1)^2 \tag{10.32}$$

should be included, where $\nabla_1 \times \mathbf{A}_1 = \mathbf{B}_1$. This latter term is generally neglected in the normal state, because in most cases the magnetic fields produced by orbital motions of electrons or spin magnetic moments are negligibly small. In superconductors, however, a supercurrent can produce a large magnetic field so that this contribution must be manifest. There is also a contribution from the *Zeeman effect*:

$$\hat{\mathscr{H}}_Z \equiv \mu_B \int d\xi_1 \hat{\psi}^\dagger(\mathbf{r}_1\alpha_1') \left(\boldsymbol{\sigma}\right)_{\alpha_1'\alpha_1} \hat{\psi}(\mathbf{r}_1\alpha_1) \cdot \mathbf{B}_1, \tag{10.33}$$

due to the spin magnetic moment, where μ_B is the Bohr magneton (6.39). However, because it is much less important in single-element superconductors than the orbital diamagnetism due to the supercurrent, we shall omit this contribution, noting that it may be easily incorporated when necessary.

The corresponding BdG equations can be derived in the same way as in Sect. 8.3. Indeed, they are still given by (8.38) by replacing $\hat{\mathbf{p}}_1 \to \hat{\mathbf{p}}_1 - e\mathbf{A}_1$ in $\hat{\mathscr{K}}_1$ of (8.36). As for the vector potential, we require that the magnetic field actually realized in the system minimizes $\Omega[\hat{\rho}] + \mathscr{H}_{\text{mag}}$. A necessary condition for this is that $\Omega[\hat{\rho}] + \mathscr{H}_{\text{mag}}$ is stationary with respect to the variation $\mathbf{A}_1 \to \mathbf{A}_1 + \delta\mathbf{A}_1$. As a preliminary to calculate the relevant first-order variation $\delta\Omega[\hat{\rho}] + \delta\mathscr{H}_{\text{mag}}$, we rewrite the kinetic energy in (8.32) using integration by parts,

$$\begin{aligned}\Omega_{\text{kin}} &= \int \left. \frac{(\hat{\mathbf{p}}_1 - e\mathbf{A}_1)^2}{2m} \rho^{(1)}(\xi_1, \xi_2) \right|_{\xi_2=\xi_1} d\xi_1 \\ &= \int \sum_{\alpha_1} \left. \frac{(-\hat{\mathbf{p}}_2 - e\mathbf{A}_2) \cdot (\hat{\mathbf{p}}_1 - e\mathbf{A}_1)}{2m} \rho^{(1)}(\xi_1, \xi_2) \right|_{\xi_2=\xi_1} d^3r_1. \end{aligned} \tag{10.34}$$

Using the latter expression, we can transform the first-order variation as

$$\delta\big(\Omega[\hat{\rho}] + \mathscr{H}_{\text{mag}}\big) = \delta\Omega_{\text{kin}} + \delta\mathscr{H}_{\text{mag}}$$

$$= -e \int \sum_{\alpha_1} \frac{\delta \mathbf{A}_2 \cdot (\hat{\mathbf{p}}_1 - e\mathbf{A}_1) + (-\hat{\mathbf{p}}_2 - e\mathbf{A}_2) \cdot \delta \mathbf{A}_1}{2m} \rho^{(1)}(\xi_1, \xi_2)\bigg|_{\xi_2=\xi_1} \mathrm{d}^3 r_1$$

$$+ \frac{2}{2\mu_0} \int (\nabla_1 \times \delta \mathbf{A}_1) \cdot (\nabla_1 \times \mathbf{A}_1) \mathrm{d}^3 r_1$$

$$= \int \delta \mathbf{A}_1 \cdot \left[-e \sum_{\alpha_1} \frac{(\hat{\mathbf{p}}_1 - e\mathbf{A}_1) + (-\hat{\mathbf{p}}_2 - e\mathbf{A}_2)}{2m} \rho^{(1)}(\xi_1, \xi_2)\bigg|_{\xi_2=\xi_1} \right.$$

$$\left. + \frac{1}{\mu_0} \nabla_1 \times (\nabla_1 \times \mathbf{A}_1) \right] \mathrm{d}^3 r_1, \tag{10.35}$$

where we have applied the mathematical identity $(\nabla \times \delta \mathbf{A}) \cdot \mathbf{B} = \nabla \cdot (\delta \mathbf{A} \times \mathbf{B}) + \delta \mathbf{A} \cdot (\nabla \times \mathbf{B})$ [1] to the magnetic energy and subsequently removed $\nabla \cdot (\delta \mathbf{A} \times \mathbf{B})$ using Gauss' theorem [1] and condition $\delta \mathbf{A} = \mathbf{0}$ on the surface. For the equality $\delta \Omega[\hat{\rho}] + \delta \mathscr{H}_{\mathrm{mag}} = 0$ to hold in terms of an arbitrary $\delta \mathbf{A}_1$, it is necessary that the coefficient of $\delta \mathbf{A}_1$ be zero. We thereby obtain Ampère's law (10.30) with a microscopic expression for the current density,

$$\mathbf{j}(\mathbf{r}_1) = e \sum_{\alpha_1} \frac{(\hat{\mathbf{p}}_1 - e\mathbf{A}_1) + (-\hat{\mathbf{p}}_2 - e\mathbf{A}_2)}{2m} \rho^{(1)}(\xi_1, \xi_2)\bigg|_{\xi_2=\xi_1}. \tag{10.36}$$

10.2.2 London Equation

The BdG equation (8.38) and Ampère's law (10.30) with current density (10.36) form a set of self-consistency equations for the quasiparticle eigenstates and magnetic field. Here, we solve them approximately to derive the London equation.

Let us generalize the argument of Sect. 10.1 by presuming that the current density spatially changes its magnitude and direction. More specifically, we replace the phase $\mathbf{q} \cdot (\mathbf{r}_1 + \mathbf{r}_2)/2$ in (10.1) with a function $[\varphi(\mathbf{r}_1) + \varphi(\mathbf{r}_2)]/2$ that varies slowly in space; the term "slow" is used here in comparison with the radius of the bound wave function ϕ. Accordingly, the phase factor in (10.3)–(10.9) is replaced by $\mathbf{q} \cdot \mathbf{r}_j \to \varphi(\mathbf{r}_j)$ $(j = 1, 2)$, as is the phase of the one-particle density matrix in (10.14). Let us substitute the density matrix into (10.36) and repeat the steps for (10.19)→(10.21). We thereby obtain an expression for the current density

$$\mathbf{j}(\mathbf{r}) = e n_{\mathrm{s}} \mathbf{v}_{\mathrm{s}}(\mathbf{r}), \tag{10.37}$$

where n_{s} denotes the superfluid density (10.22), and $\mathbf{v}_{\mathrm{s}}$ is now given by

$$\mathbf{v}_{\mathrm{s}} \equiv \frac{\hbar}{2m} \left(\nabla \varphi - \frac{2e}{\hbar} \mathbf{A} \right). \tag{10.38}$$

This $\mathbf{v}_{\mathrm{s}}$ reduces to (10.21) for neutral systems by setting $\varphi \to \mathbf{q} \cdot \mathbf{r}$ and $e \to 0$.

Substituting (10.37) into (10.30), we can express Ampère's law as

$$\nabla \times \mathbf{B} = \frac{\mu_0 e n_s \hbar}{2m} \left(\nabla \varphi - \frac{2e}{\hbar} \mathbf{A} \right). \tag{10.39}$$

We further operate with $\nabla\times$ on this equation and use identities $\nabla \times \nabla \times \mathbf{B} = \nabla\nabla \cdot \mathbf{B} - \nabla^2 \mathbf{B}$ and $\nabla \times \nabla \varphi = \mathbf{0}$ together with Gauss' law $\nabla \cdot \mathbf{B} = 0$ for magnetism. We thereby obtain the *London equation* [8] in the form

$$\nabla^2 \mathbf{B}(\mathbf{r}) = \frac{1}{\lambda_\mathrm{L}^2} \mathbf{B}(\mathbf{r}), \qquad \lambda_\mathrm{L} \equiv \sqrt{\frac{m}{\mu_0 n_s e^2}}, \tag{10.40}$$

where λ_L is called the *London penetration depth*. Note that this derivation of the London equation assumes no reduction in the energy gap in the presence of an applied magnetic field.

10.2.3 Meissner Effect

Let us solve (10.40) for a simple one-dimensional geometry to establish a theoretical basis for the Meissner effect.

We consider the case where a uniform magnetic field of flux density B_0 is applied along the z axis in the vacuum occupying the domain $x < 0$, and a superconductor is placed in the region $x \geq 0$. In this geometry, the magnetic flux density in the superconductor is expressible as $\mathbf{B}(\mathbf{r}) = (0, 0, B(x))$, and the London equation (10.40) reduces to

$$\frac{\mathrm{d}^2 B(x)}{\mathrm{d}x^2} = \frac{1}{\lambda_\mathrm{L}^2} B(x).$$

Its general solution is expressible in the form $B(x) = C_1 \mathrm{e}^{-x/\lambda_\mathrm{L}} + C_2 \mathrm{e}^{x/\lambda_\mathrm{L}}$ where C_1 and C_2 are two constants of integration. The continuity of the flux density at $x = 0$ yields $B(0) = C_1 + C_2 = B_0$, whereas the physical boundary condition $|B(x \to \infty)| < \infty$ gives $C_2 = 0$. We thereby obtain the solution

$$B(x > 0) = B_0 \mathrm{e}^{-x/\lambda_\mathrm{L}}. \tag{10.41}$$

Hence, the magnetic field decreases exponentially near the surface over the length λ_L and is excluded completely from the bulk of the superconductor. The corresponding current density is obtained from (10.30) as $\mathbf{j}(\mathbf{r}) = (0, B(x)/\mu_0 \lambda_\mathrm{L}, 0)$, which is also confined to the surface region to depth λ_L. Thus, the *Meissner effect* can be understood as a response of the superconductor in preventing an energy increase in the bulk due to the magnetic field and superflow.

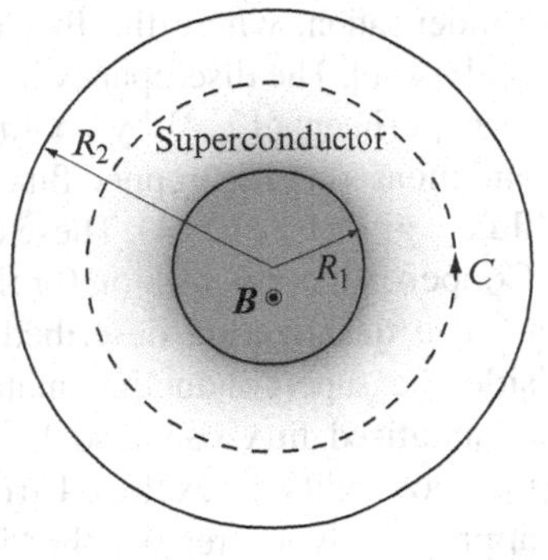

Fig. 10.2 Superconductivity in an annulus with a trapped magnetic field

10.2.4 Flux Quantization

Next, we consider superconductivity in an annulus with a weak magnetic field trapped in the central vacuum region, as depicted in Fig. 10.2. We shall show that the trapped magnetic flux is quantized in deriving (10.43) below.

If the difference between the outside diameter R_2 and inside diameter R_1 satisfies $R_2 - R_1 \gg \lambda_L$, the magnetic field is excluded from the bulk. Let us perform a line integral of (10.39) along a closed path C in the bulk region. The integral on the left-hand side yields 0 because $\mathbf{B} = \mathbf{0}$ along C. Next, the equality is transformed using Stokes' theorem [1] to give

$$\begin{aligned} 0 &= \oint_C \left(\nabla\varphi - \frac{2e}{\hbar}\mathbf{A}\right)\cdot d\mathbf{r} = \oint_C \nabla\varphi\cdot d\mathbf{r} - \frac{2e}{\hbar}\int_R (\nabla\times\mathbf{A})\cdot d\mathbf{S} \\ &= 2\pi n - \frac{2e}{\hbar}\int_R \mathbf{B}\cdot d\mathbf{S} = 2\pi n - \frac{2e}{\hbar}\Phi, \end{aligned} \tag{10.42}$$

where R denotes the region enclosed by C, and Φ is the total flux confined in the central region. Noting that (i) φ is the phase of the pair wave function $\underline{\phi}(\mathbf{r}_1, \mathbf{r}_2)$ in terms of the center-of-mass coordinate $(\mathbf{r}_1 + \mathbf{r}_2)/2$ and (ii) $\underline{\phi}(\mathbf{r}_1, \mathbf{r}_2)$ must be single-valued, we conclude that the integration constant n should be an integer. This implies that the total flux in the central region is quantized,[2]

$$\Phi = -n\Phi_0 \qquad (n = 0, \pm 1, \pm 2, \cdots), \tag{10.43}$$

where

$$\Phi_0 \equiv \frac{h}{2|e|} = 2.068 \times 10^{-15}\,\text{Wb} \tag{10.44}$$

forms the unit of magnetic flux in a superconductor called the *flux quantum*. Flux quantization was predicted by London in 1948 [7, 8] prior to the idea of pair

[2]The minus sign in (10.43) reflects $e < 0$.

condensation, where the flux quantum was predicted to be twice Φ_0 because of $2|e| \to |e|$. The discrepancy in the factor 2 was understood subsequently in a couple of experiments [2, 4] by cooling Sn or Pb in weak magnetic fields from above T_c and measuring the trapped flux. The results showed clearly that the unit of magnetic flux is given by (10.44). These experiments may also be regarded as establishing the Cooper-pair condensation for the mechanism of superconductivity.

The quantization described above is relevant to the flux trapped mostly outside the superconducting material. Nevertheless, we shall see in Chap. 15 that a quantized flux can also be trapped inside *type-II* superconductors. However, this possibility is excluded from the London equation (10.40). To incorporate it appropriately, we replace the "identity" $\nabla \times \nabla\varphi = \mathbf{0}$ with

$$\nabla \times \nabla\varphi = 2\pi n \hat{\mathbf{z}} \delta^2(\mathbf{r} - \mathbf{r}_0), \tag{10.45}$$

where $\hat{\mathbf{z}}$ is the unit vector along the z axis, $\delta^2(\mathbf{r}) \equiv \delta(x)\delta(y)$, and $\mathbf{r}_0$ denotes the singular point of $\varphi(\mathbf{r})$ around which $\varphi(\mathbf{r})$ changes by $2\pi n$ upon a counterclockwise rotation in the (x, y) plane. The validity of (10.45) is confirmed by taking its scalar product with $\hat{\mathbf{z}}$, integrating the resulting equation over a region R in the (x, y) plane that includes $\mathbf{r}_0$, and transforming the left-hand side using Stokes' theorem [1]. Noting (10.45), let us operate $\nabla\times$ on (10.39). We thereby find that (10.40) is now replaced by

$$-\lambda_L^2 \nabla^2 \mathbf{B}(\mathbf{r}) + \mathbf{B}(\mathbf{r}) = -n\Phi_0 \hat{\mathbf{z}} \delta^2(\mathbf{r} - \mathbf{r}_0). \tag{10.46}$$

We note that the singularity of $\varphi(\mathbf{r})$ at $\mathbf{r} = \mathbf{r}_0$ is superimposed by a zero of the pair potential $\Delta(\mathbf{r})$ to remove the singularity, as will be studied in detail in Sects. 15.6 and 16.3. Equation (10.46) has been useful in clarifying the flux-line structures of so-called *extreme type-II superconductors* [3]; see also (15.79) on this point.

Problems

10.1. In regard to the Yosida function $Y(T)$,

(a) Show that (10.16) can be written alternatively as (10.17) by using (9.39).
(b) Prove

$$Y(T \lesssim T_c) = 1 - 2\frac{T_c - T}{T_c}, \tag{10.47}$$

using (9.42) and (10.17).
(c) Show that $Y(T \to 0)$ is given approximately by

$$Y(T \to 0) \approx \sqrt{\frac{2\pi\Delta}{k_B T}}\, e^{-\Delta/k_B T}. \tag{10.48}$$

References

1. G.B. Arfken, H.J. Weber, *Mathematical Methods for Physicists* (Academic, New York, 2012)
2. B.S. Deaver, W.M. Fairbank, Phys. Rev. Lett. **7**, 43 (1961)
3. P.G. de Gennes, *Superconductivity of Metals and Alloys* (W.A. Benjamin, New York, 1966)
4. R. Doll, M. Näbauer, Phys. Rev. Lett. **7**, 51 (1961)
5. A.I. Larkin, A.B. Migdal, J. Exp. Theor. Phys. **44**, 1703 (1963). (Sov. Phys. JETP **17**, 1146 (1963))
6. A.J. Leggett, Phys. Rev. **140**, A1869 (1965)
7. F. London, Phys. Rev. **74**, 562 (1948)
8. F. London, *Macroscopic Theory of Superconductivity* (Dover, New York, 1961)
9. W. Meissner, R. Ochsenfeld, Naturwissenschaften **21**, 787 (1933)
10. K. Yosida, Phys. Rev. **110**, 769 (1958)

Chapter 11
Responses to External Perturbations

Abstract One basic experimental method to probe condensed matter is to subject the system to small perturbing forces, using for example electromagnetic fields, and measure responses. In this chapter, we first develop a linear-response theory for analyzing the resulting data. We then use it to obtain theoretical formulas for ultrasonic attenuation and nuclear-spin relaxation in s-wave superconductors. It is thereby shown that changes in the excitation spectrum through the superconducting transition can be captured unambiguously by these experiments.

11.1 Linear-Response Theory

We consider an arbitrary grand canonical ensemble in equilibrium at $t = -\infty$ described by Hamiltonian $\hat{\mathscr{H}}$, and apply a small time-dependent perturbation $\hat{\mathscr{H}}'(t)$ for $t > -\infty$. We formulate a linear-response theory following Kubo [7]. Key formulas are given by (11.7) with (11.5) for the response in the time domain, (11.13) for the response in the frequency domain, and (11.19) with (11.17) for the energy dissipation.

11.1.1 Response in Time Domain

First, we study responses in the time domain to derive a linear-response formula, (11.7), in terms of the operator (11.5).

Our starting point is the density-matrix operator (5.1), the ket and bra of which now obeys the time-dependent Schrödinger equation:

$$\mathrm{i}\hbar \frac{\mathrm{d}|\Phi_\nu\rangle}{\mathrm{d}t} = \left(\hat{\mathscr{H}} + \hat{\mathscr{H}}'\right)|\Phi_\nu\rangle, \qquad -\mathrm{i}\hbar \frac{\mathrm{d}\langle\Phi_\nu|}{\mathrm{d}t} = \langle\Phi_\nu|\left(\hat{\mathscr{H}} + \hat{\mathscr{H}}'\right), \tag{11.1}$$

© Springer Japan 2015

T. Kita, *Statistical Mechanics of Superconductivity*, Graduate Texts in Physics,
DOI 10.1007/978-4-431-55405-9_11

instead of the time-independent equation.[1] Using them, we differentiate (5.1) with respect to t, changing $\hat{\rho} \to \hat{\rho}_{\rm P}(t)$ to emphasize the presence of the time-dependent perturbation. We thereby obtain[2]

$$\mathrm{i}\hbar \frac{\mathrm{d}\hat{\rho}_{\rm P}(t)}{\mathrm{d}t} = \left[\hat{\mathscr{H}} + \hat{\mathscr{H}}'(t), \hat{\rho}_{\rm P}(t)\right], \tag{11.2}$$

with $[\hat{A}, \hat{B}] \equiv \hat{A}\hat{B} - \hat{B}\hat{A}$. It is convenient to express $\hat{\rho}_{\rm P}(t)$ as

$$\hat{\rho}_{\rm P}(t) = \mathrm{e}^{-\mathrm{i}\hat{\mathscr{H}}t/\hbar} \hat{\rho}_{\rm H}(t) \mathrm{e}^{\mathrm{i}\hat{\mathscr{H}}t/\hbar}. \tag{11.3}$$

Indeed, substitution of this expression into (11.2) produces a cancelation of $\hat{\mathscr{H}}$ on the right-hand side. Multiplying the resulting equation by $\mathrm{e}^{\mathrm{i}\hat{\mathscr{H}}t/\hbar}$ from the left and $\mathrm{e}^{-\mathrm{i}\hat{\mathscr{H}}t/\hbar}$ from the right, we obtain

$$\mathrm{i}\hbar \frac{\mathrm{d}\hat{\rho}_{\rm H}(t)}{\mathrm{d}t} = \left[\hat{\mathscr{H}}'_{\rm H}(t), \hat{\rho}_{\rm H}(t)\right], \tag{11.4}$$

where $\hat{\mathscr{H}}'_{\rm H}(t)$ is defined by

$$\hat{\mathscr{H}}'_{\rm H}(t) \equiv \mathrm{e}^{\mathrm{i}\hat{\mathscr{H}}t/\hbar} \hat{\mathscr{H}}'(t) \mathrm{e}^{-\mathrm{i}\hat{\mathscr{H}}t/\hbar}. \tag{11.5}$$

With the initial condition $\hat{\rho}_{\rm H}(-\infty) = \mathrm{e}^{-\mathrm{i}\hat{\mathscr{H}}t/\hbar}\big(\mathrm{e}^{-\beta\hat{\mathscr{H}}}/Z_{\rm G}\big)\mathrm{e}^{\mathrm{i}\hat{\mathscr{H}}t/\hbar} = \mathrm{e}^{-\beta\hat{\mathscr{H}}}/Z_{\rm G} = \hat{\rho}$, we integrate (11.4)

$$\hat{\rho}_{\rm H}(t) = \hat{\rho} - \frac{\mathrm{i}}{\hbar}\int_{-\infty}^{t} \mathrm{d}t' \left[\hat{\mathscr{H}}'_{\rm H}(t'), \hat{\rho}_{\rm H}(t')\right] \approx \hat{\rho} - \frac{\mathrm{i}}{\hbar}\int_{-\infty}^{t} \mathrm{d}t' \left[\hat{\mathscr{H}}'_{\rm H}(t'), \hat{\rho}\right], \tag{11.6}$$

where the last expression is valid up to first order in $\hat{\mathscr{H}}'(t)$.

Now, consider an observable $\hat{\mathscr{O}}$ with no explicit time dependence. Its expectation at time t is given by $\mathrm{Tr}\hat{\rho}_{\rm P}(t)\hat{\mathscr{O}}$. Let us substitute (11.6) into (11.3) and transform $\mathrm{Tr}\hat{\rho}_{\rm P}(t)\hat{\mathscr{O}} \equiv \mathscr{O}(t)$ using the invariance of the trace under cyclic permutation of the operators [1, 9]. We thereby obtain an expression for the expectation under the same approximation

$$\begin{aligned}\mathscr{O}(t) &\equiv \mathrm{Tr}\,\hat{\rho}_{\rm P}(t)\hat{\mathscr{O}} = \mathrm{Tr}\,\hat{\rho}_{\rm H}(t)\hat{\mathscr{O}}_{\rm H}(t) \approx \hat{\rho}\hat{\mathscr{O}} - \frac{\mathrm{i}}{\hbar}\int_{-\infty}^{t} \mathrm{d}t' \mathrm{Tr}\big[\hat{\mathscr{H}}'_{\rm H}(t'), \hat{\rho}\big]\hat{\mathscr{O}}_{\rm H}(t) \\ &= \langle\hat{\mathscr{O}}\rangle - \frac{\mathrm{i}}{\hbar}\int_{-\infty}^{t} \mathrm{d}t' \big\langle\big[\hat{\mathscr{O}}_{\rm H}(t), \hat{\mathscr{H}}'_{\rm H}(t')\big]\big\rangle,\end{aligned} \tag{11.7}$$

[1]The density matrix (5.1) in equilibrium remains invariant under the change in definition of the ket and bra, because the additional phase factor $\mathrm{e}^{-\mathrm{i}\mathscr{E}_\nu t/\hbar}$ of $|\Phi_\nu\rangle$ in the absence of $\hat{\mathscr{H}}'$ is canceled by $\mathrm{e}^{\mathrm{i}\mathscr{E}_\nu t/\hbar}$ of $\langle\Phi_\nu|$.

[2]Probabilities w_ν are assumed to have no time dependence at all, which is justified when considering linear responses.

with $\langle\hat{\mathscr{O}}\rangle \equiv \mathrm{Tr}\,\hat{\rho}\hat{\mathscr{O}}$ as defined by (5.2). Thus, we can express the linear response due to the perturbation in terms of the expectation in equilibrium as

$$\mathscr{O}^{(1)}(t) = -\frac{\mathrm{i}}{\hbar}\int_{-\infty}^{t} \mathrm{d}t' \langle[\hat{\mathscr{O}}_{\mathrm{H}}(t), \hat{\mathscr{H}}'_{\mathrm{H}}(t')]\rangle. \tag{11.8}$$

11.1.2 Response in Frequency Domain

Next, we show that (11.8) in the time domain can be Fourier-transformed into (11.13) below in the frequency domain.

Let us express $\hat{\mathscr{H}}'(t)$ in a continuous Fourier series,

$$\hat{\mathscr{H}}'(t) = \int_{-\infty}^{\infty} \frac{\mathrm{d}\omega}{2\pi} \mathrm{e}^{-\mathrm{i}\omega t} \hat{\mathscr{H}}'_{\omega}, \tag{11.9}$$

where $\hat{\mathscr{H}}'^{\dagger}_{\omega} = \hat{\mathscr{H}}'_{-\omega}$ holds because of the Hermiticity of $\hat{\mathscr{H}}'(t)$. Using this equation and introducing $\hat{\mathscr{H}}'_{\omega\mathrm{H}}(t) \equiv \mathrm{e}^{\mathrm{i}\hat{\mathscr{H}}t/\hbar}\hat{\mathscr{H}}'_{\omega}\mathrm{e}^{-\mathrm{i}\hat{\mathscr{H}}t/\hbar}$, we can rewrite (11.8) as

$$\begin{aligned}\mathscr{O}^{(1)}(t) &= -\frac{\mathrm{i}}{\hbar}\int_{-\infty}^{t} \mathrm{d}t' \int_{-\infty}^{\infty} \frac{\mathrm{d}\omega}{2\pi} \mathrm{e}^{-\mathrm{i}\omega t'} \langle[\hat{\mathscr{O}}_{\mathrm{H}}(t), \hat{\mathscr{H}}'_{\omega\mathrm{H}}(t')]\rangle \\ &= \int_{-\infty}^{\infty} \frac{\mathrm{d}\omega}{2\pi} \mathrm{e}^{-\mathrm{i}\omega t} \int_{-\infty}^{t} \mathrm{d}t' \mathrm{e}^{\mathrm{i}\omega(t-t')} \left(-\frac{\mathrm{i}}{\hbar}\right) \langle[\hat{\mathscr{O}}_{\mathrm{H}}(t), \hat{\mathscr{H}}'_{\omega\mathrm{H}}(t')]\rangle. \end{aligned} \tag{11.10}$$

Now, one can show (**Problem** 11.1)

$$\langle[\hat{\mathscr{O}}_{\mathrm{H}}(t), \hat{\mathscr{H}}'_{\omega\mathrm{H}}(t')]\rangle = \langle[\hat{\mathscr{O}}_{\mathrm{H}}(t-t'), \hat{\mathscr{H}}'_{\omega}]\rangle. \tag{11.11}$$

With this, and making a change of variable $t' \to t_1 \equiv t - t'$, we can express $\mathscr{O}^{(1)}(t)$ above as

$$\mathscr{O}^{(1)}(t) = \int_{-\infty}^{\infty} \frac{\mathrm{d}\omega}{2\pi} \mathrm{e}^{-\mathrm{i}\omega t} \mathscr{O}^{(1)}_{\omega}, \tag{11.12}$$

with

$$\mathscr{O}^{(1)}_{\omega} \equiv \int_{0}^{\infty} \left(-\frac{\mathrm{i}}{\hbar}\right) \langle[\hat{\mathscr{O}}_{\mathrm{H}}(t_1), \hat{\mathscr{H}}'_{\omega}]\rangle \mathrm{e}^{\mathrm{i}\omega_{+}t_1} \mathrm{d}t_1, \tag{11.13}$$

where $\omega_{+} \equiv \omega + \mathrm{i}0_{+}$ with 0_{+} denoting an infinitesimal positive constant. The additional factor $\mathrm{e}^{-0_{+}t_1}$ introduced here makes the integrand vanish for $t_1 \to \infty$ and thereby ensures the convergence of the t_1 integral. It physically corresponds to the fact that the perturbation is absent at $t' = -\infty$.

Equation (11.12) is given as a linear combination of $\mathscr{O}^{(1)}_{\omega}$ that represents the response of observable $\hat{\mathscr{O}}$ to perturbation $\hat{\mathscr{H}}'_{\omega}$ of the same frequency. In other words, the response to each ω can be calculated independently within the linear-response regime.

11.1.3 Energy Dissipation

Finally, we consider the perturbation[3]

$$\hat{\mathscr{H}}'(t) = \frac{\hat{\mathscr{H}}'_{\omega}\,\mathrm{e}^{-\mathrm{i}\omega t} + \hat{\mathscr{H}}'_{-\omega}\,\mathrm{e}^{\mathrm{i}\omega t}}{2} = \frac{1}{2}\sum_{\sigma=\pm 1} \hat{\mathscr{H}}'_{\sigma\omega}\,\mathrm{e}^{-\sigma\mathrm{i}\omega t}, \qquad (11.14)$$

to obtain an expression for the energy dissipation per unit time, (11.19) below, in terms of the retarded Green's function (11.17). The pair of frequencies of opposite sign are included here to make $\hat{\mathscr{H}}'(t)$ Hermitian, which is indispensable for studying dissipation.

The relevant observable is

$$\hat{\mathscr{R}}(t) \equiv \frac{\mathrm{d}\hat{\mathscr{H}}'(t)}{\mathrm{d}t} = -\frac{\mathrm{i}\omega}{2}\sum_{\sigma=\pm 1} \sigma\hat{\mathscr{H}}'_{\sigma\omega}\,\mathrm{e}^{-\sigma\mathrm{i}\omega t}. \qquad (11.15)$$

The energy injected into the system per unit time is given by the time average $\langle \mathrm{d}\hat{\mathscr{H}}'(t)/\mathrm{d}t\rangle$, which is transformed into *heat* associated with random motions within the system. It is carried to the surface by electrons or phonons to eventually disperse into the surroundings. All these processes are beyond linear order. Hence, we can choose $\hat{\mathscr{O}}$ in (11.8) as $\hat{\mathscr{R}}$ above for the evaluation of the energy *dissipation* rate, which here denotes the energy transferred to its surroundings per unit time in the steady state.

With two relevant frequencies in $\hat{\mathscr{R}}(t)$ above, (11.12) is modified for taking the time average as follows. Let us substitute (11.14) and (11.15) into (11.8) with $\hat{\mathscr{O}} \to \hat{\mathscr{R}}$. We then transform $\mathscr{R}^{(1)}$ similarly to (11.10)–(11.13), obtaining

$$\begin{aligned}
\mathscr{R}^{(1)}(t) &= -\frac{\mathrm{i}}{\hbar}\int_{-\infty}^{t}\mathrm{d}t'\frac{-\mathrm{i}\omega}{4}\sum_{\sigma,\sigma'=\pm 1}\sigma\mathrm{e}^{-\sigma\mathrm{i}\omega t-\sigma'\mathrm{i}\omega t'}\langle\!\langle[\hat{\mathscr{H}}'_{\sigma\omega,\mathrm{H}}(t),\hat{\mathscr{H}}'_{\sigma'\omega,\mathrm{H}}(t')]\rangle\!\rangle \\
&= \frac{-\mathrm{i}\omega}{4}\sum_{\sigma,\sigma'}\sigma\mathrm{e}^{-(\sigma+\sigma')\mathrm{i}\omega t}\left(-\frac{\mathrm{i}}{\hbar}\right)\int_{-\infty}^{t}\mathrm{d}t'\mathrm{e}^{\sigma'\mathrm{i}\omega(t-t')}\langle\!\langle[\hat{\mathscr{H}}'_{\sigma\omega,\mathrm{H}}(t-t'),\hat{\mathscr{H}}'_{\sigma'\omega}]\rangle\!\rangle \\
&= \frac{-\mathrm{i}\omega}{4}\sum_{\sigma,\sigma'}\sigma\mathrm{e}^{-(\sigma+\sigma')\mathrm{i}\omega t}\left(-\frac{\mathrm{i}}{\hbar}\right)\int_{0}^{\infty}\mathrm{d}t_1\mathrm{e}^{\sigma'\mathrm{i}\omega t_1}\langle\!\langle[\hat{\mathscr{H}}'_{\sigma\omega,\mathrm{H}}(t_1),\hat{\mathscr{H}}'_{\sigma'\omega}]\rangle\!\rangle.
\end{aligned}$$

Subsequently, we average $\mathscr{R}^{(1)}(t)$ over a single period, through which only terms with $\sigma = -\sigma'$ survive,

$$\bar{\mathscr{R}}^{(1)} \equiv \frac{\omega}{2\pi}\int_{t_0}^{t_0+2\pi/\omega}\mathscr{R}^{(1)}(t)\mathrm{d}t = \frac{\mathrm{i}\omega}{4}\left[K^{\mathrm{R}}(\omega) - K^{\mathrm{R}}(-\omega)\right], \qquad (11.16)$$

[3] We adopt a normalization for $\hat{\mathscr{H}}_{\omega}$ different from (11.9) for the continuous spectrum.

where $K^{\rm R}(\omega)$ is defined by

$$K^{\rm R}(\omega) \equiv -\frac{\rm i}{\hbar}\int_0^\infty {\rm d}t_1\, {\rm e}^{{\rm i}\omega + t_1}\langle[\hat{\mathscr{H}}'_{-\omega,{\rm H}}(t_1), \hat{\mathscr{H}}'_\omega]\rangle, \tag{11.17}$$

in terms of $\hat{\mathscr{H}}'_{\pm\omega}$ given in (11.14). Note that factor ${\rm e}^{-0+t_1}$ has been incorporated into the integrand once again. This is a *retarded Green's function* that represents the response of $\hat{\mathscr{H}}'_{-\omega,{\rm H}}(t_1 \geq 0)$ due to perturbation $\hat{\mathscr{H}}'_\omega$ at $t_1 = 0$ in the frequency domain; it satisfies (**Problem** 11.2)

$$[K^{\rm R}(\omega)]^* = K^{\rm R}(-\omega). \tag{11.18}$$

Using this, we can express (11.16) as

$$\bar{\mathscr{R}}^{(1)} = -\frac{\omega}{2}{\rm Im}K^{\rm R}(\omega), \tag{11.19}$$

where Im denotes the imaginary part.

11.2 Ultrasonic Attenuation

We study the longitudinal ultrasonic attenuation of s-wave superconductors based on (11.19) to derive (11.39) for the relevant energy dissipation, which is plotted in Fig. 11.1 below.

Ultrasound is an oscillating sound wave whose frequency $f \equiv \omega/2\pi$ ranges from 20 kHz up to several giga-Hertz (i.e., 10^9 Hz). When applied to superconductors, it perturbs the system with energy $2\pi\hbar f \ll \Delta$ and wave number $q \ll k_{\rm F}$.[4] The disturbance by longitudinal waves may be modeled by the time-dependent potential $\mathscr{U}(\mathbf{r}, t) = \mathscr{U}_1\cos(\mathbf{q}\cdot\mathbf{r} - \omega t)$ for electrons. The corresponding Hamiltonian is given by

$$\hat{\mathscr{H}}'(t) = \int {\rm d}\xi\, \hat{\psi}^\dagger(\xi)\mathscr{U}(\mathbf{r}, t)\hat{\psi}(\xi) = \frac{\hat{\mathscr{H}}'_\omega {\rm e}^{-{\rm i}\omega t} + \hat{\mathscr{H}}'_{-\omega}{\rm e}^{{\rm i}\omega t}}{2}, \tag{11.20}$$

where $\hat{\mathscr{H}}'_\omega$ denotes

$$\hat{\mathscr{H}}'_\omega \equiv \mathscr{U}_1\sum_\alpha\int {\rm d}^3r\, \hat{\psi}^\dagger(\xi)\hat{\psi}(\xi)\,{\rm e}^{{\rm i}\mathbf{q}\cdot\mathbf{r}} = \mathscr{U}_1\sum_{\mathbf{k}\alpha}\hat{c}^\dagger_{\mathbf{k}+\alpha}\hat{c}_{\mathbf{k}-\alpha}, \tag{11.21}$$

[4]The energy of sound in the temperature scale is of the order of $\Delta T \equiv 2\pi\hbar f/k_{\rm B} \lesssim 2\pi \times 10^{-34} \times 10^9/10^{-23} \sim 0.1$K, which is much smaller than $T_{\rm c}$ in general. The corresponding wave number q is given in terms of the speed of sound $s \sim 10^3$ m/s by $q = 2\pi f/s \lesssim 2\pi \times 10^9/10^3 \sim 10^7\,{\rm m}^{-1}$, which is also much smaller than $k_{\rm F} \sim a^{-1} \sim 10^{10}\,{\rm m}^{-1}$ with $a \sim 10^{-10}$ m denoting the lattice spacing of metals.

with $\mathbf{k}_\pm \equiv \mathbf{k} \pm \mathbf{q}/2$. In deriving the second expression, we have expanded $\hat{\psi}^\dagger(\xi)$ and $\hat{\psi}(\xi)$ in plane waves,

$$\hat{\psi}^\dagger(\mathbf{r}\alpha) = \frac{1}{\sqrt{V}} \sum_{\mathbf{k}_1} \hat{c}^\dagger_{\mathbf{k}_1\alpha} e^{-i\mathbf{k}_1\cdot\mathbf{r}}, \qquad \hat{\psi}(\mathbf{r}\alpha) = \frac{1}{\sqrt{V}} \sum_{\mathbf{k}_2} \hat{c}_{\mathbf{k}_2\alpha} e^{i\mathbf{k}_2\cdot\mathbf{r}}, \tag{11.22}$$

then used orthonormality $\langle \mathbf{k}_1 | \mathbf{k}_2 + \mathbf{q} \rangle = \delta_{\mathbf{k}_1,\mathbf{k}_2+\mathbf{q}}$ to eliminate $\mathbf{k}_1$, and set $\mathbf{k}_2 = \mathbf{k} - \mathbf{q}/2$ to change the summation variable from $\mathbf{k}_2$ to $\mathbf{k}$.

Next, we expand $\hat{c}_{\mathbf{k}\alpha}$ in terms of quasiparticle fields with unitary matrix (9.7),

$$\begin{bmatrix} \hat{c}_{\mathbf{k}\uparrow} \\ \hat{c}_{\mathbf{k}\downarrow} \\ \hat{c}^\dagger_{-\mathbf{k}\uparrow} \\ \hat{c}^\dagger_{-\mathbf{k}\downarrow} \end{bmatrix} = \begin{bmatrix} u_k & 0 & 0 & -v_k^* \\ 0 & u_k & v_k^* & 0 \\ 0 & -v_k & u_k & 0 \\ v_k & 0 & 0 & u_k \end{bmatrix} \begin{bmatrix} \hat{\gamma}_{\mathbf{k}1} \\ \hat{\gamma}_{\mathbf{k}2} \\ \hat{\gamma}^\dagger_{-\mathbf{k}1} \\ \hat{\gamma}^\dagger_{-\mathbf{k}2} \end{bmatrix}. \tag{11.23}$$

Using this, we transform (11.21) into

$$\begin{aligned} \hat{\mathscr{H}}'_\omega &= \mathscr{U}_1 \sum_{\mathbf{k}} \sum_{\tilde{\alpha}=1}^{2} \left[u_{k_+} \hat{\gamma}^\dagger_{\mathbf{k}_+\tilde{\alpha}} + (-1)^{\tilde{\alpha}} v_{k_+} \hat{\gamma}_{-\mathbf{k}_+ 3-\tilde{\alpha}} \right] \left[u_{k_-} \hat{\gamma}_{\mathbf{k}_-\tilde{\alpha}} + (-1)^{\tilde{\alpha}} v^*_{k_-} \hat{\gamma}^\dagger_{-\mathbf{k}_- 3-\tilde{\alpha}} \right] \\ &= \mathscr{U}_1 \sum_{\mathbf{k}\tilde{\alpha}} \Bigg[(u_{k_+} u_{k_-} - v_{k_-} v^*_{k_+}) \hat{\gamma}^\dagger_{\mathbf{k}_+\tilde{\alpha}} \hat{\gamma}_{\mathbf{k}_-\tilde{\alpha}} + \frac{(-1)^{\tilde{\alpha}}}{2} (u_{k_+} v^*_{k_-} + u_{k_-} v^*_{k_+}) \\ &\quad \times \hat{\gamma}^\dagger_{\mathbf{k}_+\tilde{\alpha}} \hat{\gamma}^\dagger_{-\mathbf{k}_- 3-\tilde{\alpha}} + \frac{(-1)^{\tilde{\alpha}}}{2} (u_{k_-} v_{k_+} + u_{k_+} v_{k_-}) \hat{\gamma}_{-\mathbf{k}_+ 3-\tilde{\alpha}} \hat{\gamma}_{\mathbf{k}_-\tilde{\alpha}} \Bigg], \end{aligned} \tag{11.24}$$

where the second expression has been derived by (i) changing the summation variables for the vv and uv terms as $(\mathbf{k}, \tilde{\alpha}) \to (-\mathbf{k}, 3 - \tilde{\alpha})$ and (ii) using the anticommutation relations of field operators for $\mathbf{q} \neq \mathbf{0}$.

Substituting (11.24) and $\hat{\mathscr{H}}'_{-\omega} = \hat{\mathscr{H}}'^\dagger_\omega$ into (11.17), we can express the expectation of the commutator as

$$\begin{aligned} &\langle [\hat{\mathscr{H}}'_{-\omega,\mathrm{H}}(t), \hat{\mathscr{H}}'_\omega] \rangle \\ &= \mathscr{U}_1^2 \sum_{\mathbf{k}\tilde{\alpha}} \Bigg\{ |u_{k_+} u_{k_-} - v_{k_-} v^*_{k_+}|^2 \langle [\hat{\gamma}^\dagger_{\mathbf{k}_-\tilde{\alpha}}(t) \hat{\gamma}_{\mathbf{k}_+\tilde{\alpha}}(t), \hat{\gamma}^\dagger_{\mathbf{k}_+\tilde{\alpha}} \hat{\gamma}_{\mathbf{k}_-\tilde{\alpha}}] \rangle \\ &\quad + \frac{2}{4} |u_{k_+} v_{k_-} + u_{k_-} v_{k_+}|^2 \langle [\hat{\gamma}_{-\mathbf{k}_- 3-\tilde{\alpha}}(t) \hat{\gamma}_{\mathbf{k}_+\tilde{\alpha}}(t), \hat{\gamma}^\dagger_{\mathbf{k}_+\tilde{\alpha}} \hat{\gamma}^\dagger_{-\mathbf{k}_- 3-\tilde{\alpha}}] \rangle \\ &\quad + \frac{2}{4} |u_{k_+} v_{k_-} + u_{k_-} v_{k_+}|^2 \langle [\hat{\gamma}^\dagger_{\mathbf{k}_-\tilde{\alpha}}(t) \hat{\gamma}^\dagger_{-\mathbf{k}_+ 3-\tilde{\alpha}}(t), \hat{\gamma}_{-\mathbf{k}_+ 3-\tilde{\alpha}} \hat{\gamma}_{\mathbf{k}_-\tilde{\alpha}}] \rangle \Bigg\}, \end{aligned} \tag{11.25}$$

where $\hat{\gamma}_{\mathbf{k}\tilde{\alpha}}(t) \equiv e^{i\hat{\mathscr{H}}t/\hbar}\hat{\gamma}_{\mathbf{k}\tilde{\alpha}}e^{-i\hat{\mathscr{H}}t/\hbar}$, and we have retained only terms with finite expectations. Factor 2 in the uv terms may be confirmed by choosing the summation variables of $\hat{\mathscr{H}}'_{\omega}$ and $\hat{\mathscr{H}}'_{-\omega}$ as $(\mathbf{k}, \tilde{\alpha})$ and $(\mathbf{k}', \tilde{\alpha}')$, respectively, and noting that the two combinations $(\mathbf{k}', \tilde{\alpha}') = (\mathbf{k}, \tilde{\alpha}), (-\mathbf{k}, 3 - \tilde{\alpha})$ yield the same average.

We now evaluate (11.25) based on the mean-field theory of superconductivity where Hamiltonian $\hat{\mathscr{H}}$ is approximated as (8.57). Specifically for the present homogeneous s-wave pairing, it reads

$$\hat{\mathscr{H}} \approx \sum_{\mathbf{k}\tilde{\alpha}} E_k \hat{\gamma}^{\dagger}_{\mathbf{k}\tilde{\alpha}}\hat{\gamma}_{\mathbf{k}\tilde{\alpha}} + \text{const.},$$

with E_k given by (9.5). Now, we have shown that (5.9) is expressible as (5.10). Similarly, operator $\hat{\gamma}_{\mathbf{k}\tilde{\alpha}}(t) = e^{i\hat{\mathscr{H}}t/\hbar}\hat{\gamma}_{\mathbf{k}\tilde{\alpha}}e^{-i\hat{\mathscr{H}}t/\hbar}$ and its Hermitian conjugate with the above Hamiltonian can be written as

$$\hat{\gamma}_{\mathbf{k}\tilde{\alpha}}(t) = e^{-iE_k t/\hbar}\hat{\gamma}_{\mathbf{k}\tilde{\alpha}}, \qquad \hat{\gamma}^{\dagger}_{\mathbf{k}\tilde{\alpha}}(t) = e^{iE_k t/\hbar}\hat{\gamma}^{\dagger}_{\mathbf{k}\tilde{\alpha}}. \tag{11.26}$$

Using them, we can perform the Wick decomposition of the first average in (11.25),

$$\begin{aligned}
&\langle[\hat{\gamma}^{\dagger}_{\mathbf{k}_-\tilde{\alpha}}(t)\hat{\gamma}_{\mathbf{k}_+\tilde{\alpha}}(t), \hat{\gamma}^{\dagger}_{\mathbf{k}_+\tilde{\alpha}}\hat{\gamma}_{\mathbf{k}_-\tilde{\alpha}}]\rangle \\
&= e^{-i(E_{k_+}-E_{k_-})t/\hbar}\left[\langle\hat{\gamma}_{\mathbf{k}_+\tilde{\alpha}}\hat{\gamma}^{\dagger}_{\mathbf{k}_+\tilde{\alpha}}\rangle\langle\hat{\gamma}^{\dagger}_{\mathbf{k}_-\tilde{\alpha}}\hat{\gamma}_{\mathbf{k}_-\tilde{\alpha}}\rangle - \langle\hat{\gamma}^{\dagger}_{\mathbf{k}_+\tilde{\alpha}}\hat{\gamma}_{\mathbf{k}_+\tilde{\alpha}}\rangle\langle\hat{\gamma}_{\mathbf{k}_-\tilde{\alpha}}\hat{\gamma}^{\dagger}_{\mathbf{k}_-\tilde{\alpha}}\rangle\right] \\
&= e^{-i(E_{k_+}-E_{k_-})t/\hbar}\left[(1-\bar{n}_{k_+})\bar{n}_{k_-} - \bar{n}_{k_+}(1-\bar{n}_{k_-})\right],
\end{aligned}$$

with $\bar{n}_k \equiv (e^{\beta E_k} + 1)^{-1}$. The other two expectations can be estimated similarly. Substituting them into (11.25) and summing over $\tilde{\alpha}$, we obtain

$$\begin{aligned}
&\langle[\hat{\mathscr{H}}'_{-\omega,\mathrm{H}}(t), \hat{\mathscr{H}}'_{\omega}]\rangle \\
&= 2\mathscr{U}_1^2\sum_{\mathbf{k}}\Bigg\{|u_{k_+}u_{k_-} - v_{k_-}v^*_{k_+}|^2(\bar{n}_{k_-} - \bar{n}_{k_+})e^{-i(E_{k_+}-E_{k_-})t/\hbar} \\
&+ \frac{|u_{k_+}v_{k_-} + u_{k_-}v_{k_+}|^2}{2}(1 - \bar{n}_{k_+} - \bar{n}_{k_-})\left[e^{-i(E_{k_+}+E_{k_-})t/\hbar} - e^{i(E_{k_+}+E_{k_-})t/\hbar}\right]\Bigg\}.
\end{aligned} \tag{11.27}$$

Using this expression, we can perform the integration of (11.17) to obtain

$$\begin{aligned}
K^{\mathrm{R}}(\omega) = 2\mathscr{U}_1^2\sum_{\mathbf{k}}\Bigg\{&|u_{k_+}u_{k_-} - v_{k_-}v^*_{k_+}|^2\frac{\bar{n}_{k_-} - \bar{n}_{k_+}}{\hbar\omega_+ - (E_{k_+} - E_{k_-})} \\
&+ \frac{|u_{k_+}v_{k_-} + u_{k_-}v_{k_+}|^2}{2}
\end{aligned}$$

$$\times \left[\frac{1 - \bar{n}_{k_+} - \bar{n}_{k_-}}{\hbar\omega_+ - (E_{k_+} + E_{k_-})} - \frac{1 - \bar{n}_{k_+} - \bar{n}_{k_-}}{\hbar\omega_+ + (E_{k_+} + E_{k_-})} \right] \Bigg\} . \tag{11.28}$$

Quantities $u_{k_+}u_{k_-} - v_{k_-}v^*_{k_+}$ and $u_{k_+}v_{k_-} + u_{k_-}v_{k_+}$ are called *coherence factors*, whose magnitudes are given using (9.6) by

$$|u_{k_+}u_{k_-} \mp v_{k_-}v^*_{k_+}|^2 = \frac{E_{k_+}E_{k_-} + \xi_{k_+}\xi_{k_-} \mp \Delta^2}{2E_{k_+}E_{k_-}},$$

$$|u_{k_+}v_{k_-} \pm u_{k_-}v_{k_+}|^2 = \frac{E_{k_+}E_{k_-} - \xi_{k_+}\xi_{k_-} \pm \Delta^2}{2E_{k_+}E_{k_-}}, \tag{11.29}$$

where the lower signs are incorporated for later convenience. Let us substitute (11.29) into (11.28) and use identity

$$\frac{1}{x \pm \mathrm{i}0_+} = \mathrm{P}\frac{1}{x} \mp \mathrm{i}\pi\delta(x), \tag{11.30}$$

with P denoting the principal value, which may be seen to hold by integrating both sides over an arbitrary interval on the real x axis. Equation (11.19) with (11.28) is thereby transformed into

$$\begin{aligned}\bar{\mathscr{R}}^{(1)} = \pi\omega\mathscr{U}_1^2 \sum_{\mathbf{k}} \Bigg\{ & \frac{E_{k_+}E_{k_-} + \xi_{k_+}\xi_{k_-} - \Delta^2}{2E_{k_+}E_{k_-}} (\bar{n}_{k_-} - \bar{n}_{k_+})\delta(\hbar\omega - E_{k_+} + E_{k_-}) \\ & + \frac{E_{k_+}E_{k_-} - \xi_{k_+}\xi_{k_-} + \Delta^2}{4E_{k_+}E_{k_-}} (1 - \bar{n}_{k_+} - \bar{n}_{k_-}) \big[\delta(\hbar\omega - E_{k_+} - E_{k_-}) \\ & - \delta(\hbar\omega + E_{k_+} + E_{k_-}) \big] \Bigg\} . \end{aligned} \tag{11.31}$$

Noting that $\hbar\omega \ll \Delta \leq E_{k_\pm}$, we omit the second term in the curly brackets. For convenience, we then shift the summation variable as $\mathbf{k} - \mathbf{q}/2 \to \mathbf{k}$, and subsequently convert the sum over $\mathbf{k}$ into an integral using (8.90),

$$\begin{aligned}\bar{\mathscr{R}}^{(1)} \approx \pi\omega\mathscr{U}_1^2 V N(\varepsilon_\mathrm{F}) \int_{-\infty}^{\infty} \mathrm{d}\xi_k \int_{-1}^{1} \frac{\mathrm{d}t}{2} \frac{E_{|\mathbf{k}+\mathbf{q}|}E_k + \xi_{|\mathbf{k}+\mathbf{q}|}\xi_k - \Delta^2}{2E_{|\mathbf{k}+\mathbf{q}|}E_k} \\ \times [\bar{n}(E_k) - \bar{n}(E_{|\mathbf{k}+\mathbf{q}|})]\delta(\hbar\omega - E_{|\mathbf{k}+\mathbf{q}|} + E_k), \end{aligned} \tag{11.32}$$

where $t \equiv \cos\theta_\mathbf{k}$ in the coordinate system with $\mathbf{q} \parallel \mathbf{z}$.

We first consider the normal-state limit of $\Delta \to 0$ and $E_k \to \xi_k$, where (11.32) reduces to

$$\bar{\mathscr{R}}_\mathrm{n}^{(1)} = \pi\omega\mathscr{U}_1^2 V N(\varepsilon_\mathrm{F}) \int_{-\infty}^{\infty} \mathrm{d}\xi_k \int_{-1}^{1} \frac{\mathrm{d}t}{2} [\bar{n}(\xi_k) - \bar{n}(\xi_{|\mathbf{k}+\mathbf{q}|})]\delta(\hbar\omega - \xi_{|\mathbf{k}+\mathbf{q}|} + \xi_k). \tag{11.33}$$

Noting $q \ll k_F$, we approximate $\xi_{|\mathbf{k}+\mathbf{q}|} \approx \xi_k + \hbar^2\mathbf{k}\cdot\mathbf{q}/m \approx \xi_k + (\hbar^2 k_F q/m)t$. Then, we can express the argument of the δ function in terms of the Fermi velocity $v_F \equiv \hbar k_F/m \sim 10^6$ m/s and the speed of ultrasound $s \equiv \omega/q \sim 10^3$ m/s as

$$\hbar\omega - \xi_{|\mathbf{k}+\mathbf{q}|} + \xi_k \approx \frac{\hbar^2 k_F q}{m}\left(\frac{m\omega}{\hbar k_F q} - t\right) = \frac{\hbar^2 k_F q}{m}\left(\frac{s}{v_F} - t\right). \tag{11.34}$$

Note that $s/v_F \sim 10^{-3}$ certainly lies in the interval $[-1, 1]$ of the t integral. Hence, the t integral in (11.33) only has the effect of setting $\xi_{|\mathbf{k}+\mathbf{q}|} = \xi_k + \hbar\omega$ via the delta function $\delta(\hbar\omega - \xi_{|\mathbf{k}+\mathbf{q}|} + \xi_k) = (m/\hbar^2 k_F q)\delta(t - s/v_F)$. With this observation, we can simplify (11.33),

$$\begin{aligned}\bar{\mathscr{R}}_{\mathrm{n}}^{(1)} &\approx \pi\omega\mathscr{U}_1^2 VN(\varepsilon_F)\frac{m}{2\hbar^2 k_F q}\int_{-\infty}^{\infty} d\xi_k[\bar{n}(\xi_k) - \bar{n}(\xi_k + \hbar\omega)] \\ &= \pi\mathscr{U}_1^2 VN(\varepsilon_F)\frac{m\omega^2}{2\hbar k_F q},\end{aligned} \tag{11.35}$$

where we have used

$$I(\varepsilon) \equiv \int_{-\infty}^{\infty}[\bar{n}(\xi) - \bar{n}(\xi + \varepsilon)]d\xi = \varepsilon, \tag{11.36}$$

as shown easily based on $dI(\varepsilon)/d\varepsilon = 1$ and $I(0) = 0$.

We now proceed to (11.32); the delta function transforms as

$$\delta(\hbar\omega - E_{|\mathbf{k}+\mathbf{q}|} + E_k) = \left|\frac{dE_{|\mathbf{k}+\mathbf{q}|}}{dt}\right|^{-1}\delta(t - t_0) \approx \left|\frac{dE_k}{d\xi_k}\frac{\hbar^2 k_F q}{m}\right|^{-1}\delta(t - t_0), \tag{11.37}$$

where t_0 ($|t_0| \ll 1$) denotes the value of t at which $E_{|\mathbf{k}+\mathbf{q}|} = \hbar\omega + E_k$ holds. With this and noting $q \ll k_F$, we can rewrite (11.32) as

$$\begin{aligned}\bar{\mathscr{R}}^{(1)} &\approx \pi\omega\mathscr{U}_1^2 VN(\varepsilon_F)2\int_0^{\infty} d\xi_k\frac{E_k^2 + \xi_k^2 - \Delta^2}{2E_k^2}[\bar{n}(E_k) - \bar{n}(E_k + \hbar\omega)]\frac{d\xi_k}{dE_k}\frac{m}{2\hbar^2 k_F q} \\ &\approx \pi\omega\mathscr{U}_1^2 VN(\varepsilon_F)\frac{m}{\hbar^2 k_F q}\int_{\Delta}^{\infty} dE_k\left(\frac{d\xi_k}{dE_k}\right)^2\frac{\xi_k^2}{E_k^2}\left[-\frac{d\bar{n}(E_k)}{dE_k}\right]\hbar\omega \\ &= \pi\mathscr{U}_1^2 VN(\varepsilon_F)\frac{m\omega^2}{\hbar k_F q}\bar{n}(\Delta),\end{aligned} \tag{11.38}$$

where we have used $d\xi_k/dE_k = E_k/\xi_k$ for $\xi_k \geq 0$, which follows from (9.5). Hence, we obtain the energy dissipation of ultrasound in s-wave superconductors relative to (11.35) for the normal state as ($\mathscr{R}^{(1)} \to \mathscr{R}_{\mathrm{s}}^{(1)}$)

$$\mathscr{R}_{\mathrm{s}}^{(1)}/\mathscr{R}_{\mathrm{n}}^{(1)} = 2\bar{n}(\Delta), \tag{11.39}$$

Fig. 11.1 Coefficient of attenuation for ultrasound as a function of temperature

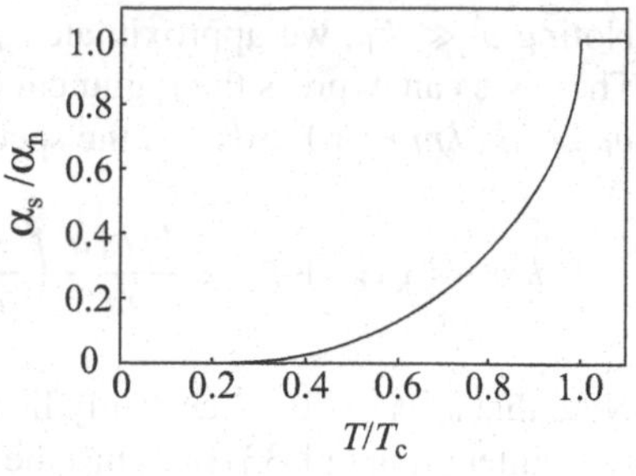

which is equal to the ratio α_s/α_n of the ultrasound attenuation rate α_s observed in experiments. Figure 11.1 plots (11.39) as a function of temperature. We see a steep decrease in the attenuation rate for $T \lesssim T_c$ resulting from the widening of the energy gap. An excellent agreement between theory and experiment was reported soon after the publication of the BCS theory [10].

11.3 Nuclear-Spin Relaxation

Next, we consider nuclear-spin relaxation in s-wave superconductors to derive an expression for the relaxation rate, (11.57), plotted in Fig. 11.2.

We have already seen in Sect. 3.2 that every particle has an internal degree of freedom called spin with a proper magnetic moment. We focus here on the nuclei in a metal that are arranged periodically at lattice sites $\{\mathbf{R}_j\}$ $(j = 1, 2, \cdots, N_a)$ with electrons moving around them. Each nucleus j has a composite spin $\hat{\mathbf{I}}_j$ of the same magnitude I. Its magnetic moment is given in terms of the nuclear magneton $\mu_N = 5.05 \times 10^{-27}$ J/T and g-factor g_I of order 1 as

$$\hat{\boldsymbol{\mu}}_{Ij} = g_I \mu_N \hat{\mathbf{I}}_j. \tag{11.40}$$

Electrons also carry spin moments, for which the density is expressible as

$$\hat{\boldsymbol{\mu}}_e(\mathbf{r}) \equiv -\mu_B \sum_{\alpha\alpha'} \hat{\psi}^\dagger(\mathbf{r}\alpha')\boldsymbol{\sigma}_{\alpha'\alpha}\hat{\psi}(\mathbf{r}\alpha) = -\frac{\mu_B}{V}\sum_{\alpha\alpha'}\sum_{\mathbf{kq}} \hat{c}^\dagger_{\mathbf{k}_+\alpha'}\boldsymbol{\sigma}_{\alpha'\alpha}\hat{c}_{\mathbf{k}_-\alpha}\mathrm{e}^{-\mathrm{i}\mathbf{q}\cdot\mathbf{r}}, \tag{11.41}$$

where μ_B is the Bohr magneton (6.39), $\underline{\boldsymbol{\sigma}} \equiv (\boldsymbol{\sigma}_{\alpha'\alpha})$ are the Pauli matrices, (8.42), and we have expanded the field operators as given in (11.22) with $\mathbf{k}_1 = \mathbf{k} + \mathbf{q}/2$ and $\mathbf{k}_2 = \mathbf{k} - \mathbf{q}/2$. The moment density $\hat{\boldsymbol{\mu}}_e(\mathbf{r})$ yields a local magnetic field at each nuclear site given by the flux density:

$$\hat{\mathbf{b}}(\mathbf{R}_j) = \frac{\mu_0}{4\pi}\int \mathrm{d}^3 r \left[\frac{3(\mathbf{R}_j - \mathbf{r})(\mathbf{R}_j - \mathbf{r})\cdot\hat{\boldsymbol{\mu}}_e(\mathbf{r})}{|\mathbf{R}_j - \mathbf{r}|^5} - \frac{\hat{\boldsymbol{\mu}}_e(\mathbf{r})}{|\mathbf{R}_j - \mathbf{r}|^3} + \frac{8\pi}{3}\hat{\boldsymbol{\mu}}_e(\mathbf{r})\delta(\mathbf{R}_j - \mathbf{r})\right], \tag{11.42}$$

where $\mu_0 = 4\pi \times 10^{-7}$ N·A^{-2} denotes the *vacuum permeability*. The first two terms in the square brackets are the dipole magnetic field, whereas the third term denotes the *Fermi contact interaction*. The corresponding molecular Zeeman energy:

$$\hat{\mathscr{H}}_{\text{hf}} \equiv -\sum_{j=1}^{N_{\text{a}}} \hat{\boldsymbol{\mu}}_{\text{I}j} \cdot \hat{\mathbf{b}}(\mathbf{R}_j) \tag{11.43}$$

is called the *hyperfine interaction*.

Now, suppose that a homogeneous magnetic field with flux density $\mathbf{B} = B\hat{\mathbf{z}}$ is present initially, in which the nuclear spins are subject to the *Zeeman effect*:

$$\hat{\mathscr{H}}_{\text{Z}} \equiv -\sum_{j=1}^{N_{\text{a}}} \hat{\boldsymbol{\mu}}_{\text{I}j} \cdot \mathbf{B}. \tag{11.44}$$

It splits each nuclear energy level into $2I + 1$ distinct sublevels all equally spaced[5]:

$$\hbar\omega \equiv g_{\text{I}}\mu_{\text{N}}B, \tag{11.45}$$

making the state $I_{jz} = I$ with the Zeeman energy $-\hbar\omega I$ the nuclear ground state. Note that $\hbar\omega/k_{\text{B}} \sim 10^{-3}B$ K in the temperature scale, which is much smaller than T_{c} even for a strong field of $B \sim 1$ T. Next, we turn off the external magnetic field at $t = -\infty$ and let the nuclear state $I_{jz} = I$ relax to one of the $2I + 1$ degenerate states. Assuming that the electronic density around each nucleus is isotropic, we can identify the term in (11.43) that is responsible for the relaxation; this is due to the last term in the square brackets of (11.42) and given explicitly by

$$\begin{aligned}\hat{\mathscr{H}}' &= -\frac{2\mu_0}{3}\sum_{j=1}^{N_{\text{a}}} \frac{\hat{\mu}_{\text{e}-}(\mathbf{R}_j)\hat{\mu}_{\text{I}j+} + \hat{\mu}_{\text{e}+}(\mathbf{R}_j)\hat{\mu}_{\text{I}j-}}{2} \\ &= \frac{2g_{\text{I}}\mu_{\text{N}}\mu_{\text{B}}\mu_0}{3V}\sum_{j=1}^{N_{\text{a}}} \text{e}^{-\text{i}\mathbf{q}\cdot\mathbf{R}_j}\sum_{\mathbf{kq}}\left(\hat{c}^\dagger_{\mathbf{k}_+\downarrow}\hat{c}_{\mathbf{k}_-\uparrow}\hat{I}_{j+} + \hat{c}^\dagger_{\mathbf{k}_+\uparrow}\hat{c}_{\mathbf{k}_-\downarrow}\hat{I}_{j-}\right),\end{aligned} \tag{11.46}$$

where $\hat{\mu}_{\text{e}\pm} \equiv \hat{\mu}_{\text{e}x} \pm \text{i}\hat{\mu}_{\text{e}y}$, $\hat{\mu}_{\text{I}j\pm} \equiv \hat{\mu}_{\text{I}jx} \pm \text{i}\hat{\mu}_{\text{I}jy}$, and we have used (11.40) and (11.41). Operator $\hat{I}_{j\pm} \equiv \hat{I}_{jx} \pm \text{i}\hat{I}_{jy}$ causes a change $I_{jz} \to I_{jz} \pm 1$ [2, 8, 11].

As the original Hamiltonian is given as a sum of the electronic Hamiltonian $\hat{\mathscr{H}}$ and nuclear Hamiltonian (11.44), we need to take additional care to construct $\hat{\mathscr{H}}'_{\text{H}}(t)$. Indeed, the nuclear-spin operators give rise to an extra time dependence due to (11.44), i.e.,

[5]There is also a molecular-field contribution from (11.43), which yields a correction $\Delta\omega \lesssim 0.01\omega$ called the *Knight shift* [4, 5].

$$\hat{I}_{j\pm}(t) \equiv \mathrm{e}^{\mathrm{i}(\hat{\mathscr{H}}+\hat{\mathscr{H}}_Z)t/\hbar}\hat{I}_{j\pm}\mathrm{e}^{-\mathrm{i}(\hat{\mathscr{H}}+\hat{\mathscr{H}}_Z)t/\hbar} = \mathrm{e}^{-\mathrm{i}\omega t\hat{I}_{jz}}\hat{I}_{j\pm}\mathrm{e}^{\mathrm{i}\omega t\hat{I}_{jz}}, \tag{11.47}$$

with ω defined by (11.45). To remove this, we differentiate (11.47) with respect to t,

$$\frac{\mathrm{d}\hat{I}_{j\pm}(t)}{\mathrm{d}t} = -\mathrm{i}\omega\mathrm{e}^{-\mathrm{i}\omega t\hat{I}_{jz}}[\hat{I}_{jz}, \hat{I}_{j\pm}]\mathrm{e}^{\mathrm{i}\omega t\hat{I}_{jz}} = \mp\mathrm{i}\omega\hat{I}_{j\pm}(t),$$

where we have used the commutation relation $[\hat{I}_{jz}, \hat{I}_{j\pm}] = \mathrm{i}\hat{I}_{jy} \pm \hat{I}_{jx} = \pm\hat{I}_{j\pm}$ for the angular momentum operators [2, 8, 11]. Integrating subject to $\hat{I}_{j\pm}(0) = \hat{I}_{j\pm}$ yields

$$\hat{I}_{j\pm}(t) = \mathrm{e}^{\mp\mathrm{i}\omega t}\hat{I}_{j\pm}. \tag{11.48}$$

Hence, using this, $\hat{\mathscr{H}}'_{\mathrm{H}}(t) \equiv \mathrm{e}^{\mathrm{i}(\hat{\mathscr{H}}+\hat{\mathscr{H}}_Z)t/\hbar}\hat{\mathscr{H}}'\mathrm{e}^{-\mathrm{i}(\hat{\mathscr{H}}+\hat{\mathscr{H}}_Z)t/\hbar}$ with (11.46) becomes

$$\hat{\mathscr{H}}'_{\mathrm{H}}(t) = \mathrm{e}^{\mathrm{i}\hat{\mathscr{H}}t/\hbar}\frac{\hat{\mathscr{H}}'_{\omega}\mathrm{e}^{-\mathrm{i}\omega t} + \hat{\mathscr{H}}'_{-\omega}\mathrm{e}^{\mathrm{i}\omega t}}{2}\mathrm{e}^{-\mathrm{i}\hat{\mathscr{H}}t/\hbar}, \tag{11.49}$$

where

$$\hat{\mathscr{H}}'_{\omega} \equiv \frac{4g_{\mathrm{I}}\mu_{\mathrm{N}}\mu_{\mathrm{B}}\mu_0}{3V}\sum_{j=1}^{N_{\mathrm{a}}}\mathrm{e}^{-\mathrm{i}\mathbf{q}\cdot\mathbf{R}_j}\sum_{\mathbf{kq}}\hat{c}^{\dagger}_{\mathbf{k}_+\downarrow}\hat{c}_{\mathbf{k}_-\uparrow}\hat{I}_{j+} \tag{11.50}$$

and $\hat{\mathscr{H}}'_{-\omega} = \hat{\mathscr{H}}'^{\dagger}_{\omega}$. The fraction in (11.49) thereby acquires the same expression as that in (11.14) for electrons alone. We subsequently substitute (11.23) into (11.50) and rewrite it as

$$\begin{aligned}
\hat{\mathscr{H}}'_{\omega} &= \frac{4g_{\mathrm{I}}\mu_{\mathrm{N}}\mu_{\mathrm{B}}\mu_0}{3V}\sum_{j=1}^{N_{\mathrm{a}}}\mathrm{e}^{-\mathrm{i}\mathbf{q}\cdot\mathbf{R}_j}\sum_{\mathbf{kq}}\big(u_{k_+}\hat{\gamma}^{\dagger}_{\mathbf{k}_+2} + v_{k_+}\hat{\gamma}_{-\mathbf{k}_+1}\big)\big(u_{k_-}\hat{\gamma}_{\mathbf{k}_-1} - v^*_{k_-}\hat{\gamma}^{\dagger}_{-\mathbf{k}_-2}\big)\hat{I}_{j+} \\
&= \frac{4g_{\mathrm{I}}\mu_{\mathrm{N}}\mu_{\mathrm{B}}\mu_0}{3V}\sum_{j=1}^{N_{\mathrm{a}}}\mathrm{e}^{-\mathrm{i}\mathbf{q}\cdot\mathbf{R}_j}\sum_{\mathbf{kq}}\bigg[\big(u_{k_+}u_{k_-} + v_{k_-}v^*_{k_+}\big)\hat{\gamma}^{\dagger}_{\mathbf{k}_+2}\hat{\gamma}_{\mathbf{k}_-1} \\
&\quad - \frac{u_{k_+}v^*_{k_-} - u_{k_-}v^*_{k_+}}{2}\hat{\gamma}^{\dagger}_{\mathbf{k}_+2}\gamma^{\dagger}_{-\mathbf{k}_-2} + \frac{u_{k_-}v_{k_+} - u_{k_+}v_{k_-}}{2}\hat{\gamma}_{-\mathbf{k}_+1}\hat{\gamma}_{\mathbf{k}_-1}\bigg]\hat{I}_{j+},
\end{aligned} \tag{11.51}$$

where the second expression has been obtained by (i) changing the summation variable for the vv and uv terms as $\mathbf{k} \to -\mathbf{k}$, and (ii) using the commutation relations of the field operators.

We substitute (11.51) and $\hat{\mathscr{H}}'_{-\omega} = \hat{\mathscr{H}}'^{\dagger}_{\omega}$ into the commutator of (11.17), take the extra expectations with respect to the nuclear initial state as $\langle I|\hat{I}_{j+}\hat{I}_{j'-}|I\rangle = 2I\delta_{jj'}$ and $\langle I|\hat{I}_{j-}\hat{I}_{j'+}|I\rangle = 0$ to find $\langle[\hat{\mathscr{H}}'_{-\omega,\mathrm{H}}(t), \hat{\mathscr{H}}'_{\omega}]\rangle = -\langle\hat{\mathscr{H}}'_{\omega}\hat{\mathscr{H}}'_{-\omega,\mathrm{H}}(t)\rangle$, and use orthogonality $\sum_{j=1}^{N_{\mathrm{a}}}\mathrm{e}^{\mathrm{i}(\mathbf{q}-\mathbf{q}')\cdot\mathbf{R}_j} = N_{\mathrm{a}}\delta_{\mathbf{qq}'}$. We thereby obtain the expectation for the commutator as

$$
\begin{aligned}
&\langle[\hat{\mathscr{H}}'_{-\omega,\mathrm{H}}(t),\hat{\mathscr{H}}'_{\omega}]\rangle \\
&= -2IN_{\mathrm{a}}\left(\frac{4g_{\mathrm{I}}\mu_{\mathrm{N}}\mu_{\mathrm{B}}\mu_0}{3V}\right)^2\sum_{\mathbf{kq}}\Big[|u_{k_+}u_{k_-}+v_{k_-}v_{k_+}^*|^2\langle\hat{\gamma}^\dagger_{\mathbf{k}_+2}\hat{\gamma}_{\mathbf{k}_-1}\hat{\gamma}^\dagger_{\mathbf{k}_-1}(t)\hat{\gamma}_{\mathbf{k}_+2}(t)\rangle \\
&\quad+\frac{2}{4}|u_{k_+}v_{k_-}-u_{k_-}v_{k_+}|^2\langle\hat{\gamma}^\dagger_{\mathbf{k}_+2}\hat{\gamma}^\dagger_{-\mathbf{k}_-2}\hat{\gamma}_{-\mathbf{k}_-2}(t)\hat{\gamma}_{\mathbf{k}_+2}(t)\rangle \\
&\quad+\frac{2}{4}|u_{k_+}v_{k_-}-u_{k_-}v_{k_+}|^2\langle\hat{\gamma}_{-\mathbf{k}_+1}\hat{\gamma}_{\mathbf{k}_-1}\hat{\gamma}^\dagger_{\mathbf{k}_-1}(t)\hat{\gamma}^\dagger_{-\mathbf{k}_+1}(t)\rangle\Big].
\end{aligned}
$$

Factor 2 in the uv terms may be confirmed by choosing the summation variables of $\hat{\mathscr{H}}'_{\omega}$ and $\hat{\mathscr{H}}'_{-\omega}$ as $(\mathbf{k},\tilde{\alpha})$ and $(\mathbf{k}',\tilde{\alpha}')$, respectively, and noting that the two combinations $(\mathbf{k}',\tilde{\alpha}') = (\pm\mathbf{k},\tilde{\alpha})$ yield the same average. Subsequently, we substitute (11.26) into the above expression and perform the Wick decompositions to obtain

$$
\begin{aligned}
&\langle[\hat{\mathscr{H}}'_{-\omega,\mathrm{H}}(t),\hat{\mathscr{H}}'_{\omega}]\rangle \\
&= -2IN_{\mathrm{a}}\left(\frac{4g_{\mathrm{I}}\mu_{\mathrm{N}}\mu_{\mathrm{B}}\mu_0}{3V}\right)^2\sum_{\mathbf{kq}}\Big[|u_{k_+}u_{k_-}+v_{k_-}v_{k_+}^*|^2\bar{n}_{k_+}(1-\bar{n}_{k_-})\mathrm{e}^{-\mathrm{i}(E_{k_+}-E_{k_-})t/\hbar} \\
&\quad+\frac{|u_{k_+}v_{k_-}-u_{k_-}v_{k_+}|^2}{2}\bar{n}_{k_+}\bar{n}_{k_-}\mathrm{e}^{-\mathrm{i}(E_{k_+}+E_{k_-})t/\hbar} \\
&\quad+\frac{|u_{k_+}v_{k_-}-u_{k_-}v_{k_+}|^2}{2}(1-\bar{n}_{k_+})(1-\bar{n}_{k_-})\mathrm{e}^{\mathrm{i}(E_{k_+}+E_{k_-})t/\hbar}\Big],
\end{aligned}
\tag{11.52}
$$

with $\bar{n}_k \equiv (\mathrm{e}^{\beta E_k}+1)^{-1}$. With this expression in the integrand, the integration of (11.17) is easily performed, yielding

$$
\begin{aligned}
K^{\mathrm{R}}(\omega) &= -2IN_{\mathrm{a}}\left(\frac{4g_{\mathrm{I}}\mu_{\mathrm{N}}\mu_{\mathrm{B}}\mu_0}{3V}\right)^2\sum_{\mathbf{kq}}\Bigg\{|u_{k_+}u_{k_-}+v_{k_-}v_{k_+}^*|^2\frac{\bar{n}_{k_+}(1-\bar{n}_{k_-})}{\hbar\omega_+-(E_{k_+}-E_{k_-})} \\
&\quad+\frac{|u_{k_+}v_{k_-}-u_{k_-}v_{k_+}|^2}{2}\left[\frac{\bar{n}_{k_+}\bar{n}_{k_-}}{\hbar\omega_+-(E_{k_+}+E_{k_-})}+\frac{(1-\bar{n}_{k_+})(1-\bar{n}_{k_-})}{\hbar\omega_++(E_{k_+}+E_{k_-})}\right]\Bigg\}.
\end{aligned}
\tag{11.53}
$$

Substituting (11.29) into (11.53) and using identity (11.30) to calculate (11.19), we thereby obtain the average energy gain of the electrons per unit time as

$$
\begin{aligned}
\bar{\mathscr{R}}^{(1)} &= -\pi\omega IN_{\mathrm{a}}\left(\frac{4g_{\mathrm{I}}\mu_{\mathrm{N}}\mu_{\mathrm{B}}\mu_0}{3V}\right)^2\sum_{\mathbf{kq}}\Bigg\{\frac{E_{k_+}E_{k_-}+\xi_{k_+}\xi_{k_-}+\Delta^2}{2E_{k_+}E_{k_-}}\bar{n}_{k_+}(1-\bar{n}_{k_-}) \\
&\quad\times\delta(\hbar\omega-E_{k_+}+E_{k_-})+\frac{E_{k_+}E_{k_-}-\xi_{k_+}\xi_{k_-}-\Delta^2}{4E_{k_+}E_{k_-}}[\bar{n}_{k_+}\bar{n}_{k_-}\delta(\hbar\omega-E_{k_+}-E_{k_-}) \\
&\quad+(1-\bar{n}_{k_+})(1-\bar{n}_{k_-})\delta(\hbar\omega+E_{k_+}+E_{k_-})]\Bigg\},
\end{aligned}
\tag{11.54}
$$

where the minus sign after the equality sign implies that the energy actually flows from electrons to nuclei to compensate the negative nuclear energy $-I\hbar\omega$. The quantity

$$T_1 \equiv \frac{N_a I\hbar\omega}{-\bar{\mathscr{R}}^{(1)}}, \tag{11.55}$$

which is called the *longitudinal magnetic relaxation time*, gives the time scale over which the nuclear-spin state $I_{jz} = I$ relaxes to the new equilibrium of equally populated sublevels $I_{jz} = I, I-1, \cdots, -I$. Noting that $\hbar\omega \ll \Delta \le E_{k\pm}$, we omit the second term within the curly brackets in (11.54). We then change the summation variables from $(\mathbf{k}, \mathbf{q})$ to $(\mathbf{k}_+, \mathbf{k}_-) \equiv (\mathbf{k}_1, \mathbf{k}_2)$ and transform the sums over $(\mathbf{k}_1, \mathbf{k}_2)$ into integrals using (8.90). The term $\xi_{k+}\xi_{k-}$, being odd in $\xi_{k\pm}$, gives a null contribution to the integration over $-\infty \le \xi_{k\pm} \le \infty$. Subsequently, setting $\xi_{k\pm} \to -\xi_{k\pm}$ for $-\infty \le \xi_{k\pm} \le 0$, we can rearrange $T_1^{-1} = -\bar{\mathscr{R}}^{(1)}/N_a I\hbar\omega$ as

$$\begin{aligned}
\frac{1}{T_1} &\approx \frac{\pi}{\hbar}\left(\frac{4g_I\mu_N\mu_B\mu_0}{3}\right)^2 4\int_0^\infty d\xi_1 N(\varepsilon_F)\int_0^\infty d\xi_2 N(\varepsilon_F) \\
&\quad\times\frac{E_1E_2+\Delta^2}{2E_1E_2}\bar{n}(E_1)[1-\bar{n}(E_2)]\delta(\hbar\omega - E_1 + E_2) \\
&= \frac{2\pi}{\hbar}\left(\frac{4g_I\mu_N\mu_B\mu_0}{3}\right)^2\int_\Delta^\infty dE_1 N(\varepsilon_F)\frac{E_1}{\xi_1}\int_\Delta^\infty dE_2 N(\varepsilon_F)\frac{E_2}{\xi_2} \\
&\quad\times\frac{E_1E_2+\Delta^2}{E_1E_2}\bar{n}(E_1)[1-\bar{n}(E_2)]\delta(\hbar\omega - E_1 + E_2),
\end{aligned} \tag{11.56}$$

where we have used $d\xi/dE = E/\xi = E/\sqrt{E^2-\Delta^2}$. Finally, we take the limit $\omega \to 0$ based on the inequality $\hbar\omega \ll k_B T_c$, as noted below (11.45), and express $\bar{n}(E)[1-\bar{n}(E)] = -k_B T[\partial\bar{n}(E)/\partial E]$. We thereby obtain[6]

$$\frac{1}{T_1} = \frac{2\pi}{\hbar}k_B T\left(\frac{4g_I\mu_N\mu_B\mu_0}{3}\right)^2\int_0^\infty dE\big[N_s(E)\big]^2\frac{E^2+\Delta^2}{E^2}\left[-\frac{\partial\bar{n}(E)}{\partial E}\right]. \tag{11.57}$$

This integral is different from (11.38) for ultrasound attenuation mainly because of the extra quasiparticle density of states $N_s(E) \equiv \theta(E-\Delta)N(\varepsilon_F)E/\sqrt{E^2-\Delta^2}$ in

[6]For electrons in solids, there appears another factor $\langle|u_{\mathbf{k}}(\mathbf{0})|^2\rangle_F^2$ on the right-hand side of (11.57) [12], where $|u_{\mathbf{k}}(\mathbf{0})|^2$ is the relative density of electrons at the nuclear site with Bloch vector $\mathbf{k}$, and $\langle\cdots\rangle_F$ denotes the Fermi-surface average.

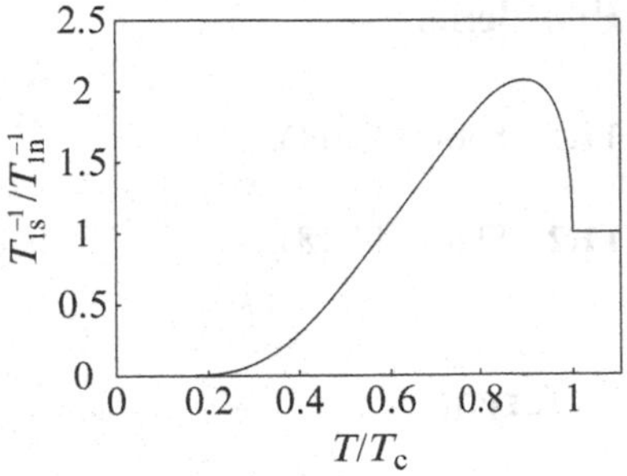

Fig. 11.2 Nuclear-spin relaxation rate as a function of temperature

the integrand, which originates from the sign difference in coherence factors (11.29). The extra factor gives rise to an enhancement of the relaxation rate T_1^{-1} for $T \lesssim T_c$ beyond the normal-state value, as seen below.

First, we consider the normal-state limit of $\Delta \to 0$ and $N_s(E) \to N(\varepsilon_F)$, where (11.57) reduces to ($T_1 \to T_{1n}$)

$$\frac{1}{T_{1n}} = \frac{\pi}{\hbar} k_B T \left[\frac{4 g_I \mu_N \mu_B \mu_0 N(\varepsilon_F)}{3} \right]^2 . \tag{11.58}$$

This formula $T_{1n}T$ = constant for the normal state is called the *Korringa relation* [6, 12]. Temperature T here reflects the number of thermally excited electrons above the Fermi surface (see Sect. 4.5) that can transfer energy to the nuclear spins.

Next, we consider (11.57), where the integral diverges because of $[N_s(E)]^2 \propto E^2/(E^2 - \Delta^2)$. However, this divergence may be regarded as unphysical, because it is removed immediately by incorporating anisotropy into the energy gap that should be present in real materials. To describe the situation, we replace the density of states in (11.57) with a smeared version,

$$\tilde{N}_s(E) = \frac{1}{2\delta} \int_{E-\delta}^{E+\delta} N_s(E') \mathrm{d}E', \tag{11.59}$$

where $0 \le \delta \ll \Delta$. Figure 11.2 plots T_{1s}^{-1}/T_{1n}^{-1} ($T_1 \to T_{1s}$) as a function of reduced temperature calculated in this way with $\delta = 0.1\Delta(T)$. As seen clearly, there is an enhancement in the relaxation rate over the normal-state value T_{1n}^{-1} for $T \lesssim T_c$, which is caused by the divergence in the quasiparticle density of states (see Fig. 9.4). The corresponding peak in T_{1s}^{-1}/T_{1n}^{-1} is called the *Hebel-Slichter peak* [3] that is characteristic of isotropic s-wave superconductors. The peak reduces gradually as the gap anisotropy is increased, vanishing completely for a gap structure with a line node, for example, which may be confirmed by substituting the rightmost density of states in Fig. 13.3 into (11.57).

Problems

11.1. Show (11.11).

11.2. Show (11.18).

References

1. A.C. Aitken, *Determinants and Matrices* (Oliver and Boyd, Edinburgh, 1956)
2. A.R. Edmonds, *Angular Momentum in Quantum Mechanics* (Princeton University Press, Princeton, 1957)
3. L.C. Hebel, C.P. Slichter, Phys. Rev. **113**, 1504 (1959)
4. C. Kittel, *Introduction to Solid State Physics* (Wiley, New York, 2005)
5. W.D. Knight, Phys. Rev. **76**, 1259 (1949); W.D. Knight, in *Solid State Physics*, vol. 2, ed. by F. Seitz, D. Turnbull (Academic, New York, 1956), pp. 93–136
6. J. Korringa, Physica **16**, 601 (1950)
7. R. Kubo, J. Phys. Soc. Jpn. **12**, 570 (1957)
8. L.D. Landau, E.M. Lifshitz, *Quantum Mechanics: Non-relativistic Theory*, 3rd edn. (Butterworth-Heinemann, Oxford, 1991)
9. S. Lang, *Linear Algebra* (Springer, New York, 1987)
10. R.W. Morse, H.V. Bohm, Phys. Rev. **108**, 1094 (1957)
11. J.J. Sakurai, *Modern Quantum Mechanics*, rev. edn. (Addison-Wesley, Reading, 1994)
12. C.P. Slichter, *Principles of Magnetic Resonance* (Springer, Berlin, 2010)

Chapter 12
Tunneling, Density of States, and Josephson Effect

Abstract The tunneling current through a superconducting-normal (SN) junction or a superconducting-superconducting (SS) junction provides rich information about the quasiparticle density of states and condensate wave function. On the basis of the linear response theory developed in the previous chapter, we first derive a general expression for the tunneling current applicable to both junctions as (12.31). It is subsequently applied to SN junctions to show that the current-voltage characteristics directly reflect the quasiparticle density of states as Fig. 12.2. Next, we consider SS junctions to clarify that, besides extra structures caused by two kinds of the quasiparticle density of states, there appears a new feature at zero bias due to the Josephson effect, as seen in Fig. 12.4, that depends on the phase difference of the two coupled superconductors. Thus, a weak contact between two superconductors provides a unique means to detect the phase of the condensate wave function.

12.1 Formula for Tunneling Current

Consider two superconductors separated by an insulating layer, as depicted in Fig. 12.1, with a chemical potential difference between superconductors L and R given by

$$\mu_L - \mu_R = eV, \tag{12.1}$$

where $e < 0$ and V are the electron charge and voltage across the barrier, respectively. Setting $|\Delta(0)|/k_B \lesssim 10$ K, we can estimate the relevant voltage as $|V| \sim |\Delta(0)|/|e| \sim 10k_B/|e| \lesssim 10^{-3}$ V. We develop an expression for the tunneling current, given in (12.31), where

$$\varphi \equiv \varphi_L - \varphi_R \tag{12.2}$$

denotes the relative phase between $\Delta_L = |\Delta_L|e^{i\varphi_L}$ and $\Delta_R = |\Delta_R|e^{i\varphi_R}$. Note that (12.31) can also describe SN junctions by letting $\Delta_R \to 0$. Those who are familiar with the microscopic derivation may skip to formula (12.31) to proceed.

© Springer Japan 2015

T. Kita, *Statistical Mechanics of Superconductivity*, Graduate Texts in Physics,
DOI 10.1007/978-4-431-55405-9_12

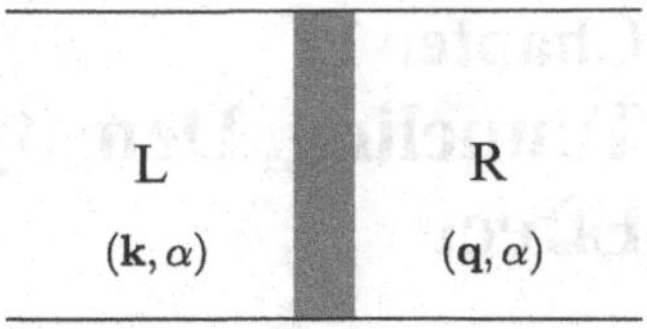

Fig. 12.1 Two superconductors L and R separated by an insulating layer

We assume that the layer in Fig. 12.1 is sufficiently thin for electrons to tunnel through but also thick enough for the tunneling process to be regarded as a small perturbation. The whole system may be described by the Hamiltonian [6]:

$$\hat{\mathscr{H}}_{\rm tot} = \hat{\mathscr{H}}_{\rm L} + \hat{\mathscr{H}}_{\rm R} + \hat{\mathscr{H}}'. \tag{12.3}$$

Here $\hat{\mathscr{H}}_{\rm L}$ ($\hat{\mathscr{H}}_{\rm R}$) is the Hamiltonian of the L (R) side without the perturbation, and $\hat{\mathscr{H}}'$ is the tunneling Hamiltonian given by

$$\hat{\mathscr{H}}' = \frac{1}{\sqrt{V_{\rm L}V_{\rm R}}}\sum_{\mathbf{kq}\alpha}\left(T_{\mathbf{kq}}\hat{c}^{\dagger}_{\mathbf{k}\alpha}\hat{d}_{\mathbf{q}\alpha} + T_{\mathbf{qk}}\hat{d}^{\dagger}_{\mathbf{q}\alpha}\hat{c}_{\mathbf{k}\alpha}\right), \tag{12.4}$$

where $V_{\rm L}$ ($V_{\rm R}$) is the volume of the L (R) side, $\hat{c}_{\mathbf{k}\alpha}$ ($\hat{d}_{\mathbf{q}\alpha}$) is the field operator of L (R) with wave vector $\mathbf{k}$ ($\mathbf{q}$) and spin α, and $T_{\mathbf{kq}} = T^{*}_{\mathbf{qk}}$ is the tunneling matrix, independent of spin, for which we also assume the time-reversal symmetry $T_{\mathbf{kq}} = T^{*}_{-\mathbf{q},-\mathbf{k}}$ [8]. This Hamiltonian adequately describes the electron transfer from one side to the other to the lowest order in $\hat{\mathscr{H}}'$ [4, 10].

The quantity of interest is the current between the superconductors. To derive the relevant operator $\hat{I}$, we consider the charge operator on the L side:

$$\hat{Q}_{\rm L} = e\sum_{\mathbf{k}'\alpha'}\hat{c}^{\dagger}_{\mathbf{k}'\alpha'}\hat{c}_{\mathbf{k}'\alpha'}. \tag{12.5}$$

Operator $\hat{Q}_{\rm L}(t) = {\rm e}^{{\rm i}\hat{\mathscr{H}}_{\rm tot}t/\hbar}\hat{Q}_{\rm L}{\rm e}^{-{\rm i}\hat{\mathscr{H}}_{\rm tot}t/\hbar}$ satisfies the Heisenberg equation of motion,

$$\frac{{\rm d}\hat{Q}_{\rm L}(t)}{{\rm d}t} = {\rm e}^{{\rm i}\hat{\mathscr{H}}_{\rm tot}t/\hbar}\frac{\rm i}{\hbar}\left[\hat{\mathscr{H}}_{\rm tot},\hat{Q}_{\rm L}\right]{\rm e}^{-{\rm i}\hat{\mathscr{H}}_{\rm tot}t/\hbar} = {\rm e}^{{\rm i}\hat{\mathscr{H}}_{\rm tot}t/\hbar}\frac{\rm i}{\hbar}\left[\hat{\mathscr{H}}',\hat{Q}_{\rm L}\right]{\rm e}^{-{\rm i}\hat{\mathscr{H}}_{\rm tot}t/\hbar},$$

where we have used $\left[\hat{\mathscr{H}}_{\rm L},\hat{Q}_{\rm L}\right] = \left[\hat{\mathscr{H}}_{\rm R},\hat{Q}_{\rm L}\right] = 0$. Hence, we can identify the current operator for L→R from ${\rm d}\hat{Q}_{\rm L}(t)/{\rm d}t$ as[1]

[1]Quantity ${\rm d}\hat{Q}_{\rm L}(t)/{\rm d}t$ denotes the gain of negative charges (i.e., the loss of positive charges) per unit time on the L side.

$$\hat{I} = \frac{\mathrm{i}}{\hbar}[\hat{\mathscr{H}}', \hat{Q}_{\mathrm{L}}] = \frac{\mathrm{i}}{\hbar}\frac{e}{\sqrt{V_{\mathrm{L}}V_{\mathrm{R}}}}\sum_{\mathbf{k}'\mathbf{q}'\alpha'}\left(T_{\mathbf{q}'\mathbf{k}'}\hat{d}^{\dagger}_{\mathbf{q}'\alpha'}\hat{c}_{\mathbf{k}'\alpha'} - T_{\mathbf{k}'\mathbf{q}'}\hat{c}^{\dagger}_{\mathbf{k}'\alpha'}\hat{d}_{\mathbf{q}'\alpha'}\right), \tag{12.6}$$

where we have used the commutation relation (3.62) for $\sigma = -1$ and also set $\mathbf{q} \to \mathbf{q}'$ for the second expression.

Let us express (12.4) and (12.6) concisely in matrix form. To this end, we transform the second term in the round brackets of (12.4) as

$$\sum_{\mathbf{kq}} T_{\mathbf{qk}}\hat{d}^{\dagger}_{\mathbf{q}\alpha}\hat{c}_{\mathbf{k}\alpha} = \sum_{\mathbf{kq}} T_{-\mathbf{q},-\mathbf{k}}\hat{d}^{\dagger}_{-\mathbf{q}\alpha}\hat{c}_{-\mathbf{k}\alpha} = -\sum_{\mathbf{kq}} T^{*}_{\mathbf{kq}}\hat{c}_{-\mathbf{k}\alpha}\hat{d}^{\dagger}_{-\mathbf{q}\alpha}, \tag{12.7}$$

where we have made a change of summation variables, $(\mathbf{k},\mathbf{q}) \to (-\mathbf{k},-\mathbf{q})$, and subsequently used $\hat{d}^{\dagger}_{-\mathbf{q}\alpha}\hat{c}_{-\mathbf{k}\alpha} = -\hat{c}_{-\mathbf{k}\alpha}\hat{d}^{\dagger}_{-\mathbf{q}\alpha}$ and $T_{-\mathbf{q},-\mathbf{k}} = T^{*}_{\mathbf{kq}}$. We also introduce

$$\hat{\mathbf{c}}_{\mathbf{k}} \equiv \begin{bmatrix} \hat{c}_{\mathbf{k}\uparrow} \\ \hat{c}_{\mathbf{k}\downarrow} \\ \hat{c}^{\dagger}_{-\mathbf{k}\uparrow} \\ \hat{c}^{\dagger}_{-\mathbf{k}\downarrow} \end{bmatrix}, \qquad \hat{\mathbf{c}}^{\dagger}_{\mathbf{k}} \equiv \begin{bmatrix} \hat{c}^{\dagger}_{\mathbf{k}\uparrow} & \hat{c}^{\dagger}_{\mathbf{k}\downarrow} & \hat{c}_{-\mathbf{k}\uparrow} & \hat{c}_{-\mathbf{k}\downarrow} \end{bmatrix}, \tag{12.8}$$

$$\hat{T}_{\mathbf{kq}} \equiv \begin{bmatrix} T_{\mathbf{kq}}\underline{\sigma}_0 & \underline{0} \\ \underline{0} & T^{*}_{\mathbf{kq}}\underline{\sigma}_0 \end{bmatrix}, \qquad \hat{\sigma}_z \equiv \begin{bmatrix} \underline{\sigma}_0 & \underline{0} \\ \underline{0} & -\underline{\sigma}_0 \end{bmatrix}, \tag{12.9}$$

where $\underline{\sigma}_0$ and $\underline{0}$ are the 2×2 unit and zero matrices, respectively. Using (12.7)–(12.9), we can express (12.4) concisely as

$$\hat{\mathscr{H}}' = \frac{1}{\sqrt{V_{\mathrm{L}}V_{\mathrm{R}}}}\sum_{\mathbf{kq}} \hat{\mathbf{c}}^{\dagger}_{\mathbf{k}}\hat{T}_{\mathbf{kq}}\hat{\sigma}_z\hat{\mathbf{d}}_{\mathbf{q}}. \tag{12.10}$$

Similarly, (12.6) is transformed into

$$\hat{I} = \frac{\mathrm{i}}{\hbar}\frac{e}{\sqrt{V_{\mathrm{L}}V_{\mathrm{R}}}}\sum_{\mathbf{k}'\mathbf{q}'} \hat{\mathbf{d}}^{\dagger}_{\mathbf{q}'}\hat{T}_{\mathbf{q}'\mathbf{k}'}\hat{\mathbf{c}}_{\mathbf{k}'}. \tag{12.11}$$

With these preliminaries, we now estimate the current through the barrier by regarding $\hat{\mathscr{H}}'$ as a perturbation that is applied adiabatically from $t = -\infty$. Hence, the current is absent at $t = -\infty$, and its expectation at time t is given by (11.8) with $\hat{\mathscr{O}} \to \hat{I}$ and $\hat{\mathscr{H}}'(t') \to \hat{\mathscr{H}}'\mathrm{e}^{0_+t'}$,

$$I(t) = -\frac{\mathrm{i}}{\hbar}\int_{-\infty}^{t} \mathrm{d}t' \langle [\hat{I}_{\mathrm{H}}(t), \hat{\mathscr{H}}'_{\mathrm{H}}(t')] \rangle \mathrm{e}^{0_+t'}, \tag{12.12}$$

where $\hat{I}_{\rm H}(t) \equiv {\rm e}^{{\rm i}\hat{\mathscr{H}}t/\hbar}\hat{I}{\rm e}^{-{\rm i}\hat{\mathscr{H}}t/\hbar}$ and $\hat{\mathscr{H}}'_{\rm H}(t) \equiv {\rm e}^{{\rm i}\hat{\mathscr{H}}t/\hbar}\hat{\mathscr{H}}'{\rm e}^{-{\rm i}\hat{\mathscr{H}}t/\hbar}$ with $\hat{\mathscr{H}} \equiv \hat{\mathscr{H}}_{\rm L} + \hat{\mathscr{H}}_{\rm R}$. The first term in the expectation becomes, using (12.10) and (12.11),

$$\begin{aligned}\langle \hat{I}_{\rm H}(t)\hat{\mathscr{H}}'_{\rm H}(t')\rangle &= \frac{\rm i}{\hbar}\frac{e}{V_{\rm L}V_{\rm R}}\sum_{\mathbf{kq}}\sum_{\mathbf{k'q'}}\langle \hat{\mathbf{d}}^{\dagger}_{\mathbf{q'}}(t)\hat{T}_{\mathbf{q'k'}}\hat{\mathbf{c}}_{\mathbf{k'}}(t)\hat{\mathbf{c}}^{\dagger}_{\mathbf{k}}(t')\hat{T}_{\mathbf{kq}}\hat{\sigma}_z\hat{\mathbf{d}}_{\mathbf{q}}(t')\rangle \\ &= \frac{\rm i}{\hbar}\frac{e}{V_{\rm L}V_{\rm R}}\sum_{\mathbf{kq}}\sum_{\mathbf{k'q'}}(-1)^2{\rm Tr}\langle \hat{\mathbf{d}}^{*}_{\mathbf{q'}}(t)\hat{\mathbf{d}}^{\rm T}_{\mathbf{q}}(t')\rangle^{\rm T}\hat{T}_{\mathbf{q'k'}}\langle \hat{\mathbf{c}}_{\mathbf{k'}}(t)\hat{\mathbf{c}}^{\dagger}_{\mathbf{k}}(t')\rangle\hat{T}_{\mathbf{kq}}\hat{\sigma}_z,\end{aligned}$$

where $\hat{\mathbf{c}}_{\mathbf{k}}(t) \equiv {\rm e}^{{\rm i}\hat{\mathscr{H}}_{\rm L}t/\hbar}\hat{\mathbf{c}}_{\mathbf{k}}{\rm e}^{-{\rm i}\hat{\mathscr{H}}_{\rm L}t/\hbar}$, $\hat{\mathbf{d}}_{\mathbf{q}}(t) \equiv {\rm e}^{{\rm i}\hat{\mathscr{H}}_{\rm R}t/\hbar}\hat{\mathbf{d}}_{\mathbf{q}}{\rm e}^{-{\rm i}\hat{\mathscr{H}}_{\rm R}t/\hbar}$, and $\hat{\mathbf{d}}^{*}_{\mathbf{q'}} \equiv \left(\hat{\mathbf{d}}^{\dagger}_{\mathbf{q'}}\right)^{\rm T}$ are column vectors, and Tr denotes trace. In the second equality, we have used $A_i B_i C_j D_j = (A_i D_j)(B_i C_j) = (\mathbf{AD}^{\rm T})_{ij}(\mathbf{BC}^{\rm T})_{ij}$ for arbitrary column vectors $\mathbf{A}, \cdots, \mathbf{D}$ in performing the Wick decomposition. Note that the two expectations in the second line are finite only for $\mathbf{k}' = \mathbf{k}$ and $\mathbf{q}' = \mathbf{q}$. The other term $-\langle \hat{\mathscr{H}}'_{\rm H}(t')\hat{I}_{\rm H}(t)\rangle$ can be expanded similarly. Substituting them into (12.12) and using the invariance of the trace under cyclic permutations, we can express the current through the barrier as

$$\begin{aligned}I(t) = \frac{e}{\hbar^2}\frac{1}{V_{\rm L}V_{\rm R}}\sum_{\mathbf{kq}}\int_{-\infty}^{t}{\rm d}t'{\rm Tr}\big[&\langle \hat{\mathbf{c}}_{\mathbf{k}}(t)\hat{\mathbf{c}}^{\dagger}_{\mathbf{k}}(t')\rangle\hat{T}_{\mathbf{kq}}\hat{\sigma}_z\langle \hat{\mathbf{d}}^{*}_{\mathbf{q}}(t)\hat{\mathbf{d}}^{\rm T}_{\mathbf{q}}(t')\rangle^{\rm T}\hat{T}_{\mathbf{qk}} \\ &-\langle \hat{\mathbf{c}}^{*}_{\mathbf{k}}(t')\hat{\mathbf{c}}^{\rm T}_{\mathbf{k}}(t)\rangle^{\rm T}\hat{T}_{\mathbf{kq}}\hat{\sigma}_z\langle \hat{\mathbf{d}}_{\mathbf{q}}(t')\hat{\mathbf{d}}^{\dagger}_{\mathbf{q}}(t)\rangle\hat{T}_{\mathbf{qk}}\big]{\rm e}^{0_+t'},\end{aligned} \tag{12.13}$$

with $\hat{\mathbf{d}}^{*}_{\mathbf{q}} \equiv \left(\hat{\mathbf{d}}^{\dagger}_{\mathbf{q}}\right)^{\rm T}$.

Next, we show that in the Heisenberg representation with respect to $\hat{\mathscr{H}} \equiv \hat{\mathscr{H}}_{\rm L} + \hat{\mathscr{H}}_{\rm R}$, the field operators in (12.13) acquire extra phases because of the potential difference (12.1). To see this, let us express $\mu_{\rm L} = \mu_{\rm R} + eV$ and write Hamiltonian $\hat{\mathscr{H}}_{\rm L} \equiv \hat{H}_{\rm L} - \mu_{\rm L}\hat{N}_{\rm L}$ on the L side explicitly as a function of V, $\hat{\mathscr{H}}_{\rm L}(V) = \hat{\mathscr{H}}_{\rm L}(0) - V\hat{Q}_L$, where $\hat{Q}_{\rm L} \equiv e\hat{N}_{\rm L}$ is given by (12.5). Then the equation of motion for $\hat{c}_{\mathbf{k}\alpha}(t) = {\rm e}^{{\rm i}\hat{\mathscr{H}}_{\rm L}(V)t/\hbar}\hat{c}_{\mathbf{k}\alpha}{\rm e}^{-{\rm i}\hat{\mathscr{H}}_{\rm L}(V)t/\hbar}$ can be written as

$$\frac{{\rm d}\hat{c}_{\mathbf{k}\alpha}(t)}{{\rm d}t} = \frac{\rm i}{\hbar}[\hat{\mathscr{H}}_{\rm L}(0), \hat{c}_{\mathbf{k}\alpha}(t)] + {\rm i}\frac{eV}{\hbar}\hat{c}_{\mathbf{k}\alpha}(t), \tag{12.14}$$

where we have used $-{\rm e}^{{\rm i}\hat{\mathscr{H}}_{\rm L}(V)t/\hbar}\big[\hat{Q}_{\rm L}, \hat{c}_{\mathbf{k}\alpha}\big]{\rm e}^{-{\rm i}\hat{\mathscr{H}}_{\rm L}(V)t/\hbar} = e\hat{c}_{\mathbf{k}\alpha}(t)$. This yields $\hat{c}_{\mathbf{k}\alpha}(t) = {\rm e}^{{\rm i}eVt/\hbar}\,\hat{c}^{0}_{\mathbf{k}\alpha}(t)$, where $\hat{c}^{0}_{\mathbf{k}\alpha}(t)$ is the solution for $V = 0$. Accordingly, the Heisenberg representation of the field operators in (12.8) is expressible as

$$\hat{\mathbf{c}}_{\mathbf{k}}(t) = \hat{\Gamma}(t)\hat{\mathbf{c}}^{0}_{\mathbf{k}}(t), \qquad \hat{\mathbf{c}}^{\dagger}_{\mathbf{k}}(t) = \hat{\mathbf{c}}^{0\dagger}_{\mathbf{k}}(t)\hat{\Gamma}^{*}(t), \tag{12.15}$$

where $\hat{\Gamma}(t)$ is defined by

$$\hat{\Gamma}(t) \equiv \begin{bmatrix} \underline{\sigma}_0\,{\rm e}^{{\rm i}eVt/\hbar} & \underline{0} \\ \underline{0} & \underline{\sigma}_0\,{\rm e}^{-{\rm i}eVt/\hbar}\end{bmatrix}. \tag{12.16}$$

Let us substitute (12.15) into (12.13), set $\hat{\mathbf{c}}^0_{\mathbf{k}\alpha}(t) \to \hat{\mathbf{c}}_{\mathbf{k}\alpha}(t)$ to simplify the notation in the following, and use $(\hat{A}\hat{B})^{\mathrm{T}} = \hat{B}^{\mathrm{T}}\hat{A}^{\mathrm{T}}$ on the transpose of the matrix product. We can thereby express the current as

$$I(t) = \frac{e}{\hbar^2}\frac{1}{V_{\mathrm{L}}V_{\mathrm{R}}}\sum_{\mathbf{kq}}\int_{-\infty}^{t} \mathrm{d}t' \mathrm{Tr}\big[\hat{\Gamma}(t)\langle\hat{\mathbf{c}}_{\mathbf{k}}(t)\hat{\mathbf{c}}^{\dagger}_{\mathbf{k}}(t')\rangle\hat{\Gamma}^*(t')\hat{T}_{\mathbf{kq}}\hat{\sigma}_z\langle\hat{\mathbf{d}}^*_{\mathbf{q}}(t)\hat{\mathbf{d}}^{\mathrm{T}}_{\mathbf{q}}(t')\rangle^{\mathrm{T}}\hat{T}_{\mathbf{qk}}$$

$$-\hat{\Gamma}(t)\langle\hat{\mathbf{c}}^*_{\mathbf{k}}(t')\hat{\mathbf{c}}^{\mathrm{T}}_{\mathbf{k}}(t)\rangle^{\mathrm{T}}\hat{\Gamma}^*(t')\hat{T}_{\mathbf{kq}}\hat{\sigma}_z\langle\hat{\mathbf{d}}_{\mathbf{q}}(t')\hat{\mathbf{d}}^{\dagger}_{\mathbf{q}}(t)\rangle\hat{T}_{\mathbf{qk}}\big]\mathrm{e}^{0_+t'}, \tag{12.17}$$

where the expectations should be calculated at $V = 0$.

Next, we focus on the expectations in (12.17). Expressing (11.23) as $\hat{\mathbf{c}}_{\mathbf{k}} = \hat{U}_k\hat{\boldsymbol{\gamma}}_{\mathbf{k}}$, we can write them as $\langle\hat{\mathbf{c}}_{\mathbf{k}}(t)\hat{\mathbf{c}}^{\dagger}_{\mathbf{k}}(t')\rangle = \hat{U}_k\langle\hat{\boldsymbol{\gamma}}_{\mathbf{k}}(t)\hat{\boldsymbol{\gamma}}^{\dagger}_{\mathbf{k}}(t')\rangle\hat{U}^{\dagger}_k$ and $\langle\hat{\mathbf{c}}^*_{\mathbf{k}}(t)\hat{\mathbf{c}}^{\mathrm{T}}_{\mathbf{k}}(t')\rangle^{\mathrm{T}} = \left[\hat{U}^*_k\langle\hat{\boldsymbol{\gamma}}^*_{\mathbf{k}}(t)\hat{\boldsymbol{\gamma}}^{\mathrm{T}}_{\mathbf{k}}(t')\rangle\hat{U}^{\mathrm{T}}_k\right]^{\mathrm{T}} = \hat{U}_k\langle\hat{\boldsymbol{\gamma}}^*_{\mathbf{k}}(t)\hat{\boldsymbol{\gamma}}^{\mathrm{T}}_{\mathbf{k}}(t')\rangle^{\mathrm{T}}\hat{U}^{\dagger}_k$. They can be reduced further using (9.7), (11.26), $\langle\hat{\gamma}^{\dagger}_{\mathbf{k}\tilde{\alpha}}\hat{\gamma}_{\mathbf{k}\tilde{\alpha}}\rangle = \bar{n}(E_k)$, and $\langle\hat{\gamma}_{\mathbf{k}\tilde{\alpha}}\hat{\gamma}^{\dagger}_{\mathbf{k}\tilde{\alpha}}\rangle = \bar{n}(-E_k)$ into

$$\langle\hat{\mathbf{c}}_{\mathbf{k}}(t)\hat{\mathbf{c}}^{\dagger}_{\mathbf{k}}(t')\rangle = \begin{bmatrix} \underline{\sigma}_0 g_{\mathrm{L}}(-E_k,t_1) & -\mathrm{i}\underline{\sigma}_y f^*_{\mathrm{L}}(E_k,-t_1) \\ \mathrm{i}\underline{\sigma}_y f_{\mathrm{L}}(E_k,t_1) & \underline{\sigma}_0 g_{\mathrm{L}}(E_k,t_1) \end{bmatrix}, \tag{12.18}$$

$$\langle\hat{\mathbf{c}}^*_{\mathbf{k}}(t)\hat{\mathbf{c}}^{\mathrm{T}}_{\mathbf{k}}(t')\rangle^{\mathrm{T}} = \begin{bmatrix} \underline{\sigma}_0 g_{\mathrm{L}}(E_k,t_1) & \mathrm{i}\underline{\sigma}_y f^*_{\mathrm{L}}(E_k,-t_1) \\ -\mathrm{i}\underline{\sigma}_y f_{\mathrm{L}}(E_k,t_1) & \underline{\sigma}_0 g_{\mathrm{L}}(-E_k,t_1) \end{bmatrix}, \tag{12.19}$$

where $t_1 \equiv t - t'$, and $g_{\mathrm{L}}(E_k,t)$ and $f_{\mathrm{L}}(E_k,t) = -f_{\mathrm{L}}(-E_k,t)$ are defined by[2]

$$g_{\mathrm{L}}(E_k,t) \equiv u_k^2\bar{n}(E_k)\mathrm{e}^{\mathrm{i}E_kt/\hbar} + |v_k|^2\bar{n}(-E_k)\mathrm{e}^{-\mathrm{i}E_kt/\hbar}, \tag{12.20}$$

$$f_{\mathrm{L}}(E_k,t) \equiv u_kv_k\left[\bar{n}(E_k)\mathrm{e}^{\mathrm{i}E_kt/\hbar} - \bar{n}(-E_k)\mathrm{e}^{-\mathrm{i}E_kt/\hbar}\right], \tag{12.21}$$

with $\bar{n}(E_k) \equiv (\mathrm{e}^{\beta E_k}+1)^{-1} = 1 - \bar{n}(-\bar{E}_k)$.

As the matrices in (12.17) are mostly diagonal except those in (12.18) and (12.19), we can calculate its trace easily. We subsequently make a change of variable with $t_1 = t - t'$ and set $\mathrm{e}^{0_+t} \to 1$ safely to obtain

$$I(t) = \frac{2e}{\hbar^2}\sum_{\mathbf{kq}}\frac{|T_{\mathbf{kq}}|^2}{V_{\mathrm{L}}V_{\mathrm{R}}}\int_0^{\infty}\mathrm{d}t_1\left[\mathrm{e}^{\mathrm{i}eVt_1/\hbar}g_2(E_k,E_q;t_1) + \mathrm{e}^{-\mathrm{i}eVt_1/\hbar}g_2(E_k,E_q;-t_1)\right.$$

$$\left.+\mathrm{e}^{\mathrm{i}eV(2t-t_1)/\hbar}f_2(E_k,E_q;t_1) + \mathrm{e}^{-\mathrm{i}eV(2t-t_1)/\hbar}f_2^*(E_k,E_q;t_1)\right]\mathrm{e}^{-0_+t_1}, \tag{12.22}$$

[2]We here regard u_k and v_k as independent of E_k by definition but omit them from the arguments of g_{L} and f_{L} for simplicity.

where $g_2(E_k, E_q; t)$ and $f_2(E_k, E_q; t)$ are defined in terms of (12.21) as

$$\begin{aligned} g_2(E_k, E_q; t) &\equiv g_{\rm L}(-E_k, t) g_{\rm R}(E_q, t) - g_{\rm L}(E_k, -t) g_{\rm R}(-E_q, -t) \\ &= \big[\bar{n}(E_q) - \bar{n}(E_k)\big]\big[u_k^2 u_q^2 {\rm e}^{-{\rm i}(E_k - E_q)t/\hbar} - |v_k|^2 |v_q|^2 {\rm e}^{{\rm i}(E_k - E_q)t/\hbar}\big] \\ &\quad + \big[1 - \bar{n}(E_q) - \bar{n}(E_k)\big]\big[u_k^2 |v_q|^2 {\rm e}^{-{\rm i}(E_k + E_q)t/\hbar} - |v_k|^2 u_q^2 {\rm e}^{{\rm i}(E_k + E_q)t/\hbar}\big], \end{aligned} \tag{12.23}$$

$$\begin{aligned} f_2(E_k, E_q; t) &\equiv f_{\rm L}^*(E_k, -t) f_{\rm R}(E_q, t) - f_{\rm L}^*(E_k, t) f_{\rm R}(E_q, -t) \\ &= u_k v_k^* u_q v_q \big\{\big[\bar{n}(E_q) - \bar{n}(E_k)\big]\big[{\rm e}^{{\rm i}(E_k - E_q)t/\hbar} - {\rm e}^{-{\rm i}(E_k - E_q)t/\hbar}\big] \\ &\quad - \big[1 - \bar{n}(E_k) - \bar{n}(E_q)\big]\big[{\rm e}^{{\rm i}(E_k + E_q)t/\hbar} - {\rm e}^{-{\rm i}(E_k + E_q)t/\hbar}\big]\big\}. \end{aligned} \tag{12.24}$$

Integration in (12.22) can be performed easily yielding

$$\begin{aligned} I(t) &= \frac{2e}{\hbar^2} \sum_{\mathbf{kq}} \frac{|T_{\mathbf{kq}}|^2}{V_{\rm L} V_{\rm R}} \big[\tilde{g}_2(E_k, E_q; \omega) \\ &\quad + {\rm e}^{2{\rm i}\omega} \tilde{f}_2(E_k, E_q; \omega) + {\rm e}^{-2{\rm i}\omega} \tilde{f}_2^*(E_k, E_q; \omega)\big]_{\omega = \frac{eV}{\hbar}}, \end{aligned} \tag{12.25}$$

where $\tilde{g}_2$ is the Fourier transform of g_2 defined by

$$\begin{aligned} \tilde{g}_2(E_k, E_q; \omega) &\equiv \int_{-\infty}^{\infty} g_2(E_k, E_q; t) {\rm e}^{{\rm i}\omega t} {\rm d}t \\ &= 2\pi\hbar \big\{\big[\bar{n}(E_q) - \bar{n}(E_k)\big]\big[u_k^2 u_q^2 \delta(\hbar\omega - E_k + E_q) - |v_k|^2 |v_q|^2 \delta(\hbar\omega + E_k - E_q)\big] \\ &\quad + \big[1 - \bar{n}(E_q) - \bar{n}(E_k)\big]\big[u_k^2 |v_q|^2 \delta(\hbar\omega - E_k - E_q) - |v_k|^2 u_q^2 \delta(\hbar\omega + E_k + E_q)\big]\big\}, \end{aligned} \tag{12.26}$$

and $\tilde{f}_2$ denotes $(\omega_- \equiv \omega - {\rm i}0_+)$

$$\begin{aligned} \tilde{f}_2(E_k, E_q; \omega) &\equiv \int_0^{\infty} f_2(E_k, E_q; t) {\rm e}^{-{\rm i}\omega_- t} {\rm d}t \\ &= u_k v_k^* u_q v_q \left\{\big[\bar{n}(E_q) - \bar{n}(E_k)\big] \left(\frac{-{\rm i}\hbar}{\hbar\omega_- - E_k + E_q} - \frac{-{\rm i}\hbar}{\hbar\omega_- + E_k - E_q}\right) \right. \\ &\quad \left. - \big[1 - \bar{n}(E_k) - \bar{n}(E_q)\big] \left(\frac{-{\rm i}\hbar}{\hbar\omega_- - E_k - E_q} - \frac{-{\rm i}\hbar}{\hbar\omega_- + E_k + E_q}\right)\right\}. \end{aligned} \tag{12.27}$$

We first focus on the $\tilde{g}_2$ term in (12.25) and transform the sums over **k** and **q** into integrals as (8.90) with $N_{\rm L}(\varepsilon_k) \approx N_{\rm L}(\varepsilon_{\rm F})$ and $N_{\rm R}(\varepsilon_q) \approx N_{\rm R}(\varepsilon_{\rm F})$. We may also approximate $|T_{\mathbf{kq}}|^2$ by its Fermi surface average $\langle |T_{\mathbf{kq}}|^2 \rangle_{\rm F}$ to take it outside the integral. We can thereby express the $\tilde{g}_2$ contribution in (12.25) as

$$I_g \equiv \frac{2e}{\hbar^2}\langle|T_{\mathbf{kq}}|^2\rangle_{\rm F}\int_{-\infty}^{\infty}{\rm d}\xi_k N_{\rm L}(\varepsilon_{\rm F})\int_{-\infty}^{\infty}{\rm d}\xi_q N_{\rm R}(\varepsilon_{\rm F})\tilde{g}_2(E_k,E_q;eV/\hbar). \qquad (12.28)$$

Let us substitute (12.26) into (12.28) and express $u_{k,q}^2 = \frac{1}{2}(1+\xi_{k,q}/E_{k,q})$ and $|v_{k,q}|^2 = \frac{1}{2}(1-\xi_{k,q}/E_{k,q})$ based on (9.6). We then see that the odd functions $\xi_{k,q}/E_{k,q}$ give a null contribution to the integral so that both $u_{k,q}^2$ and $|v_{k,q}|^2$ in its integrand can be replaced by 1/2. Subsequently, we set $\xi_{k,q} \to -\xi_{k,q}$ for $-\infty \le \xi_{k,q} \le 0$ to map the regions onto $0 \le \xi_{k,q} \le \infty$. We thereby obtain

$$\begin{aligned} I_g = {} & \frac{2e}{\hbar^2}\langle|T_{\mathbf{kq}}|^2\rangle_{\rm F}\,2^2\int_0^{\infty}{\rm d}\xi_k N_{\rm L}(\varepsilon_{\rm F})\int_0^{\infty}{\rm d}\xi_q N_{\rm R}(\varepsilon_{\rm F})\frac{2\pi\hbar}{2^2} \\ & \times\{[\bar{n}(E_q)-\bar{n}(E_k)]\delta(eV-E_k+E_q)+[\bar{n}(-E_q)-\bar{n}(-E_k)]\delta(eV+E_k-E_q) \\ & +[\bar{n}(-E_q)-\bar{n}(E_k)]\delta(eV-E_k-E_q)+[\bar{n}(E_q)-\bar{n}(-E_k)]\delta(eV+E_k+E_q)]\}, \end{aligned}$$

where we have used $\bar{n}(-E) = 1-\bar{n}(E)$. Next, we make a change of variable ${\rm d}\xi_k = {\rm d}E_k/({\rm d}E_k/{\rm d}\xi_k) = {\rm d}E_k\theta(E_k-|\Delta_{\rm L}|)E_k/\sqrt{E_k^2-|\Delta_{\rm L}|^2}$ based on (9.5). We then notice that the four terms in the curly brackets above can be expressed as a single integral of the first term over $-\infty \le E_{k,q} \le \infty$. Performing the E_q integraion, wc find that the resulting $I_g(V)$ is expressible in terms of the superconducting density of states,

$$N_{\rm Ls}(E) = N_{\rm L}(\varepsilon_{\rm F})\frac{|E|}{\sqrt{E^2-|\Delta_{\rm L}|^2}}\theta(|E|-|\Delta_{\rm L}|) \qquad (12.29)$$

concisely as

$$I_g(V) = \frac{4\pi e}{\hbar}\langle|T_{\mathbf{kq}}|^2\rangle_{\rm F}\int_{-\infty}^{\infty}{\rm d}E\, N_{\rm Ls}(E)N_{\rm Rs}(E-eV)[\bar{n}(E-eV)-\bar{n}(E)].$$

The $\tilde{f}_2$ term in (12.25) can be transformed similarly by expressing $u_k v_k^* u_q v_q = {\rm e}^{{\rm i}\varphi}|\Delta_{\rm L}||\Delta_{\rm R}|/4E_k E_q$ in (12.27) based on (9.6) and (12.2). Indeed, differences from the I_g case lie only in (i) $M_{\rm Ls}(E) \equiv N_{\rm Ls}(E)|\Delta_{\rm L}|/E$ in place of (12.29) and (ii) function (11.30) instead of the delta function. The M function can also be written as

$$M_{\rm Ls}(E) \equiv N_{\rm L}(\varepsilon_{\rm F})\frac{|\Delta_{\rm L}|{\rm sgn}(E)}{\sqrt{E^2-|\Delta_{\rm L}|^2}}\theta(|E|-|\Delta_{\rm L}|), \qquad (12.30)$$

with ${\rm sgn}(x) \equiv x/|x|$. We thereby obtain the total current through the barrier as

$$I(V,t) = I_g(V) + I_{fc}(V)\cos\left(\frac{2eV}{\hbar}t+\varphi\right) + I_{fs}(V)\sin\left(\frac{2eV}{\hbar}t+\varphi\right), \qquad (12.31)$$

where $I_g(V)$, $I_{fc}(V)$, and $I_{fs}(V)$ are defined in terms of (12.29) and (12.30) by

$$I_g(V) = \frac{4\pi e}{\hbar}\langle|T_{\mathbf{kq}}|^2\rangle_\mathrm{F}\int_{-\infty}^{\infty} \mathrm{d}E\, N_{\mathrm{Ls}}(E)N_{\mathrm{Rs}}(E-eV)\big[\bar{n}(E-eV)-\bar{n}(E)\big], \tag{12.32}$$

$$I_{fc}(V) = \frac{4\pi e}{\hbar}\langle|T_{\mathbf{kq}}|^2\rangle_\mathrm{F}\int_{-\infty}^{\infty} \mathrm{d}E\, M_{\mathrm{Ls}}(E)M_{\mathrm{Rs}}(E-eV)\big[\bar{n}(E-eV)-\bar{n}(E)\big], \tag{12.33}$$

$$I_{fs}(V) = \frac{4e}{\hbar}\langle|T_{\mathbf{kq}}|^2\rangle_\mathrm{F}\,\mathrm{P}\int_{-\infty}^{\infty} \mathrm{d}E\int_{-\infty}^{\infty} \mathrm{d}E'\, M_{\mathrm{Ls}}(E)M_{\mathrm{Rs}}(E')\frac{\bar{n}(E')-\bar{n}(E)}{E'+eV-E}, \tag{12.34}$$

recalling that P denotes the principal value.

12.2 NN Junction

First, we consider the situation where L and R are both normal. The corresponding current I_{NN} is obtained from (12.31) by setting $I_{fc} = I_{fs} = 0$, replacing $N_{\mathrm{Ls,Rs}}(E) \to N_{\mathrm{L,R}}(\varepsilon_\mathrm{F})$ in (12.32), and using (11.36). We thereby obtain Ohm's law for the junction:

$$I_{\mathrm{NN}}(V) = R_\mathrm{N}^{-1}V, \tag{12.35}$$

with resistance

$$R_\mathrm{N} \equiv \left[\frac{4\pi e^2}{\hbar}\langle|T_{\mathbf{kq}}|^2\rangle_\mathrm{F} N_\mathrm{L}(\varepsilon_\mathrm{F})N_\mathrm{R}(\varepsilon_\mathrm{F})\right]^{-1}. \tag{12.36}$$

12.3 SN Junction and Density of States

Next, we consider a superconducting L and normal R. The corresponding current I_{SN} is obtained from (12.31) by setting $I_{fc} = I_{fs} = 0$ and replacing $N_{\mathrm{Rs}}(E) \to N_\mathrm{R}(\varepsilon_\mathrm{F})$ in (12.32) as [6]

$$I_{\mathrm{SN}}(V) = \frac{1}{eR_\mathrm{N}N_\mathrm{L}(\varepsilon_\mathrm{F})}\int_{-\infty}^{\infty} N_{\mathrm{Ls}}(E)\big[\bar{n}(E-eV)-\bar{n}(E)\big]\mathrm{d}E, \tag{12.37}$$

where we have used (12.36). Its derivative with respect to V is obtained using $\partial\bar{n}(E-eV)/\partial V = -e\partial\bar{n}(E-eV)/\partial E$ as

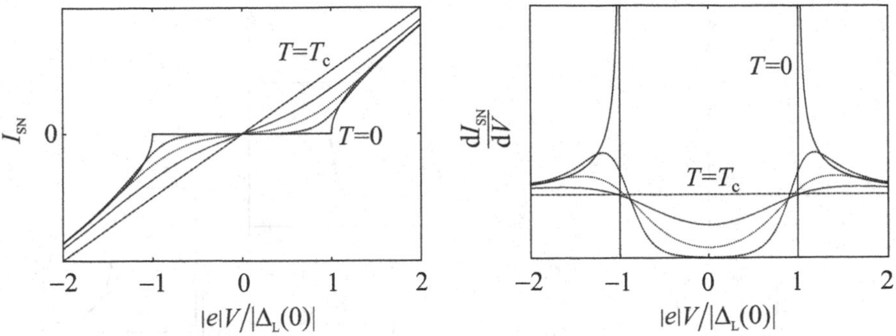

Fig. 12.2 Tunneling current I_{SN} and its derivative $\mathrm{d}I_{\mathrm{SN}}/\mathrm{d}V$ given by (12.37) and (12.38) for $T/T_{\mathrm{c}} = 0.0, 0.25, 0.5, 0.75, 1$

$$\frac{\mathrm{d}I_{\mathrm{SN}}(V)}{\mathrm{d}V} = \frac{1}{R_{\mathrm{N}}N_{\mathrm{L}}(\varepsilon_{\mathrm{F}})}\int_{-\infty}^{\infty} N_{\mathrm{Ls}}(E)\left[-\frac{\partial \bar{n}(E-eV)}{\partial E}\right]\mathrm{d}E \quad \overset{T\to 0}{\longrightarrow} \quad \frac{N_{\mathrm{Ls}}(eV)}{R_{\mathrm{N}}N_{\mathrm{L}}(\varepsilon_{\mathrm{F}})}. \tag{12.38}$$

Thus, derivative $\mathrm{d}I_{\mathrm{SN}}(V)/\mathrm{d}V$ at low temperatures directly measures the superconducting density of states, as first shown by Giaever experimentally [7].

Figure 12.2 presents graphs of (12.37) and (12.38) as functions of V for $T/T_{\mathrm{c}} = 0.0, 0.25, 0.5, 0.75, 1$. The $\mathrm{d}I_{\mathrm{SN}}/\mathrm{d}V$ curves clearly show a transition from the normal density of states at $T = T_{\mathrm{c}}$ to the superconducting density of states at $T = 0$. The behavior of $I_{\mathrm{SN}} = I_{\mathrm{SN}}(V)$ at $T = 0$ can be realized schematically by drawing the densities of states on both sides as in Fig. 12.3a; the current flows for $eV \geq |\Delta|$ ($eV \leq -|\Delta|$) where there are filled states on the left-hand (right-hand) side that can move horizontally to the empty states on the right-hand (left-hand) side.

12.4 SS Junction and Josephson Effect

We now focus on the direct current through superconducting-superconducting (SS) junctions.

Equation (12.31) implies that, besides the quasiparticle current I_g, there are extra contributions, i.e., the second and third terms on its right-hand side, which for $V \neq 0$ form an alternating current with frequency $2eV/h$ (the *AC Josephson effect*). For the special case with $V = 0$, however, the contribution converts to direct current whose magnitude depends on the phase difference (12.2); this is called the *DC Josephson effect* [9]. Noting $I_{fc}(0) = 0$ from (12.33), we can express the total direct current as

$$I_{\mathrm{DC}}(V) = I_g(V) - \delta_{V0} I_{\mathrm{c}} \sin\varphi, \tag{12.39}$$

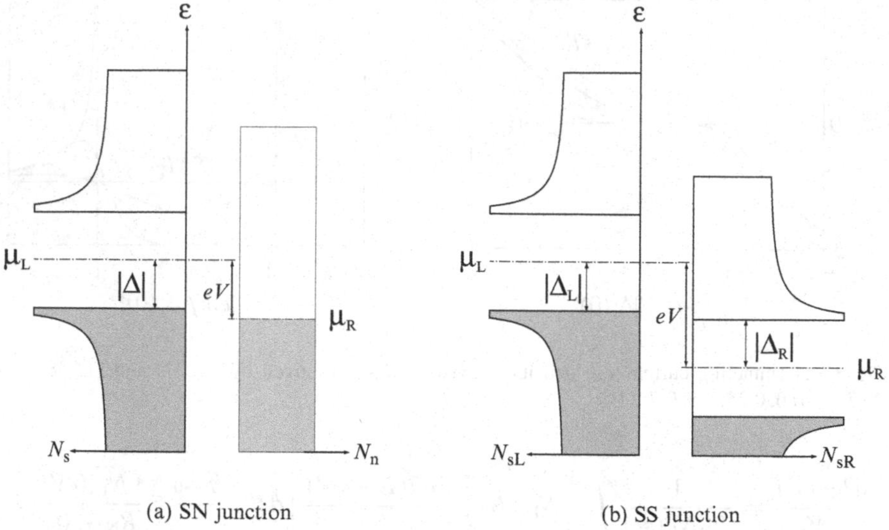

Fig. 12.3 Schematics illustrating the density of states in electron tunneling for (**a**) an SN junction and (**b**) an SS junction ($e < 0$). Shaded areas denote filled states at $T = 0$

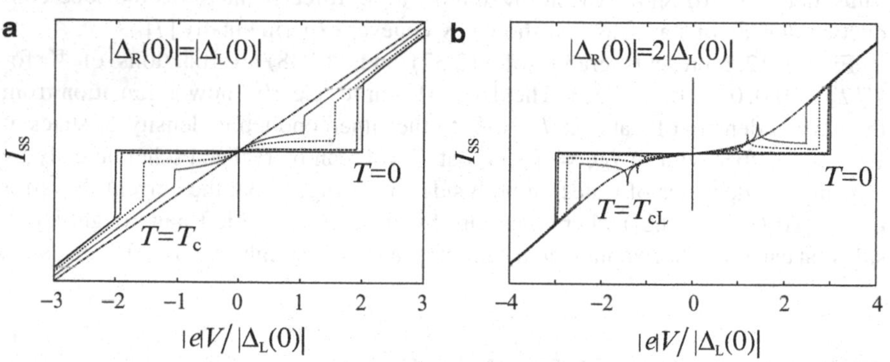

Fig. 12.4 Tunneling current I_{SS} of SS junctions given by (12.39) for (**a**) $|\Delta_R(0)| = |\Delta_L(0)|$ and (**b**) $|\Delta_R(0)| = 2|\Delta_L(0)|$ at $T/T_{cL} = 0.0, 0.5, 0.75, 0.9, 1$

where $I_g(V)$ is given by (12.32), and $I_c \equiv -I_{fs}(0)$ denotes the maximum of the extra direct current at $V = 0$. Figure 12.4 plots $I_{SS} = I_{SS}(V)$ for $|\Delta_R| = |\Delta_L|$ and $|\Delta_R| = 2|\Delta_L|$ at five different temperatures.

We first focus on the quasiparticle current $I_g(V)$. It develops from $eV = \pm(|\Delta_L| + |\Delta_R|)$ at $T = 0$ discontinuously, which may be understood schematically by graphing the superconducting densities of states (Fig. 12.3b). The temperature variation of I_{SS} for $|\Delta_R| = |\Delta_L|$ is much slower at low temperatures than that of I_{SN} in Fig. 12.2 because the threshold $2|\Delta_L|$ is larger by factor 2 than for the SN junction. For $|\Delta_R| \neq |\Delta_L|$, extra peaks develop at finite temperatures near

$eV = \pm(|\Delta_\mathrm{L}| - |\Delta_\mathrm{R}|)$, arising from thermally excited quasiparticles around the edges of the excitation thresholds with a smaller energy gap. As the temperature increases, these peaks near $eV = \pm(|\Delta_\mathrm{L}|-|\Delta_\mathrm{R}|)$ move towards the steeper sections at $eV = \pm(|\Delta_\mathrm{L}| + |\Delta_\mathrm{R}|)$ and eventually merge into the latter at $T = T_\mathrm{cL}$ where $|\Delta_\mathrm{L}| = 0$. The curve at $T = T_\mathrm{cL}$ ($T = 0.5T_\mathrm{cR}$) is identical in form to the I_SN curve at $T/T_\mathrm{c} = 0.5$ in Fig. 12.2.

Next, we consider the Josephson current at $V = 0$ in (12.39). Its magnitude directly depends on the phase difference $\varphi = \varphi_\mathrm{L} - \varphi_\mathrm{R}$ as $\sin\varphi$ so that it differs every time the SS junction is cooled during its work cycle. Thus, the phenomenon can be regarded as an experimental manifestation of the *spontaneously broken gauge symmetry*. A finite current for $\varphi_\mathrm{L} \neq \varphi_\mathrm{R}$ may be regarded as a response of the coupled system in attaining total phase coherence of $\varphi_\mathrm{L} = \varphi_\mathrm{R}$. The critical current $I_\mathrm{c} \equiv -I_{fs}(0)$ corresponds to $\varphi_\mathrm{R} - \varphi_\mathrm{L} = \pi/2$, whose magnitude depends on both temperature and T_c ratio $T_\mathrm{cR}/T_\mathrm{cL}$ between the two superconductors. Its precise expression is obtained from (12.34) (see **Problem** 12.1),

$$I_\mathrm{c} = \frac{|\Delta_\mathrm{L}||\Delta_\mathrm{R}|}{|e|R_\mathrm{N}}\frac{2\pi}{\beta}\sum_{n=0}^{\infty}\frac{1}{\sqrt{(\varepsilon_n^2 + |\Delta_\mathrm{L}|^2)(\varepsilon_n^2 + |\Delta_\mathrm{R}|^2)}}, \tag{12.40}$$

where R_N is defined by (12.36) and $\varepsilon_n \equiv (2n+1)\pi k_\mathrm{B}T$. The critical current I_c is plotted as a function of T in Fig. 12.5 for two different cases, specifically $|\Delta_\mathrm{R}(0)| = |\Delta_\mathrm{L}(0)|$ and $|\Delta_\mathrm{R}(0)| = 2|\Delta_\mathrm{L}(0)|$. There are two limits where the sum over n in (12.40) can be performed analytically. One is $|\Delta_\mathrm{L}| = |\Delta_\mathrm{R}| \equiv |\Delta|$, where we can use (9.39) to obtain

$$I_\mathrm{c} = \frac{\pi|\Delta|}{2|e|R_\mathrm{N}}\tanh\frac{\beta|\Delta|}{2}, \qquad (|\Delta_\mathrm{L}| = |\Delta_\mathrm{R}| \equiv |\Delta|). \tag{12.41}$$

The other is $T = 0$, where $2\pi\beta^{-1}\sum_n$ reduces to the integral of ε_n over $0 \le \varepsilon_n \le \infty$ to yield (**Problem** 12.2)

$$I_\mathrm{c} = \frac{2}{|e|R_\mathrm{N}}\frac{|\Delta_\mathrm{L}||\Delta_\mathrm{R}|}{|\Delta_\mathrm{L}| + |\Delta_\mathrm{R}|}K\left(\frac{||\Delta_\mathrm{L}| - |\Delta_\mathrm{R}||}{|\Delta_\mathrm{L}| + |\Delta_\mathrm{R}|}\right), \qquad (T = 0), \tag{12.42}$$

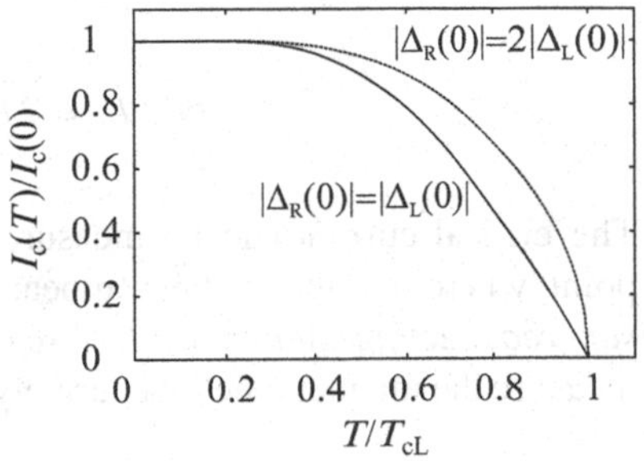

Fig. 12.5 Normalized critical current $I_\mathrm{c}(T)$ as a function of reduced temperature for $|\Delta_\mathrm{R}(0)| = |\Delta_\mathrm{L}(0)|$ and $|\Delta_\mathrm{R}(0)| = 2|\Delta_\mathrm{L}(0)|$

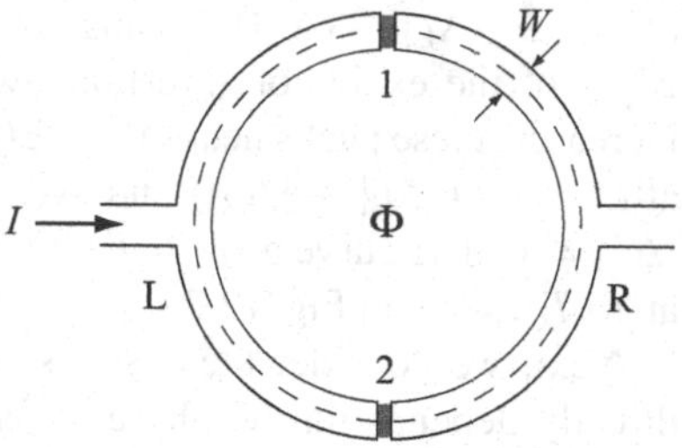

Fig. 12.6 A superconducting loop with two Josephson junctions in a magnetic field

where K denotes the complete elliptic integral [1, 3]. Expressions (12.41) and (12.42) are due to Ambegaokar and Baratoff [2].

Finally, we consider a superconducting loop with two Josephson junctions, depicted in Fig. 12.6, where the width is assumed much larger than the London penetration depth, $W \gg \lambda_{\rm L}$. We perform a line integral of (10.39) along the broken line far inside the loop in Fig. 12.6 in the counterclockwise direction. The left-hand side yields 0 from the Meissner effect; the contribution from the junction regions can be neglected. Proceeding similarly with (10.42) for the right-hand side, we obtain

$$0 = \oint \left(\nabla\varphi - \frac{2e}{\hbar}\mathbf{A} \right) \cdot d\mathbf{r} = \varphi_{\rm L2} - \varphi_{\rm L1} + \varphi_{\rm R1} - \varphi_{\rm R2} + 2\pi \frac{\Phi}{\Phi_0},$$

where Φ and Φ_0 are the total flux in the loop and flux quantum (10.44), respectively. Thus, we obtain

$$\varphi_2 = \varphi_1 - 2\pi \frac{\Phi}{\Phi_0}, \tag{12.43}$$

with $\varphi_j \equiv \varphi_{{\rm L}j} - \varphi_{{\rm R}j}$ $(j = 1, 2)$. We thereby obtain an expression for the DC Josephson current in the loop,

$$I = -I_{\rm c1} \sin\varphi_1 - I_{\rm c2} \sin\left(\varphi_1 - 2\pi \frac{\Phi}{\Phi_0} \right) \overset{I_{\rm c2}=I_{\rm c1}}{\longrightarrow} -2I_{\rm c1} \cos\left(\pi \frac{\Phi}{\Phi_0} \right) \sin\left(\varphi_1 - \pi \frac{\Phi}{\Phi_0} \right). \tag{12.44}$$

The last expression states that the critical current $I_{\rm c}$ of the loop for $I_{\rm c1} = I_{\rm c2}$ varies as a function of Φ,

$$I_{\rm c} = 2I_{\rm c1} \left| \cos\left(\pi \frac{\Phi}{\Phi_0} \right) \right|. \tag{12.45}$$

The critical current can be measured easily by increasing I and identifying the point where a finite voltage appears across the junction. The device, called the *superconducting quantum interference device (SQUID)*, has been used widely to measure the magnetic field accurately [5].

Problems

12.1. Derive (12.40) from (12.34) at $V = 0$.

12.2. Show that (12.40) at $T = 0$ reduces to (12.42).

References

1. M. Abramowitz, I.A. Stegun (eds.), *Handbook of Mathematical Functions: With Formulas, Graphs, and Mathematical Tables* (Dover, New York, 1965)
2. V. Ambegaokar, A. Baratoff, Phys. Rev. Lett. **10**, 486 (1963); **11**, 104 (1963)
3. G.B. Arfken, H.J. Weber, *Mathematical Methods for Physicists* (Academic, New York, 2012)
4. J. Bardeen, Phys. Rev. Lett. **6**, 57 (1961)
5. A. Barone, G. Paternò, *Physics and Applications of the Josephson Effect* (John Wiley, New York, 1982)
6. M.H. Cohen, L.M. Falicov, J.C. Phillips, Phys. Rev. Lett. **8**, 316 (1962)
7. I. Giaever, Phys. Rev. Lett. **5**, 147 (1960)
8. T. Inui, Y. Tanabe, Y. Onodera, *Group Theory and Its Applications in Physics* (Springer, Berlin, 1990)
9. B.D. Josephson, Phys. Lett. **1**, 251 (1962)
10. R.E. Prange, Phys. Rev. **131**, 1083 (1963)

Chapter 13
P-Wave Superfluidity

Abstract Superfluidity in liquid ^{3}He was discovered at ultra-low temperatures below around 3 mK by Osheroff, Richardson, and Lee in 1972 (Osheroff et al., Phys Rev Lett 28:885, 1972; Phys Rev Lett 29:920, 1972). The ^{3}He atom is composed of two protons, one neutron, and two electrons, each of which carries a spin of magnitude 1/2, and hence is classified as a fermion according to the spin-statistics theorem. Quantum effects in liquid ^{3}He is expected to emerge below $T_Q \sim 3\,\text{K}$ according to Table 4.1, and the superfluid transition occurs at about $10^{-3}T_Q$ to be attributed to the Cooper-pair condensation. As the atom can be regarded roughly as a rigid sphere, it is clearly impossible for a pair of ^{3}He atoms to make up an s-wave bound state that has a high probability of occupying the same position in space. However, they may form a bound state while being separated through a higher ($\ell \geq 1$) channel of expansion (8.83) to overcome repulsion. Among various theoretical predictions, superfluidity was soon identified to be associated with p-wave ($\ell = 1$) pairing with total spin $s = 1$. Hence, the bound state has a total of $(2\ell + 1)(2s + 1) = 9$ internal degrees of freedom, which brings unique features to the p-wave superfluidity including two distinct phases A and B observed in the bulk (see Fig. 13.1). Here, we survey the fundamentals of this superfluidity (Leggett, Rev Mod Phys 47:331, 1975; Vollhardt and Wölfle, The superfluid phases of helium 3. Taylor & Francis, London, 1990, p 31).

13.1 Effective Pairing Interaction

The p-wave superfluidity exploits the $\ell = 1$ channel of the expansion (8.83). It follows from (8.84) that the corresponding spherical harmonic functions are linear in $\hat{\mathbf{k}} \equiv \mathbf{k}/|\mathbf{k}|$,

$$Y_{1,\pm 1}(\hat{\mathbf{k}}) = \mp\sqrt{\frac{3}{8\pi}}(\hat{k}_x \pm \mathrm{i}\hat{k}_y), \qquad Y_{10}(\hat{\mathbf{k}}) = \sqrt{\frac{3}{4\pi}}\hat{k}_z. \tag{13.1}$$

To understand the pairing interaction microscopically, we need a treatment beyond the mean-field level [2] that is outside the scope of this book. Nevertheless, it has been established that the effective interaction near the Fermi surface can also be expressed to a first approximation as

© Springer Japan 2015

T. Kita, *Statistical Mechanics of Superconductivity*, Graduate Texts in Physics,
DOI 10.1007/978-4-431-55405-9_13

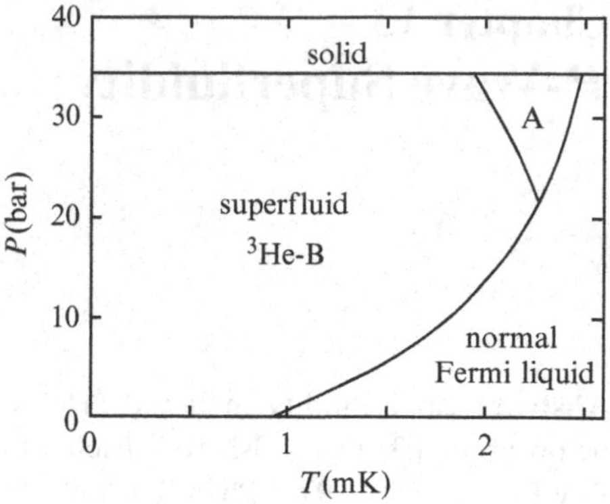

Fig. 13.1 Schematic T-P phase diagram for ^{3}He at ultra-low temperatures [5, 16]

$$\mathscr{V}_1^{(\text{eff})}(k,k') = \mathscr{V}_1^{(\text{eff})}\theta(\varepsilon_c - |\xi_k|)\theta(\varepsilon_c - |\xi_{k'}|) \tag{13.2}$$

with $k_B T_c \ll \varepsilon_c \ll \varepsilon_F$, in exactly the same way as (9.26) for s-wave pairing. Hence, we retain only the $\ell = 1$ element in (8.83), set $\mathscr{V}_1(k,k') \to \mathscr{V}_1^{(\text{eff})}(k,k')$, and then substitute (13.1). We thereby obtain the p-wave pairing interaction as

$$\mathscr{V}_{|\mathbf{k}-\mathbf{k}'|} = 4\pi \mathscr{V}_1^{(\text{eff})}(k,k') \sum_{m=-1}^{1} Y_{1m}(\hat{\mathbf{k}})Y_{1m}^*(\hat{\mathbf{k}}') = 3\mathscr{V}_1^{(\text{eff})}(k,k')\hat{\mathbf{k}}\cdot\hat{\mathbf{k}}'. \tag{13.3}$$

13.2 Gap Matrix

The pair potential for homogeneous systems within the mean-field theory is defined generally by (8.76) as a 2×2 gap matrix. Let us express it in another way that has been used widely especially for p-wave pairing. To this end, we notice that any 2×2 matrix $\underline{M} = (M_{ij})$ can be expanded in terms of the complete set (8.42),

$$\underline{M} \equiv [M_0\underline{\sigma}_0 + \mathbf{M}\cdot\underline{\boldsymbol{\sigma}}]\mathrm{i}\underline{\sigma}_y = \begin{bmatrix} -M_x + \mathrm{i}M_y & M_0 + M_z \\ -M_0 + M_z & M_x + \mathrm{i}M_y \end{bmatrix}, \tag{13.4}$$

where $\mathrm{i}\underline{\sigma}_y$ has been introduced for convenience. Thus, the four matrix elements M_{ij} $(i,j = 1,2)$ can be expressed alternatively in terms of the four coefficients $(M_0, \mathbf{M})$. The gap matrix $\underline{\Delta}(\mathbf{k})$ is also a 2×2 matrix, which for p-wave $(\ell = 1)$ pairing should obey the symmetry relation $\underline{\Delta}^{\mathrm{T}}(\mathbf{k}) = \underline{\Delta}(\mathbf{k})$ from (8.87) and (8.89). Hence, the coefficient of $\underline{\sigma}_0$ is 0 in the expansion of $\underline{\Delta}(\mathbf{k})$ as (13.4). We can also neglect the $k \equiv |\mathbf{k}|$ dependence of $\underline{\Delta}(\mathbf{k})$ near the Fermi surface for the pairing interaction (13.2). For $|\xi_k| \le \varepsilon_c$, we therefore expand the gap matrix for p-wave pairing to within an overall constant $\Delta \ge 0$,

$$\underline{\Delta}(\mathbf{k}) = \Delta\, \mathbf{d}(\hat{\mathbf{k}}) \cdot \underline{\sigma}\, \mathrm{i}\underline{\sigma}_y. \tag{13.5}$$

Explicitly, we have

$$\begin{bmatrix} \Delta_{\uparrow\uparrow}(\mathbf{k}) & \Delta_{\uparrow\downarrow}(\mathbf{k}) \\ \Delta_{\downarrow\uparrow}(\mathbf{k}) & \Delta_{\downarrow\downarrow}(\mathbf{k}) \end{bmatrix} = \Delta \begin{bmatrix} -d_x(\hat{\mathbf{k}}) + \mathrm{i}d_y(\hat{\mathbf{k}}) & d_z(\hat{\mathbf{k}}) \\ d_z(\hat{\mathbf{k}}) & d_x(\hat{\mathbf{k}}) + \mathrm{i}d_y(\hat{\mathbf{k}}) \end{bmatrix}. \tag{13.6}$$

With amplitude Δ extracted in the expansion, vector $\mathbf{d}$ should obey some normalization condition in terms of angular integration (8.85); we choose for convenience

$$\int \frac{\mathrm{d}\Omega_{\mathbf{k}}}{4\pi} |\mathbf{d}(\hat{\mathbf{k}})|^2 = 1. \tag{13.7}$$

It follows from (13.6) that vector $\mathbf{d}(\hat{\mathbf{k}})$ lies in a plane perpendicular to the quantization axis of spin.

As the $Y_{1m}(\hat{\mathbf{k}})$'s are linear in $\hat{\mathbf{k}}$, we can in general expand $\mathbf{d}$ for p-wave pairing as

$$d_\eta(\hat{\mathbf{k}}) = \sum_{\eta'} A_{\eta\eta'} \hat{k}_{\eta'} \qquad (\eta, \eta' = x, y, z), \tag{13.8}$$

where $\{A_{\eta\eta'}\}$ are the expansion coefficients obeying constraint (13.7). To find the equilibrium state, we need to minimize the free energy in terms of not only Δ but also $\{A_{\eta\eta'}\}$.

It can be shown that the transition temperature T_c is the same for any configuration of $\{A_{\eta\eta'}\}$ within the p-wave pairing (**Problem** 13.1). This degeneracy in the free energy is lifted eventually for $T < T_c$.

13.3 Two Bulk Phases

As already mentioned, liquid ^{3}He realizes two bulk superfluid phases distinguished as "A" and "B" in Fig. 13.1, whose gap structures have been confirmed unambiguously (Fig. 13.2). We discuss each of them separately in this and following sections.

13.3.1 B Phase

Although discovered later [11, 12], the *B phase* has now been established as occupying a large domain in the P-T phase diagram except for the high-pressure region near T_c [15]. Its gap structure is identified as the *Balian-Werthamer (BW) state* [4].

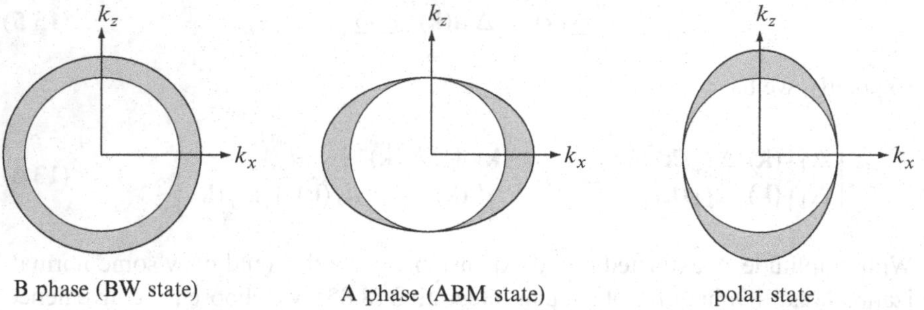

Fig. 13.2 Energy gaps of the B and A phases (*shaded areas*) above the spherical Fermi surface, which correspond to the lowest excitation energies given in (13.11) and (13.20), respectively. The A-phase gap has point nodes at the north and south poles. The polar state with a line node along the equator, which has never been observed as a bulk phase, is also plotted for later convenience in Sect. 13.4

The $\mathbf{d}$ vector and gap matrix are expressible, for example, as

$$\mathbf{d}(\hat{\mathbf{k}}) = \hat{\mathbf{k}} \quad \longleftrightarrow \quad \underline{\Delta}(\mathbf{k}) = \Delta \begin{bmatrix} -\hat{k}_\perp^* & \hat{k}_z \\ \hat{k}_z & \hat{k}_\perp \end{bmatrix}, \tag{13.9}$$

where $\hat{k}_\perp \equiv \hat{k}_x + \mathrm{i}\hat{k}_y$, and we have used (13.6) and (13.7). Substituting (9.2) and (13.9) into (8.79), we find the matrix to be diagonalized,

$$\hat{\mathscr{H}}_{\mathrm{BdG}} \equiv \begin{bmatrix} \xi_k & 0 & -\Delta\hat{k}_\perp^* & \Delta\hat{k}_z \\ 0 & \xi_k & \Delta\hat{k}_z & \Delta\hat{k}_\perp \\ -\Delta\hat{k}_\perp & \Delta\hat{k}_z & -\xi_k & 0 \\ \Delta\hat{k}_z & \Delta\hat{k}_\perp^* & 0 & -\xi_k \end{bmatrix}. \tag{13.10}$$

The eigenvalues of this 4×4 matrix are easily calculated based on Laplace's expansion of the determinant in terms of cofactors [3, 9]. However, we obtain them more readily here by noting that $\hat{\mathscr{H}}_{\mathrm{BdG}}^2$ is proportional to the 4×4 unit matrix $\hat{1}$,

$$\hat{\mathscr{H}}_{\mathrm{BdG}}^2 = E_k^2 \hat{1}, \qquad E_k \equiv \sqrt{\xi_k^2 + \Delta^2}. \tag{13.11}$$

It follows from this equality and (8.43) that the eigenvalues of $\hat{\mathscr{H}}_{\mathrm{BdG}}$ are given by $\pm E_k$ with double degeneracy for each. The eigenvectors corresponding to E_k are obtained from the first and second rows of the eigenvalue equation:

$$\left(\hat{\mathscr{H}}_{\mathrm{BdG}} - E_k \hat{1}\right) \begin{bmatrix} \mathbf{u}_{\tilde{\alpha}}(\mathbf{k}) \\ \mathbf{v}_{\tilde{\alpha}}(\mathbf{k}) \end{bmatrix} = \begin{bmatrix} \mathbf{0} \\ \mathbf{0} \end{bmatrix} \qquad (\tilde{\alpha} = 1, 2),$$

by comparing them with the first and second rows of (9.8) for $\tilde{\alpha} = 1, 2$. We obtain

$$\begin{bmatrix} \mathbf{u}_1(\mathbf{k}) \\ \mathbf{v}_1(\mathbf{k}) \end{bmatrix} = \begin{bmatrix} u_k \\ 0 \\ -v_k \hat{k}_\perp \\ v_k \hat{k}_z \end{bmatrix}, \quad \begin{bmatrix} \mathbf{u}_2(\mathbf{k}) \\ \mathbf{v}_2(\mathbf{k}) \end{bmatrix} = \begin{bmatrix} 0 \\ u_k \\ v_k \hat{k}_z \\ v_k \hat{k}_\perp^* \end{bmatrix}, \tag{13.12}$$

where u_k and v_k are also given by (9.6) for s-wave pairing. Substitution of (13.12) into (8.72) yields $\underline{\tilde{\rho}}^{(1)}(\mathbf{k})$ in terms of $\underline{\Delta}(\mathbf{k})$ in (13.9),

$$\underline{\tilde{\rho}}^{(1)}(\mathbf{k}) = \frac{\Delta}{2E_k} \begin{bmatrix} -\hat{k}_\perp^* & \hat{k}_z \\ \hat{k}_z & \hat{k}_\perp \end{bmatrix} (1 - 2\bar{n}_k) = \underline{\Delta}(\mathbf{k}) \frac{1}{2E_k} \tanh \frac{\beta E_k}{2}. \tag{13.13}$$

Next we substitute (13.3) and (13.13) into (8.76), transform the sum over $\mathbf{k}'$ into an integral as (8.90), approximate $N(\varepsilon_{k'}) \approx N(\varepsilon_\mathrm{F})$, make the change of variable, $\varepsilon_{k'} \to \xi_{k'}$, and perform the angular integration over $\hat{\mathbf{k}}'$ using

$$\int \frac{\mathrm{d}\Omega_\mathbf{k}}{4\pi} \hat{k}_\eta \hat{k}_{\eta'} = \delta_{\eta\eta'} \frac{1}{3} \qquad (\eta, \eta' = x, y, z). \tag{13.14}$$

We thereby obtain the gap equation

$$\underline{\Delta}(\mathbf{k}) = -N(\varepsilon_\mathrm{F}) \mathscr{V}_1^{(\mathrm{eff})} \int_{-\varepsilon_\mathrm{c}}^{\varepsilon_\mathrm{c}} \mathrm{d}\xi \, \underline{\Delta}(\mathbf{k}) \frac{1}{2E} \tanh \frac{\beta E}{2}. \tag{13.15}$$

Further, we remove a common "factor" $\underline{\Delta}(\mathbf{k})$ and introduce a dimensionless coupling constant:

$$g_1 \equiv -N(\varepsilon_\mathrm{F}) \mathscr{V}_1^{(\mathrm{eff})}. \tag{13.16}$$

We then find that (13.15) is identical to (9.30) for s-wave pairing with $g_0 \to g_1$. In particular, the transition temperature can be expressed in terms of g_1 above,

$$k_\mathrm{B} T_\mathrm{c} = \frac{2\mathrm{e}^\gamma}{\pi} \varepsilon_\mathrm{c} \mathrm{e}^{-1/g_1} \approx 1.13 \varepsilon_\mathrm{c} \mathrm{e}^{-1/g_1}. \tag{13.17}$$

It also follows from (13.11) that the thermodynamic properties of the BW state in terms of T/T_c are completely identical to those for s-wave pairing.

Finally, of note is the fact that the state $\underline{R}\mathbf{d}(\hat{\mathbf{k}})$ obtained from (13.9) by a three-dimensional rotation $\underline{R}$ is also the BW state. The degeneracy is partially spontaneously broken by the dipole interaction between nuclear spins, which is not included here [10, 15]. The remaining degeneracy is also lifted by other factors such as surface effects and initial fluctuations.

13.3.2 A Phase

The superfluid phase of ^{3}He that was first discovered was the high-temperature high-pressure phase called the *A phase* [11, 12], which is now identified as the *Anderson-Brinkmann-Morel (ABM) state*. Having an anisotropic energy gap [10, 15], this state was found as a theoretical candidate for the stable p-wave pairing by Anderson and Morel in 1961 [2], and studied in more detail by Anderson and Brinkmann in regard to its stabilization [1].

Given the normalization condition of (13.7), the **d** vector of the ABM state may be written

$$\mathbf{d}(\hat{\mathbf{k}}) = \sqrt{\frac{3}{2}}\hat{k}_\perp \hat{\mathbf{z}} \quad \longleftrightarrow \quad \underline{\Delta}(\mathbf{k}) = \sqrt{\frac{3}{2}}\Delta \begin{bmatrix} 0 & \hat{k}_\perp \\ \hat{k}_\perp & 0 \end{bmatrix} \tag{13.18}$$

with $\hat{k}_\perp \equiv \hat{k}_x + i\hat{k}_y$, for example. Let us substitute (9.2) and (13.18) into (8.79). The resulting BdG equation can be solved easily by the procedure used in obtaining (9.8) for s-wave pairing,

$$\begin{bmatrix} \xi_k & 0 & 0 & \sqrt{\frac{3}{2}}\Delta\hat{k}_\perp \\ 0 & \xi_k & \sqrt{\frac{3}{2}}\Delta\hat{k}_\perp & 0 \\ 0 & \sqrt{\frac{3}{2}}\Delta\hat{k}_\perp^* & -\xi_k & 0 \\ \sqrt{\frac{3}{2}}\Delta\hat{k}_\perp^* & 0 & 0 & -\xi_k \end{bmatrix} \hat{U}_\mathbf{k} = \hat{U}_\mathbf{k} \begin{bmatrix} E_\mathbf{k} & 0 & 0 & 0 \\ 0 & E_\mathbf{k} & 0 & 0 \\ 0 & 0 & -E_\mathbf{k} & 0 \\ 0 & 0 & 0 & -E_\mathbf{k} \end{bmatrix}. \tag{13.19}$$

Here, the excitation energies are anisotropic,

$$E_\mathbf{k} = \sqrt{\xi_k^2 + \frac{3}{2}\Delta^2|\hat{k}_\perp|^2}, \tag{13.20}$$

for which the energy gap vanishes at the north and south poles of the Fermi surface where $\hat{k}_\perp = 0$ (Fig. 13.2). The unitary matrix $\hat{U}_\mathbf{k}$ takes the form

$$\hat{U}_\mathbf{k} = \begin{bmatrix} u_\mathbf{k} & 0 & 0 & v_\mathbf{k}^* \\ 0 & u_\mathbf{k} & v_\mathbf{k}^* & 0 \\ 0 & v_\mathbf{k} & u_\mathbf{k} & 0 \\ v_\mathbf{k} & 0 & 0 & u_\mathbf{k} \end{bmatrix}, \qquad \begin{cases} u_\mathbf{k} \equiv \sqrt{\dfrac{E_\mathbf{k} + \xi_k}{2E_\mathbf{k}}} \\ v_\mathbf{k} \equiv \dfrac{\sqrt{\frac{3}{2}}\Delta\hat{k}_\perp^*}{\sqrt{2E_\mathbf{k}(E_\mathbf{k} + \xi_k)}} \end{cases}. \tag{13.21}$$

Hence, the eigenvectors belonging to $E_\mathbf{k}$ are given by

$$\mathbf{u}_1(\mathbf{k}) = \begin{bmatrix} u_\mathbf{k} \\ 0 \end{bmatrix}, \quad \mathbf{v}_1(\mathbf{k}) = \begin{bmatrix} 0 \\ v_\mathbf{k} \end{bmatrix}, \quad \mathbf{u}_2(\mathbf{k}) = \begin{bmatrix} 0 \\ u_\mathbf{k} \end{bmatrix}, \quad \mathbf{v}_2(\mathbf{k}) = \begin{bmatrix} v_\mathbf{k} \\ 0 \end{bmatrix}. \tag{13.22}$$

Substitution of (13.22) into (8.72) yields $\underline{\tilde{\rho}}^{(1)}(\mathbf{k})$ as

$$\underline{\tilde{\rho}}^{(1)}(\mathbf{k}) = \begin{bmatrix} 0 & u_{\mathbf{k}}v_{\mathbf{k}}^*(1-2\bar{n}_{\mathbf{k}}) \\ u_{\mathbf{k}}v_{\mathbf{k}}^*(1-2\bar{n}_{\mathbf{k}}) & 0 \end{bmatrix} = \underline{\Delta}(\mathbf{k})\frac{1}{2E_{\mathbf{k}}}\tanh\frac{\beta E_{\mathbf{k}}}{2}. \tag{13.23}$$

Thus, $\underline{\tilde{\rho}}^{(1)}(\mathbf{k})$ is expressible in terms of the gap matrix also for the ABM state. Let us substitute (13.3) and (13.23) into (8.76), change the sum over $\mathbf{k}'$ to an integral as (8.90), approximate $N(\varepsilon_{k'}) \approx N(\varepsilon_{\mathrm{F}})$, make a change of variable $\varepsilon_{k'} \to \xi_{k'}$, and perform the angular integrations over $\hat{\mathbf{k}}'$ using (13.14). Gap equation (8.76) is thereby transformed, using g_1 in (13.16),

$$\begin{aligned}
\hat{k}_\perp &= -N(\varepsilon_{\mathrm{F}})\mathscr{V}_1^{(\mathrm{eff})}\int_{-\varepsilon_c}^{\varepsilon_c} \mathrm{d}\xi_{k'}\int\frac{\mathrm{d}\Omega_{\mathbf{k}'}}{4\pi}\frac{3(\hat{k}_x\hat{k}'_x+\hat{k}_y\hat{k}'_y+\hat{k}_z\hat{k}'_z)\hat{k}'_\perp}{2E_{\mathbf{k}'}}\tanh\frac{\beta E_{\mathbf{k}'}}{2} \\
&= g_1\int_{-\varepsilon_c}^{\varepsilon_c} \mathrm{d}\xi_{k'}\int\frac{\mathrm{d}\Omega_{\mathbf{k}'}}{4\pi}\frac{3(\hat{k}_x\hat{k}'^2_x+\mathrm{i}\hat{k}_y\hat{k}'^2_y)}{2E_{\mathbf{k}'}}\tanh\frac{\beta E_{\mathbf{k}'}}{2} \\
&= g_1\int_{-\varepsilon_c}^{\varepsilon_c} \mathrm{d}\xi_{k'}\int\frac{\mathrm{d}\Omega_{\mathbf{k}'}}{4\pi}\frac{\frac{3}{2}(\hat{k}_x+\mathrm{i}\hat{k}_y)|\hat{k}'_\perp|^2}{2E_{\mathbf{k}'}}\tanh\frac{\beta E_{\mathbf{k}'}}{2},
\end{aligned}$$

where we have expressed $\hat{\mathbf{k}}' = (\sin\theta_{\mathbf{k}'}\cos\varphi_{\mathbf{k}'}, \sin\theta_{\mathbf{k}'}\sin\varphi_{\mathbf{k}'}, \cos\theta_{\mathbf{k}'})$ in polar coordinates and used identities:

$$\int_0^{2\pi}\frac{\mathrm{d}\varphi_{\mathbf{k}'}}{2\pi}\hat{k}'_\perp\hat{k}'_z = \int_0^{2\pi}\frac{\mathrm{d}\varphi_{\mathbf{k}'}}{2\pi}\hat{k}'_x\hat{k}'_y = 0, \qquad \int_0^{2\pi}\frac{\mathrm{d}\varphi_{\mathbf{k}'}}{2\pi}\hat{k}'^2_x = \int_0^{2\pi}\frac{\mathrm{d}\varphi_{\mathbf{k}'}}{2\pi}\hat{k}'^2_y = \frac{|\hat{k}'_\perp|^2}{2}.$$

We thereby obtain the gap equation for the ABM state as

$$1 = g_1\int_0^{\varepsilon_c}\mathrm{d}\xi_k\int\frac{\mathrm{d}\Omega_{\mathbf{k}}}{4\pi}\frac{\frac{3}{2}|\hat{k}_\perp|^2}{E_{\mathbf{k}}}\tanh\frac{\beta E_{\mathbf{k}}}{2}. \tag{13.24}$$

The transition temperature T_c is determined by setting $T \to T_c$ and $E_{\mathbf{k}} \to \xi_k$ in (13.24). One confirms easily that it is also given by (13.17). As mentioned earlier, all the p-wave states are thermodynamically degenerate at $T = T_c$ (**Problem** 13.1). This degeneracy is lifted as the temperature is lowered below T_c.

Proceeding similarly as in (9.34) for the energy integral, the gap equation (13.24) at $T = 0$ can be solved analytically by changing the order of the energy and angular integrals,

$$\frac{1}{g_1} = \int\frac{\mathrm{d}\Omega_{\mathbf{k}}}{4\pi}\frac{3}{2}|\hat{k}_\perp|^2\int_0^{\varepsilon_c}\frac{\mathrm{d}\xi_k}{E_{\mathbf{k}}}$$

$$= \int_0^{2\pi} \frac{d\varphi_{\mathbf{k}}}{2\pi} \int_0^{\pi} \frac{d\theta_{\mathbf{k}} \sin\theta_{\mathbf{k}}}{2} \frac{3}{2} \sin^2\theta_{\mathbf{k}} \ln \frac{2\varepsilon_c}{\sqrt{\frac{3}{2}}\Delta_0 \sin\theta_{\mathbf{k}}} \qquad (t \equiv \cos\theta_{\mathbf{k}})$$

$$= \frac{3}{4}\int_{-1}^{1} dt\, (1-t^2) \left[\ln\frac{2\varepsilon_c}{\Delta_0} - \frac{1}{2}\ln\frac{3}{2} - \frac{1}{2}\ln(1-t^2) \right]$$

$$= \ln\frac{2\varepsilon_c}{\Delta_0} + \frac{5}{6} - \frac{1}{2}\ln 6,$$

where the last integral over t has been performed using

$$\int (x+a)^n \ln(x+a) dx = \frac{(x+a)^{n+1}}{n+1}\left[\ln(x+a) - \frac{1}{n+1}\right].$$

Hence, we obtain the angular average of the energy gap for the ABM state as ($\Delta_0 \to \Delta_0^{\mathrm{ABM}}$)

$$\Delta_0^{\mathrm{ABM}} = 2\varepsilon_c e^{-1/g_1 + 5/6 - \frac{1}{2}\ln 6} = \frac{e^{5/6}}{\sqrt{6}}\Delta_0^{\mathrm{BW}} = 0.94\Delta_0^{\mathrm{BW}}, \tag{13.25}$$

where $\Delta_0^{\mathrm{BW}} \equiv 2\varepsilon_c e^{-1/g_1}$ is the energy gap of the BW state. The condensation energy of the ABM state at $T = 0$ is obtained from (9.52) by changing $\Delta_0^2 \to \frac{3}{2}\left|\Delta_0^{\mathrm{ABM}}\hat{k}_\perp\right|^2$ and subsequently performing the angular integral $\int d\Omega_{\mathbf{k}}/4\pi$,

$$F_{\mathrm{sn}}^{\mathrm{ABM}} = -\frac{D(\varepsilon_F)}{4}\left(\Delta_0^{\mathrm{ABM}}\right)^2 \int \frac{d\Omega_{\mathbf{k}}}{4\pi}\frac{3}{2}|\hat{k}_\perp|^2 = -\frac{D(\varepsilon_F)}{4}\left(\Delta_0^{\mathrm{ABM}}\right)^2 = 0.88 F_{\mathrm{sn}}^{\mathrm{BW}}, \tag{13.26}$$

where $F_{\mathrm{sn}}^{\mathrm{BW}} \equiv -\frac{1}{4}D(\varepsilon_F)\left(\Delta_0^{\mathrm{BW}}\right)^2$ is the condensation energy of the BW state. Thus, our mean-field theory predicts that the ABM state is metastable with a higher free energy than the BW state. It turns out that this conclusion holds for $0 \le T \le T_c$ within the mean-field theory. Indeed, the prediction agrees with the phase diagram at low pressures in Fig. 13.1. However, the stabilization of the ABM state in the high-pressure region cannot be explained within the mean-field theory; its explanation is attributed to spin-fluctuation effects that are describable only by a treatment beyond the mean-field level [1, 8, 13].

A couple of comments are in order before closing the section. First, the ABM state also has a large degeneracy. To be specific, consider (13.18) and rotate both $\hat{\mathbf{z}}$ in the spin space and $\hat{k}_\perp$ in the orbital space independently around arbitrary axes by arbitrary angles. The state thereby obtained is also an ABM state with the same free energy. Part of the degeneracy is spontaneously broken through dipole interactions between nuclear spins that are not included here [10, 15]. The remaining degeneracies are also lifted by other effects such as surface effects and initial fluctuations. Second, the complex **d** vector of (13.18) implicitly describes a Cooper pair with an orbital angular momentum $\hbar$, which is expected to produce a total angular momentum of $(N/2)\hbar$ at $T = 0$ due to the macroscopic condensation into identical two-particle bound states [6, 7]. This effect remains to be confirmed experimentally.

13.4 Gap Anisotropy and Quasiparticle Density of States

The gap structures are different between the A and B phases; whereas it is isotropic for the B phase, the gap closes at a couple of points in the A phase. We now see that this difference manifests itself in the quasiparticle density of states at low energies, which are observable by low-temperature thermodynamic experiments. Thus, our main purpose here is to find general means to identify this gap anisotropy in experiments [14].

The quasiparticle density of states is defined by (9.46), which is expressible as (9.48) for *s*-wave pairing. Its extension to *p*-wave pairing is easily performed by incorporating the angular dependence into the energy gap as $\Delta \to |\Delta_{\mathbf{k}}|$ and averaging the resulting expression over the solid angle,

$$D_{\rm s}(E) = D(\varepsilon_{\rm F}) \int \frac{d\Omega_{\mathbf{k}}}{4\pi} \frac{|E|}{(E^2 - |\Delta_{\mathbf{k}}|^2)^{1/2}} \theta(|E| - |\Delta_{\mathbf{k}}|). \tag{13.27}$$

We consider three $|\Delta_{\mathbf{k}}|$'s given in terms of $\hat{\mathbf{k}} = (\sin\theta_{\mathbf{k}}\cos\varphi_{\mathbf{k}}, \sin\theta_{\mathbf{k}}\sin\varphi_{\mathbf{k}}, \cos\theta_{\mathbf{k}})$ as

$$|\Delta_{\mathbf{k}}| = \begin{cases} \Delta & : \text{BW state} \\ \sqrt{3/2}\,\Delta\sin\theta_{\mathbf{k}} & : \text{ABM state} \\ \sqrt{3}\Delta\cos\theta_{\mathbf{k}} & : \text{polar state} \end{cases}. \tag{13.28}$$

Thus, besides the BW and ABM states given by (13.11) and (13.20), respectively, a novel *polar state* $\mathbf{d}(\mathbf{k}) = \sqrt{3}\hat{k}_z$ [10, 15] is included here. Although it is not stabilized as a bulk phase, it has a distinct gap structure in that it closes on the equator of the Fermi surface (Fig. 13.2) to make it worth including in the present considerations.

Integral (13.27) for each gap structure can be performed easily, and we obtain (**Problem** 13.2)

$$\frac{D_{\rm s}(E)}{D(\varepsilon_{\rm F})} = \begin{cases} \dfrac{|E|}{(E^2-\Delta^2)^{1/2}}\theta(|E|-\Delta) & : \text{BW state} \\[2ex] \dfrac{|E|}{2\Delta_{\rm max}} \ln\left|\dfrac{|E|+\Delta_{\rm max}}{|E|-\Delta_{\rm max}}\right| & : \text{ABM state} \\[2ex] \dfrac{\pi|E|}{2\Delta_{\rm max}}\left[\theta(\Delta_{\rm max}-|E|)+\theta(|E|-\Delta_{\rm max})\dfrac{2}{\pi}\arcsin\dfrac{\Delta_{\rm max}}{|E|}\right] & : \text{polar state} \end{cases}, \tag{13.29}$$

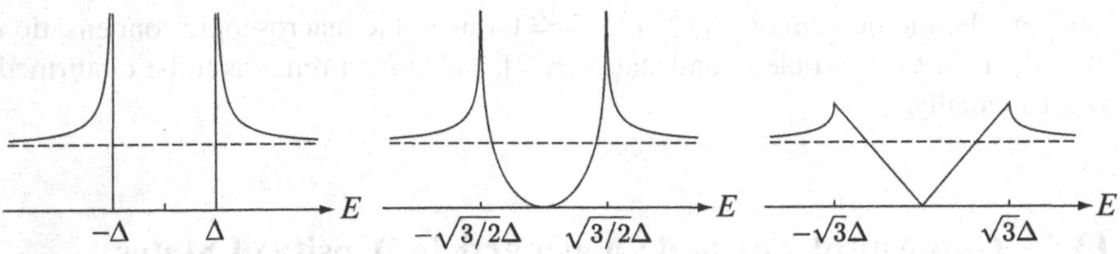

Fig. 13.3 Quasiparticle density of states given by (13.29). Broken lines denote the normal density of states

with $\Delta_{\max} = \sqrt{3/2}\,\Delta$ ($\Delta_{\max} = \sqrt{3}\Delta$) for the ABM (polar) state; each is plotted in Fig. 13.3. At low energies of $|E/\Delta_{\max}| \ll 1$, $D_s(E)$ can be approximated as

$$\frac{D_s(E)}{D(\varepsilon_F)} = \begin{cases} 0 & : \text{BW state (full gap)} \\ (|E|/\Delta_{\max})^2 & : \text{ABM state (point nodes)} \\ \pi|E|/2\Delta_{\max} & : \text{polar state (line node)} \end{cases} . \tag{13.30}$$

These relations give the general connection between the low-energy density of states and the dimension of nodes in the gap; they are also valid for anisotropic Fermi surfaces except for the prefactor.

The low-energy density of states manifests itself in temperature dependences of various thermodynamic quantities. First, we consider heat capacity. Let us substitute the latter two expressions of (13.30) into (9.47), rewrite $\partial\bar{n}(E)/\partial T = -(E/T)\partial\bar{n}(E)/\partial E$, perform an integration by parts, and express C in terms of the dimensionless integral (4.39). We thereby find that the low-temperature heat capacity behaves as

$$C(T \to 0) \propto \begin{cases} e^{-\Delta/k_B T} & : \text{full gap} \\ T^3 & : \text{point nodes} \\ T^2 & : \text{line nodes} \end{cases} , \tag{13.31}$$

where we have used (9.49) for the full gap.

Second, we consider an anisotropic extension of the Yosida function (10.16) given by

$$Y(\hat{\mathbf{k}}, T) \equiv -2\int_0^\infty d\xi \frac{\partial\bar{n}(E_{\mathbf{k}})}{\partial E_{\mathbf{k}}}, \qquad E_{\mathbf{k}} \equiv \sqrt{\xi^2 + |\Delta_{\mathbf{k}}|^2}. \tag{13.32}$$

Its angular average can be transformed by writing $d\xi = (d\xi/dE_{\mathbf{k}})dE_{\mathbf{k}}$, substituting $d\xi/dE_{\mathbf{k}} = E_{\mathbf{k}}\theta(E_{\mathbf{k}} - |\Delta_{\mathbf{k}}|)/\sqrt{E_{\mathbf{k}}^2 - |\Delta_{\mathbf{k}}|^2}$, exchanging the order of integrations over $d\Omega_{\mathbf{k}}$ and $dE_{\mathbf{k}}$, and using (13.27) as

$$\bar{Y}(T) \equiv \int \frac{d\Omega_{\mathbf{k}}}{4\pi} Y(\hat{\mathbf{k}}, T) = -2\int_0^\infty dE \frac{D_s(E)}{D(\varepsilon_F)} \frac{\partial \bar{n}(E)}{\partial E}. \tag{13.33}$$

Let us substitute the latter two expressions of (13.30) into (13.33), perform an integration by parts in terms of E, and express $\bar{Y}(T)$ using dimensionless integral (4.39). We thereby obtain

$$\bar{Y}(T \to 0) \propto \begin{cases} e^{-\Delta/k_B T} & : \text{full gap} \\ T^2 & : \text{point nodes} \\ T & : \text{line nodes} \end{cases}, \tag{13.34}$$

where we have used (10.48) for the full gap. The temperature dependence is observable by measuring the spin susceptibility and penetration depth, as described by (10.18) and (10.40) with (10.22).

Problems

13.1. Follow the procedure below to show that every homogeneous p-wave ($\ell = 1$) state has an identical transition temperature T_c.

(a) Let us consider the cases where submatrix $\underline{\mathscr{K}}_{HF}$ in the homogeneous BdG equation (8.79) is given by $\underline{\mathscr{K}}_{HF} = \xi_k \underline{\sigma}_0$ in terms of ξ_k in (9.17) and $\underline{\sigma}_0$ in (8.42). Solve the BdG equation by a perturbation expansion in terms of $\underline{\Delta}(\mathbf{k})$ to show that (8.72) to lowest order becomes

$$\underline{\tilde{\rho}}^{(1)}(\mathbf{k}) \approx \underline{\Delta}(\mathbf{k}) \frac{1}{2\xi_k} \tanh \frac{\beta \xi_k}{2}.$$

(b) Substitute the result of (a) and (13.3) into (8.76) to show that the resulting equation at $T = T_c$ is given for any internal state as

$$\frac{-1}{N(\varepsilon_F)\mathscr{V}_1^{(\text{eff})}} = \int_{-\varepsilon_c}^{\varepsilon_c} d\xi \frac{1}{2\xi} \tanh \frac{\xi}{2k_B T_c} \tag{13.35}$$

that yields (13.17) for the transition temperature.

13.2. Show (13.29) for the ABM and polar states.

References

1. P.W. Anderson, W.F. Brinkman, Phys. Rev. Lett. **30**, 1108 (1973)
2. P.W. Anderson, P. Morel, Phys. Rev. **123**, 1911 (1961)
3. G.B. Arfken, H.J. Weber, *Mathematical Methods for Physicists* (Academic, New York, 2012)

4. R. Balian, N.R. Werthamer, Phys. Rev. **131**, 1553 (1963)
5. D.S. Greywall, Phys. Rev. B **33**, 7520 (1986)
6. M. Ishikawa, Prog. Theor. Phys. **57**, 1836 (1977)
7. T. Kita, J. Phys. Soc. Jpn. **67**, 216 (1998)
8. Y. Kuroda, Prog. Theor. Phys. **53**, 349 (1975)
9. S. Lang, *Linear Algebra* (Springer, New York, 1987)
10. A.J. Leggett, Rev. Mod. Phys. **47**, 331 (1975)
11. D.D. Osheroff, W.J. Gully, R.C. Richardson, D.M. Lee, Phys. Rev. Lett. **29**, 920 (1972)
12. D.D. Osheroff, R.C. Richardson, D.M. Lee, Phys. Rev. Lett. **28**, 885 (1972)
13. J.W. Serene, D. Rainer, Phys. Rep. **101**, 221 (1983)
14. M. Sigrist, K. Ueda, Rev. Mod. Phys. **63**, 239 (1991)
15. D. Vollhardt, P. Wölfle, *The Superfluid Phases of Helium 3* (Taylor & Francis, London, 1990), p. 31
16. J.C. Wheatley, Rev. Mod. Phys. **47**, 415 (1975)

Chapter 14
Gor'kov, Eilenberger, and Ginzburg–Landau Equations

Abstract One of the most outstanding features of superconductivity is that there can be various stable structures with quasimacroscopic inhomogeneity, such as the flux-line lattice realized in certain superconductors under an applied magnetic field. To describe these structures concisely, we here simplify the BdG equations in three steps. First, we derive the Gor'kov equations (14.26) for the Matsubara Green's functions, which is equivalent to the BdG equations. Second, we integrate out an independent variable from the Gor'kov equations to derive the Eilenberger equations (14.61) and (14.62) for the quasiclassical Green's function (14.59). Third, we focus on the region near T_c to simplify the Eilenberger equations further into the Ginzburg–Landau (GL) equations (14.89) and (14.94). Those who are interested mainly in the physical phenomena rather than the microscopic derivations of the standard equations may skip this chapter.

14.1 Matsubara Green's Function

Introduced in 1955 [14], the Matsubara Green's function is now regarded as one of the most fundamental tools in equilibrium statistical mechanics. We introduce it here and enumerate its basic properties.

Let us distinguish the creation and annihilation operators with integer subscripts [11],

$$\hat{\psi}_1(\xi) \equiv \hat{\psi}(\xi), \qquad \hat{\psi}_2(\xi) \equiv \hat{\psi}^\dagger(\xi), \tag{14.1}$$

so that $\hat{\psi}_i^\dagger(\xi) = \hat{\psi}_{3-i}(\xi)$ holds ($i = 1, 2$). Next, we introduce the field operators in the Heisenberg representation in terms of a new variable $\tau_1 \in [0, \beta]$,

$$\hat{\psi}_i(1) \equiv \mathrm{e}^{\tau_1 \mathscr{H}} \hat{\psi}_i(\xi_1) \mathrm{e}^{-\tau_1 \mathscr{H}}, \tag{14.2}$$

with $\hat{\psi}_i(1) \equiv \hat{\psi}_i(\xi_1, \tau_1)$. Replacing $\tau_1 \to it_1/\hbar$ yields the standard Heisenberg representation with respect to time t_1.

© Springer Japan 2015

T. Kita, *Statistical Mechanics of Superconductivity*, Graduate Texts in Physics,
DOI 10.1007/978-4-431-55405-9_14

We now define the *Matsubara Green's function* using the field operators and step function (4.11) as

$$G_{ij}(1,2) \equiv -\theta(\tau_1-\tau_2)\langle\hat{\psi}_i(1)\hat{\psi}_{3-j}(2)\rangle + \theta(\tau_2-\tau_1)\langle\hat{\psi}_{3-j}(2)\hat{\psi}_i(1)\rangle$$
$$\equiv -\langle\hat{T}_\tau\hat{\psi}_i(1)\hat{\psi}_{3-j}(2)\rangle, \tag{14.3}$$

where the second expression is to be regarded as defining the $\hat{T}_\tau$ operator. Thus, $\hat{T}_\tau$ rearranges field operators to its right in descending order of τ, multiplying the result by $\sigma = -1$ for each exchange of adjacent operators. Diagonal elements $G_{11}(1,2)$ and $G_{22}(1,2)$ are composed of a pair of creation and annihilation operators, which remain finite even for normal states. In contrast, the off-diagonal elements are characteristic of superconductivity; for example, $G_{12}(1,2)$ is made up of two annihilation operators and sometimes called the "anomalous" Green's function.

The Matsubara Green's function has the following properties.

(a) $G_{ij}(1,2)$ is a function of only $\tau_1-\tau_2$.

The proof proceeds using Tr$\hat{A}\hat{B}$ =Tr$\hat{B}\hat{A}$ and the commutativity of $e^{-\beta\mathscr{H}}$ and $e^{-\tau_1\mathscr{H}}$ as follows:

$$G_{ij}(1,2) = -\mathrm{Tr}\,\hat{T}_\tau e^{\beta(\Omega-\mathscr{H})}e^{\tau_1\mathscr{H}}\hat{\psi}_i(\xi_1)e^{-\tau_1\mathscr{H}}e^{\tau_2\mathscr{H}}\hat{\psi}_{3-j}(\xi_2)e^{-\tau_2\mathscr{H}}$$
$$= -\mathrm{Tr}\,\hat{T}_\tau e^{\beta(\Omega-\mathscr{H})}e^{(\tau_1-\tau_2)\mathscr{H}}\hat{\psi}_i(\xi_1)e^{-(\tau_1-\tau_2)\mathscr{H}}\hat{\psi}_{3-j}(\xi_2)$$
$$\equiv G_{ij}(\xi_1,\xi_2;\tau_1-\tau_2). \tag{14.4}$$

Operator $\hat{T}_\tau$ does not affect the proof at all, as may be confirmed by performing it separately for $\tau_1>\tau_2$ and $\tau_1<\tau_2$ without $\hat{T}_\tau$. It follows from $0\le\tau_1,\tau_2\le\beta$ that $-\beta\le\tau_1-\tau_2\le\beta$.

(b) $G_{ij}(\xi_1,\xi_2;\tau+\beta) = -G_{ij}(\xi_1,\xi_2;\tau)$ for $\tau\in[-\beta,0]$.

This is shown as follows. First, the right-hand side is given explicitly by

$$-G_{ij}(\xi_1,\xi_2;\tau) = \langle\hat{T}_\tau\hat{\psi}_i(\xi_1,\tau)\hat{\psi}_{3-j}(\xi_2)\rangle$$
$$= -e^{\beta\Omega}\,\mathrm{Tr}\,e^{-\beta\mathscr{H}}\hat{\psi}_{3-j}(\xi_2)e^{\tau\mathscr{H}}\hat{\psi}_i(\xi_1)e^{-\tau\mathscr{H}}.$$

The left-hand side can be expanded as

$$G_{ij}(\xi_1,\xi_2;\tau+\beta) = -e^{\beta\Omega}\,\mathrm{Tr}\,e^{-\beta\mathscr{H}}e^{(\tau+\beta)\mathscr{H}}\hat{\psi}_i(\xi_1)e^{-(\tau+\beta)\mathscr{H}}\hat{\psi}_{3-j}(\xi_2)$$
$$= -e^{\beta\Omega}\,\mathrm{Tr}\,e^{\tau\mathscr{H}}\hat{\psi}_i(\xi_1)e^{-\tau\mathscr{H}}e^{-\beta\mathscr{H}}\hat{\psi}_{3-j}(\xi_2)$$
$$= -e^{\beta\Omega}\,\mathrm{Tr}\,e^{-\beta\mathscr{H}}\hat{\psi}_{3-j}(\xi_2)e^{\tau\mathscr{H}}\hat{\psi}_i(\xi_1)e^{-\tau\mathscr{H}}.$$

Hence, we conclude $G_{ij}(\xi_1, \xi_2; \tau + \beta) = -G_{ij}(\xi_1, \xi_2; \tau)$. It is convenient to expand G_{ij} in a complete set that satisfies boundary condition $g(\tau + \beta) = -g(\tau)$. With this aim, we consider the first-order differential equation:

$$\frac{\mathrm{d}g(\tau)}{\mathrm{d}\tau} = -\mathrm{i}\varepsilon g(\tau), \qquad g(\tau + \beta) = -g(\tau), \tag{14.5}$$

where factor $-\mathrm{i}$ has been introduced for convenience. Its general solution is $g(\tau) \propto \mathrm{e}^{-\mathrm{i}\varepsilon\tau}$. Imposing the boundary condition yields eigenvalues ε (with units of energy)

$$\varepsilon_n = (2n + 1)\pi/\beta \qquad (n = 0, \pm 1, \pm 2, \cdots). \tag{14.6}$$

The quantity $\varepsilon_n/\hbar$ is called the *Matsubara frequency for fermions*. The basis functions $\{\mathrm{e}^{-\mathrm{i}\varepsilon_n\tau}/\sqrt{\beta}\}$ with $\tau \in [0, \beta]$ form a complete orthonormal set for an arbitrary function $f(\tau)$ that satisfies $f(\tau + \beta) = -f(\tau)$. Their completeness relation reads

$$\delta(\tau_1 - \tau_2) = \frac{1}{\beta} \sum_{n=-\infty}^{\infty} \mathrm{e}^{-\mathrm{i}\varepsilon_n(\tau_1 - \tau_2)}. \tag{14.7}$$

Expanding in the basis functions, the Matsubara Green's function becomes

$$G_{ij}(\xi_1, \xi_2; \tau) = \frac{1}{\beta} \sum_{n=-\infty}^{\infty} G_{ij}(\xi_1, \xi_2; \varepsilon_n)\mathrm{e}^{-\mathrm{i}\varepsilon_n\tau}, \tag{14.8}$$

where we distinguish the G_{ij}'s on each side by their arguments. The inverse transform is obtained by multiplying the equation by $\mathrm{e}^{\mathrm{i}\varepsilon_{n'}\tau}$ and performing integration over $\tau \in [0, \beta]$. The result is given with $n' \to n$ as

$$G_{ij}(\xi_1, \xi_2; \varepsilon_n) = \int_0^\beta G_{ij}(\xi_1, \xi_2; \tau)\mathrm{e}^{\mathrm{i}\varepsilon_n\tau}\, \mathrm{d}\tau. \tag{14.9}$$

(c) $G_{ij}(1, 2) = G^*_{ji}(\xi_2\tau_1, \xi_1\tau_2) = -G_{3-j,3-i}(2, 1)$.

The first equality for $\tau \equiv \tau_1 - \tau_2 \geq 0$ can be proved using $\langle \hat{A}\hat{B}\rangle^* = \langle \hat{B}^\dagger \hat{A}^\dagger\rangle$ and $\hat{\psi}^\dagger_j(\xi) = \hat{\psi}_{3-j}(\xi)$,

$$\begin{aligned} G^*_{ij}(\xi_1, \xi_2; \tau) &= -\langle \mathrm{e}^{\tau\hat{\mathscr{H}}}\hat{\psi}_i(\xi_1)\mathrm{e}^{-\tau\hat{\mathscr{H}}}\hat{\psi}_{3-j}(\xi_2)\rangle^* \\ &= -\langle \hat{\psi}_j(\xi_2)\mathrm{e}^{-\tau\hat{\mathscr{H}}}\hat{\psi}_{3-i}(\xi_1)\mathrm{e}^{\tau\hat{\mathscr{H}}}\rangle \\ &= -\langle \mathrm{e}^{\tau\hat{\mathscr{H}}}\hat{\psi}_j(\xi_2)\mathrm{e}^{-\tau\hat{\mathscr{H}}}\hat{\psi}_{3-i}(\xi_1)\rangle \\ &= G_{ji}(\xi_2, \xi_1; \tau). \end{aligned} \tag{14.10}$$

It follows from (b) above that the equality also holds for $\tau \in [-\beta, 0]$. The second equality obtains using the anticommutation relation of the fermion field operators under the influence of $\hat{T}_\tau$ and noting $i = 3 - (3 - i)$,

$$\begin{aligned} G_{ij}(1,2) &= -\langle \hat{T}_\tau \hat{\psi}_i(1) \hat{\psi}_{3-j}(2) \rangle = \langle \hat{T}_\tau \hat{\psi}_{3-j}(2) \psi_i(1) \rangle \\ &= -G_{3-j,3-i}(2,1). \end{aligned} \tag{14.11}$$

The two symmetry relations are expressible in terms of the Fourier coefficients in (14.9), specifically

$$G_{ij}(\xi_1, \xi_2; \varepsilon_n) = G_{ji}^*(\xi_2, \xi_1; -\varepsilon_n) = -G_{3-j,3-i}(\xi_2, \xi_1; -\varepsilon_n). \tag{14.12}$$

14.2 Gor'kov Equations

We consider a system described by the Hamiltonian:

$$\begin{aligned} \hat{\mathscr{H}} \equiv & \int d\xi_2 \hat{\psi}^\dagger(\xi_2) \hat{\mathscr{K}}_2 \hat{\psi}(\xi_2) \\ & + \frac{1}{2} \int d\xi_2 \int d\xi_2' \mathscr{V}(|\mathbf{r}_2 - \mathbf{r}_2'|) \hat{\psi}^\dagger(\xi_2) \hat{\psi}^\dagger(\xi_2') \hat{\psi}(\xi_2') \hat{\psi}(\xi_2), \end{aligned} \tag{14.13}$$

to derive the Gor'kov equations within the mean-field approximation. To describe phenomena in magnetic fields, the one-particle operator $\hat{\mathscr{K}}_2$ now contains the vector potential $\mathbf{A}_2 \equiv \mathbf{A}(\mathbf{r}_2)$,

$$\hat{\mathscr{K}}_2 \equiv \frac{(\hat{\mathbf{p}}_2 - e\mathbf{A}_2)^2}{2m} - \mu, \tag{14.14}$$

with m and $e < 0$ denoting the electron mass and charge, respectively.[1]

14.2.1 *Equation of Motion for Field Operators*

As a preliminary, we obtain an equation of motion obeyed by $\hat{\psi}_i(1)$.

Differentiation of (14.2) with respect to τ_1 yields

$$\frac{\partial \hat{\psi}_i(1)}{\partial \tau_1} = e^{\tau_1 \hat{\mathscr{H}}} [\hat{\mathscr{H}} \hat{\psi}_i(\xi_1) - \hat{\psi}_i(\xi_1) \hat{\mathscr{H}}] e^{-\tau_1 \hat{\mathscr{H}}}. \tag{14.15}$$

[1] We omit the Zeeman coupling once again. See the comment below (10.33) on this point.

Using commutation relation (3.17) for $\sigma = -1$, we then move $\hat{\psi}_i(\xi_1)$ in the term $\hat{\mathscr{H}}\hat{\psi}_i(\xi_1)$ above successively to the left to cancel $-\hat{\psi}_i(\xi_1)\hat{\mathscr{H}}$. The result for $i = 1$ is given by[2]

$$\frac{\partial\hat{\psi}(1)}{\partial\tau_1} = e^{\tau_1\hat{\mathscr{H}}}\left[-\hat{\mathscr{K}}_1\hat{\psi}(\xi_1) + \frac{1}{2}\int d\xi_2 \mathscr{V}(|\mathbf{r}_2 - \mathbf{r}_1|)\hat{\psi}^\dagger(\xi_2)\hat{\psi}(\xi_1)\hat{\psi}(\xi_2)\right.$$
$$\left. - \frac{1}{2}\int d\xi_2' \mathscr{V}(|\mathbf{r}_1 - \mathbf{r}_2'|)\hat{\psi}^\dagger(\xi_2')\hat{\psi}(\xi_2')\hat{\psi}(\xi_1)\right]e^{-\tau_1\hat{\mathscr{H}}}.$$

The second and third terms in the square brackets yield the same contribution, as seen easily using commutation relation $\hat{\psi}(\xi_1)\hat{\psi}(\xi_2) = -\hat{\psi}(\xi_2)\hat{\psi}(\xi_1)$ and $\xi_2, \xi_2' \to \xi_1'$. Finally, we exchange the order of $\hat{\mathscr{K}}_1$ and $e^{\tau_1\hat{\mathscr{H}}}$,[3] insert identity operator $e^{-\tau_1\hat{\mathscr{H}}}e^{\tau_1\hat{\mathscr{H}}}$ between every pair of adjacent field operators, and express the result in terms of the operators in (14.2). We thereby obtain

$$\frac{\partial\hat{\psi}_1(1)}{\partial\tau_1} = -\hat{\mathscr{K}}_1\hat{\psi}_1(1) - \int d\xi_1' \mathscr{V}(|\mathbf{r}_1 - \mathbf{r}_1'|)\hat{\psi}_2(1')\hat{\psi}_1(1')\hat{\psi}_1(1), \tag{14.16}$$

where τ_1 in $1' \equiv (\xi_1', \tau_1)$ is equal to that in $1 \equiv (\xi_1, \tau_1)$. For $i = 2$, it is convenient as a preliminary to apply integration by parts to the first term in (14.13),

$$\int d\xi_2 \hat{\psi}^\dagger(\xi_2)\hat{\mathscr{K}}_2\hat{\psi}(\xi_2) - \int d\xi_2 \left[\hat{\mathscr{K}}_2^*\hat{\psi}^\dagger(\xi_2)\right]\hat{\psi}(\xi_2),$$

and use it to calculate the commutator of (14.15). The final result is given by

$$\frac{\partial\hat{\psi}_2(1)}{\partial\tau_1} = \hat{\mathscr{K}}_1^*\hat{\psi}_2(1) + \int d\xi_1' \mathscr{V}(|\mathbf{r}_1 - \mathbf{r}_1'|)\hat{\psi}_2(1)\hat{\psi}_2(1')\hat{\psi}_1(1'). \tag{14.17}$$

The two differential equations (14.16) and (14.17) can be expressed in a unified way as

$$\frac{\partial\hat{\psi}_i(1)}{\partial\tau_1} = (-1)^i\left[\hat{\mathscr{K}}_1^i\hat{\psi}_i(1) + \int d\xi_1' \mathscr{V}(|\mathbf{r}_1 - \mathbf{r}_1'|)\hat{N}\hat{\psi}_2(1')\hat{\psi}_1(1')\hat{\psi}_i(1)\right], \tag{14.18}$$

---

[2]The calculation of the commutator $[\hat{\psi}(\xi_1), \hat{\mathscr{H}}]_+ \equiv \hat{\psi}(\xi_1)\hat{\mathscr{H}} - \hat{\mathscr{H}}\hat{\psi}(\xi_1)$ is equivalent to the functional derivative $\delta\hat{\mathscr{H}}/\delta\hat{\psi}^\dagger(\xi_1)$ by incorporating the anticommutation relation (3.17). Similarly, $[\hat{\psi}^\dagger(\xi_1), \hat{\mathscr{H}}]_+$ is equal to $-\delta\hat{\mathscr{H}}/\delta\hat{\psi}(\xi_1)$.

[3]There is only a single point $\xi_2 = \xi_1$ in (14.13) that does not commute with $\hat{\mathscr{K}}_1$, its measure being zero in the integration. Therefore, the procedure is justified.

where $\hat{\mathscr{K}}_1^i$ denotes $\hat{\mathscr{K}}_1^1 = \hat{\mathscr{K}}_1$ and $\hat{\mathscr{K}}_1^2 = \hat{\mathscr{K}}_1^*$, and $\hat{N}$ is the so-called *normal-ordering operator* that places the creation operators to the left of the annihilation operators, multiplying by $\sigma = -1$ for each exchange of field operators; specifically,

$$\hat{N}\hat{\psi}_2(1')\hat{\psi}_1(1')\hat{\psi}_i(1) \equiv \begin{cases} \hat{\psi}_2(1')\hat{\psi}_1(1')\hat{\psi}_1(1) & : i = 1 \\ (-1)^2\hat{\psi}_2(1)\hat{\psi}_2(1')\hat{\psi}_1(1') & : i = 2 \end{cases}. \tag{14.19}$$

The case $i = 2$ can also be expressed as $(-1)^1\hat{\psi}_2(1')\hat{\psi}_2(1)\hat{\psi}_1(1')$.

14.2.2 Derivation of the Gor'kov Equations

We are now ready to derive the Gor'kov equations by differentiating $G_{ij}(1,2)$ with respect to τ_1 and adopting the mean-field approximation with the Wick-decomposition procedure.

The τ_1 dependence of (14.3) lies in the step function and $\hat{\psi}_i(1)$. Using $\delta(x) = \theta'(x) = -\theta'(-x)$ as noted below (4.11), we differentiate Green's function with respect to τ_1 to obtain

$$\frac{\partial G_{ij}(1,2)}{\partial \tau_1} = -\delta(\tau_1 - \tau_2)\Big[\langle\hat{\psi}_i(1)\hat{\psi}_{3-j}(2)\rangle + \langle\hat{\psi}_{3-j}(2)\hat{\psi}_i(1)\rangle\Big] - \left\langle \hat{T}_\tau \frac{\partial\hat{\psi}_i(1)}{\partial\tau_1}\hat{\psi}_{3-j}(2)\right\rangle.$$

The first term on the right-hand side contains expectations at $\tau_1 = \tau_2$, which do not depend on τ_1, as seen in (14.4). Hence, it can be simplified with the help of (3.17) for $\sigma = -1$ to $-\delta(\tau_1 - \tau_2)\delta_{ij}\delta(\xi_1, \xi_2) \equiv -\delta_{ij}\delta(1,2)$. The second term can also be expanded by substituting (14.18). We thereby obtain

$$\frac{\partial G_{ij}(1,2)}{\partial \tau_1} = -\delta_{ij}\delta(1,2) - (-1)^i\Big[\hat{\mathscr{K}}_1^i\langle\hat{T}_\tau\hat{\psi}_i(1)\hat{\psi}_{3-j}(2)\rangle + \int d\xi_1'\mathscr{V}(|\mathbf{r}_1 - \mathbf{r}_1'|) \times \langle\hat{T}_\tau[\hat{N}\hat{\psi}_2(1')\hat{\psi}_1(1')\hat{\psi}_i(1)]\hat{\psi}_{3-j}(2)\rangle\Big]. \tag{14.20}$$

Subsequently, we adopt the mean-field approximation as for (8.26). A couple of key points in the process are summarized as follows: (i) The first condition in (5.13) implies that the $\hat{T}_\tau$ operator should be incorporated in the expectations with different "times" after the Wick decomposition. (ii) The expectations of "equal-time" operators do not depend on τ.

Keeping these points in mind, let us write the interaction term of (14.20) for $i = 1$ without $\hat{N}$ and rearrange it using (8.29), (8.30), (8.34), and (8.35),

$$
\begin{aligned}
&\int \mathrm{d}\xi_1' \mathscr{V}(|\mathbf{r}_1-\mathbf{r}_1'|)\langle \hat{T}_\tau \hat{\psi}_2(1')\hat{\psi}_1(1')\hat{\psi}_1(1)\hat{\psi}_{3-j}(2)\rangle \\
&\approx \int \mathrm{d}\xi_1' \mathscr{V}(|\mathbf{r}_1-\mathbf{r}_1'|)\Big[\langle \hat{\psi}_2(1')\hat{\psi}_1(1')\rangle\langle \hat{T}_\tau \hat{\psi}_1(1)\hat{\psi}_{3-j}(2)\rangle \\
&\quad - \langle \hat{\psi}_2(1')\hat{\psi}_1(1)\rangle\langle \hat{T}_\tau \hat{\psi}_1(1')\hat{\psi}_{3-j}(2)\rangle + \langle \hat{\psi}_1(1')\hat{\psi}_1(1)\rangle\langle \hat{T}_\tau \hat{\psi}_2(1')\hat{\psi}_{3-j}(2)\rangle\Big] \\
&= \int \mathrm{d}\xi_1' \mathscr{V}(|\mathbf{r}_1-\mathbf{r}_1'|)\Big[-\rho^{(1)}(\xi_1',\xi_1')G_{1j}(1,2) + \rho^{(1)}(\xi_1,\xi_1')G_{1j}(1',2) \\
&\quad + \tilde{\rho}^{(1)}(\xi_1,\xi_1')G_{2j}(1',2)\Big] \\
&= -\int \mathrm{d}\xi_1' \Big[\mathscr{U}_{\mathrm{HF}}(\xi_1,\xi_1')G_{1j}(1',2) + \Delta(\xi_1,\xi_1')G_{2j}(1',2)\Big].
\end{aligned}
$$

Substituting this result into (14.20) for $i = 1$ gives

$$
\begin{aligned}
&\left(-\frac{\partial}{\partial \tau_1} - \hat{\mathscr{K}}_1\right) G_{1j}(1,2) - \int \mathrm{d}\xi_1' \left[\mathscr{U}_{\mathrm{HF}}(\xi_1,\xi_1')G_{1j}(1',2) + \Delta(\xi_1,\xi_1')G_{2j}(1',2)\right] \\
&\quad = \delta_{1j}\delta(1,2). \qquad (14.21)
\end{aligned}
$$

The interaction term of (14.20) for $i = 2$ similarly becomes

$$
\begin{aligned}
&\int \mathrm{d}\xi_1' \mathscr{V}(|\mathbf{r}_1-\mathbf{r}_1'|)\langle \hat{T}_\tau \hat{\psi}_2(1)\hat{\psi}_2(1')\hat{\psi}_1(1')\hat{\psi}_{3-j}(2)\rangle \\
&\approx \int \mathrm{d}\xi_1' \mathscr{V}(|\mathbf{r}_1-\mathbf{r}_1'|)\Big[\langle \hat{\psi}_2(1')\hat{\psi}_1(1')\rangle\langle \hat{T}_\tau \hat{\psi}_2(1)\hat{\psi}_{3-j}(2)\rangle \\
&\quad - \langle \hat{\psi}_2(1)\hat{\psi}_1(1')\rangle\langle \hat{T}_\tau \hat{\psi}_2(1')\hat{\psi}_{3-j}(2)\rangle + \langle \hat{\psi}_2(1)\hat{\psi}_2(1')\rangle\langle \hat{T}_\tau \hat{\psi}_1(1')\hat{\psi}_{3-j}(2)\rangle\Big] \\
&= \int \mathrm{d}\xi_1' \mathscr{V}(|\mathbf{r}_1-\mathbf{r}_1'|)\Big[\langle \hat{\psi}_2(1')\hat{\psi}_1(1')\rangle^*\langle \hat{T}_\tau \hat{\psi}_2(1)\hat{\psi}_{3-j}(2)\rangle \\
&\quad - \langle \hat{\psi}_2(1')\hat{\psi}_1(1)\rangle^*\langle \hat{T}_\tau \hat{\psi}_2(1')\hat{\psi}_{3-j}(2)\rangle - \langle \hat{\psi}_1(1)\hat{\psi}_1(1')\rangle^*\langle \hat{T}_\tau \hat{\psi}_1(1')\hat{\psi}_{3-j}(2)\rangle\Big] \\
&= \int \mathrm{d}\xi_1' \mathscr{V}(|\mathbf{r}_1-\mathbf{r}_1'|)\Big[-\rho^{(1)*}(\xi_1',\xi_1')G_{2j}(1,2) + \rho^{(1)*}(\xi_1,\xi_1')G_{2j}(1',2) \\
&\quad + \tilde{\rho}^{(1)*}(\xi_1,\xi_1')G_{1j}(1',2)\Big] \\
&= -\int \mathrm{d}\xi_1' \Big[\mathscr{U}^*_{\mathrm{HF}}(\xi_1,\xi_1')G_{2j}(1',2) + \Delta^*(\xi_1,\xi_1')G_{1j}(1',2)\Big].
\end{aligned}
$$

Substituting into (14.20) for $i = 2$, we obtain

$$\left(-\frac{\partial}{\partial \tau_1}+\hat{\mathscr{K}}_1^*\right) G_{2j}(1,2)+\int \mathrm{d}\xi_1' \left[\mathscr{U}_{\mathrm{HF}}^*(\xi_1,\xi_1')G_{2j}(1',2)+\Delta^*(\xi_1,\xi_1')G_{1j}(1',2)\right]$$
$$= \delta_{2j}\delta(1,2). \tag{14.22}$$

Next, we expand $\delta(\tau_1 - \tau_2)$ and $G_{ij}(1,2)$ as (14.7) and (14.8). The corresponding coupled equations for the Fourier coefficients can be calculated easily. Indeed, they are derivable from (14.21) and (14.22) using the three replacements, $G_{ij}(1,2) \to G_{ij}(\xi_1,\xi_2;\varepsilon_n)$, $-\partial/\partial\tau_1 \to \mathrm{i}\varepsilon_n$, and $\delta(1,2) \to \delta(\xi_1,\xi_2)$. To express them concisely, let us introduce matrices in the particle-hole space, i.e., the *Nambu matrces*[4]:

$$\hat{G}(\xi_1,\xi_2;\varepsilon_n) \equiv \begin{bmatrix} G_{11}(\xi_1,\xi_2;\varepsilon_n) & G_{12}(\xi_1,\xi_2;\varepsilon_n) \\ G_{21}(\xi_1,\xi_2;\varepsilon_n) & G_{22}(\xi_1,\xi_2;\varepsilon_n) \end{bmatrix}, \tag{14.23}$$

$$\hat{\mathscr{U}}_{\mathrm{BdG}}(\xi_1,\xi_2) \equiv \begin{bmatrix} \mathscr{U}_{\mathrm{HF}}(\xi_1,\xi_2) & \Delta(\xi_1,\xi_2) \\ -\Delta^*(\xi_1,\xi_2) & -\mathscr{U}_{\mathrm{HF}}^*(\xi_1,\xi_2) \end{bmatrix}, \tag{14.24}$$

$$\hat{\delta}(\xi_1,\xi_2) \equiv \begin{bmatrix} \delta(\xi_1,\xi_2) & 0 \\ 0 & \delta(\xi_1,\xi_2) \end{bmatrix}. \tag{14.25}$$

Using (14.23)–(14.25), we can combine all four equations into a single matrix equation:

$$\begin{bmatrix} \mathrm{i}\varepsilon_n - \hat{\mathscr{K}}_1 & 0 \\ 0 & \mathrm{i}\varepsilon_n + \hat{\mathscr{K}}_1^* \end{bmatrix} \hat{G}(\xi_1,\xi_2;\varepsilon_n) - \int \mathrm{d}\xi_3 \hat{\mathscr{U}}_{\mathrm{BdG}}(\xi_1,\xi_3)\hat{G}(\xi_3,\xi_2;\varepsilon_n) = \hat{\delta}(\xi_1,\xi_2). \tag{14.26}$$

This is called the *Gor'kov equations* [6]. More specifically, it is an extension of the equations that Gor'kov derived for s-wave pairing to an arbitrary pairing symmetry expressed in the concise representational form of Nambu [17]. The Gor'kov equations in the mean-field approximation are equivalent in content to the BdG equations. This may be seen by observing that the matrices that operate on the Nambu Green's function in (14.26) are identical as a whole for $\mathrm{i}\varepsilon_n \to 0$ with the matrix in (8.38) to be diagonalized.

[4]Like operators, we denote the Nambu matrices in the particle-hole space also by a caret ^ above each, but they are easily distinguished from operators by context.

14.2.3 Matrix Representation of Spin Variables

For later purposes, we shall express the spin degrees of freedom explicitly as matrices.

Let us introduce a new notation for each of G_{ij} as $G_{11} \to G$, $G_{12} \to F$, $G_{21} \to -\bar{F}$, and $G_{22} \to -\bar{G}$. We also separate the spin variable $\alpha = \uparrow, \downarrow$ from $\xi = \mathbf{r}\alpha$ to write the four new functions as

$$\begin{cases} G_{11}(\xi_1, \xi_2; \varepsilon_n) = G_{\alpha_1\alpha_2}(\mathbf{r}_1, \mathbf{r}_2; \varepsilon_n) \\ G_{12}(\xi_1, \xi_2; \varepsilon_n) = F_{\alpha_1\alpha_2}(\mathbf{r}_1, \mathbf{r}_2; \varepsilon_n) \\ G_{21}(\xi_1, \xi_2; \varepsilon_n) = -\bar{F}_{\alpha_1\alpha_2}(\mathbf{r}_1, \mathbf{r}_2; \varepsilon_n) \\ G_{22}(\xi_1, \xi_2; \varepsilon_n) = -\bar{G}_{\alpha_1\alpha_2}(\mathbf{r}_1, \mathbf{r}_2; \varepsilon_n) \end{cases}. \tag{14.27}$$

Subsequently, the spin degrees of freedom is made evident by constructing the 2×2 matrix

$$\underline{G}(\mathbf{r}_1, \mathbf{r}_2; \varepsilon_n) \equiv \begin{bmatrix} G_{\uparrow\uparrow}(\mathbf{r}_1, \mathbf{r}_2; \varepsilon_n) & G_{\uparrow\downarrow}(\mathbf{r}_1, \mathbf{r}_2; \varepsilon_n) \\ G_{\downarrow\uparrow}(\mathbf{r}_1, \mathbf{r}_2; \varepsilon_n) & G_{\downarrow\downarrow}(\mathbf{r}_1, \mathbf{r}_2; \varepsilon_n) \end{bmatrix}. \tag{14.28}$$

In the matrix notation, symmetry (14.12) now reads

$$\underline{G}(\mathbf{r}_1, \mathbf{r}_2; \varepsilon_n) = \underline{G}^{\dagger}(\mathbf{r}_2, \mathbf{r}_1; -\varepsilon_n) = \underline{\bar{G}}^{\mathrm{T}}(\mathbf{r}_2, \mathbf{r}_1; -\varepsilon_n), \tag{14.29}$$

$$\underline{F}(\mathbf{r}_1, \mathbf{r}_2; \varepsilon_n) = -\underline{\bar{F}}^{\dagger}(\mathbf{r}_2, \mathbf{r}_1; -\varepsilon_n) = -\underline{F}^{\mathrm{T}}(\mathbf{r}_2, \mathbf{r}_1; -\varepsilon_n), \tag{14.30}$$

where † and ^{T} denote Hermitian conjugate and transpose, respectively. From these symmetry relations, $\underline{\bar{G}}(\mathbf{r}_1, \mathbf{r}_2; \varepsilon_n) = \underline{G}^*(\mathbf{r}_1, \mathbf{r}_2; \varepsilon_n)$ and $\underline{\bar{F}}(\mathbf{r}_1, \mathbf{r}_2; \varepsilon_n) = \underline{F}^*(\mathbf{r}_1, \mathbf{r}_2; \varepsilon_n)$.

The Nambu matrix (14.23) in this notation can be expressed as a 4×4 matrix,

$$\hat{G}(\mathbf{r}_1, \mathbf{r}_2; \varepsilon_n) = \begin{bmatrix} \underline{G}(\mathbf{r}_1, \mathbf{r}_2; \varepsilon_n) & \underline{F}(\mathbf{r}_1, \mathbf{r}_2; \varepsilon_n) \\ -\underline{F}^*(\mathbf{r}_1, \mathbf{r}_2; \varepsilon_n) & -\underline{G}^*(\mathbf{r}_1, \mathbf{r}_2; \varepsilon_n) \end{bmatrix}. \tag{14.31}$$

Expressing (14.24) and (14.25) similarly, we have

$$\hat{\mathscr{U}}_{\mathrm{BdG}}(\mathbf{r}_1, \mathbf{r}_2) \equiv \begin{bmatrix} \underline{\mathscr{U}}_{\mathrm{HF}}(\mathbf{r}_1, \mathbf{r}_2) & \underline{\Delta}(\mathbf{r}_1, \mathbf{r}_2) \\ -\underline{\Delta}^*(\mathbf{r}_1, \mathbf{r}_2) & -\underline{\mathscr{U}}^*_{\mathrm{HF}}(\mathbf{r}_1, \mathbf{r}_2) \end{bmatrix}, \tag{14.32}$$

$$\hat{\delta}(\mathbf{r}_1, \mathbf{r}_2) \equiv \begin{bmatrix} \delta(\mathbf{r}_1, \mathbf{r}_2)\underline{\sigma}_0 & \underline{0} \\ \underline{0} & \delta(\mathbf{r}_1, \mathbf{r}_2)\underline{\sigma}_0 \end{bmatrix}. \tag{14.33}$$

where $\underline{\sigma}_0$ and $\underline{0}$ are the 2×2 unit and zero matrices, respectively. Using them, we can rewrite (14.26) as

$$\begin{bmatrix} (\mathrm{i}\varepsilon_n - \hat{\mathscr{K}}_1)\underline{\sigma}_0 & \underline{0} \\ \underline{0} & (\mathrm{i}\varepsilon_n + \hat{\mathscr{K}}_1^*)\underline{\sigma}_0 \end{bmatrix} \hat{G}(\mathbf{r}_1, \mathbf{r}_2; \varepsilon_n) - \int \mathrm{d}^3 r_3\, \hat{\mathscr{U}}_{\mathrm{BdG}}(\mathbf{r}_1, \mathbf{r}_3)\hat{G}(\mathbf{r}_3, \mathbf{r}_2; \varepsilon_n)$$
$$= \hat{\delta}(\mathbf{r}_1, \mathbf{r}_2). \tag{14.34}$$

14.2.4 Gauge Invariance

Equation (14.34) has an important property called *gauge invariance*.

Let us introduce the gauge transformation in terms of a continuously differentiable function $\chi(\mathbf{r})$ by

$$\begin{cases} \mathbf{A}(\mathbf{r}_1) = \mathbf{A}'(\mathbf{r}_1) + \boldsymbol{\nabla}_1 \chi(\mathbf{r}_1) \\ \hat{\psi}_1(1) = \hat{\psi}_1'(1)\mathrm{e}^{\mathrm{i}e\chi(\mathbf{r}_1)/\hbar} \\ \hat{\psi}_2(1) = \hat{\psi}_2'(1)\mathrm{e}^{-\mathrm{i}e\chi(\mathbf{r}_1)/\hbar} \end{cases}, \tag{14.35}$$

where a prime $'$ distinguishes f' from f as a different function. The corresponding variations of Green's function (14.31) and potential (14.32) are expressible in terms of the matrix

$$\hat{\Theta}(\mathbf{r}_1) \equiv \begin{bmatrix} \underline{\sigma}_0 \mathrm{e}^{\mathrm{i}e\chi(\mathbf{r}_1)/\hbar} & \underline{0} \\ \underline{0} & \underline{\sigma}_0 \mathrm{e}^{-\mathrm{i}e\chi(\mathbf{r}_1)/\hbar} \end{bmatrix} \tag{14.36}$$

as

$$\hat{G}(\mathbf{r}_1, \mathbf{r}_2; \varepsilon_n) = \hat{\Theta}(\mathbf{r}_1)\hat{G}'(\mathbf{r}_1, \mathbf{r}_2; \varepsilon_n)\hat{\Theta}^*(\mathbf{r}_2), \tag{14.37}$$

$$\hat{\mathscr{U}}_{\mathrm{BdG}}(\mathbf{r}_1, \mathbf{r}_2) = \hat{\Theta}(\mathbf{r}_1)\hat{\mathscr{U}}'_{\mathrm{BdG}}(\mathbf{r}_1, \mathbf{r}_2)\hat{\Theta}^*(\mathbf{r}_2). \tag{14.38}$$

Property (14.38) follows by recalling (8.29), (8.30), (8.34), and (8.35). Moreover, operation $(\mp \mathrm{i}\hbar \boldsymbol{\nabla}_1 - e\mathbf{A}_1)^2$ on $\mathrm{e}^{\pm \mathrm{i}e\chi_1/\hbar}$ yields

$$\begin{aligned}(\mp \mathrm{i}\hbar \boldsymbol{\nabla}_1 - e\mathbf{A}_1)^2 \mathrm{e}^{\pm \mathrm{i}e\chi_1/\hbar} &= \mathrm{e}^{\pm \mathrm{i}e\chi_1/\hbar}\left[\mp \mathrm{i}\hbar \boldsymbol{\nabla}_1 - e(\mathbf{A}_1 - \boldsymbol{\nabla}_1\chi_1)\right]^2 \\ &= \mathrm{e}^{\pm \mathrm{i}e\chi_1/\hbar}\left(\mp \mathrm{i}\hbar \boldsymbol{\nabla}_1 - e\mathbf{A}_1'\right)^2,\end{aligned}$$

so that

$$\begin{bmatrix} -\hat{\mathscr{K}}_1\underline{\sigma}_0 & \underline{0} \\ \underline{0} & \hat{\mathscr{K}}_1^*\underline{\sigma}_0 \end{bmatrix} \hat{\Theta}(\mathbf{r}_1) = \hat{\Theta}(\mathbf{r}_1) \begin{bmatrix} -\hat{\mathscr{K}}_1'\underline{\sigma}_0 & \underline{0} \\ \underline{0} & \hat{\mathscr{K}}_1'^*\underline{\sigma}_0 \end{bmatrix} \tag{14.39}$$

holds for the $\hat{\mathscr{K}}_1$ term of (14.34). Let us substitute (14.37) and (14.38) into (14.34), then use (14.39), and multiply the resulting equation by $\hat{\Theta}^*(\mathbf{r}_1)$ and $\hat{\Theta}(\mathbf{r}_2)$ from the left and right, respectively. We then realize that the resulting equation in terms of $\mathbf{A}'$, $\hat{G}'(\mathbf{r}_1, \mathbf{r}_2; \varepsilon_n)$, and $\hat{\mathscr{U}}'_{\mathrm{BdG}}(\mathbf{r}_1, \mathbf{r}_2)$ is identical in form to (14.34). This is *gauge invariance*, implying that there is an arbitrariness in the choice of vector potential. Incidentally, we sometimes encounter a phrase like "spontaneously broken gauge symmetry" in superfluids and superconductors [13]. What is meant by this is that superconductivity is a state described by a macroscopic wave function with a fixed phase. It should be emphasized that gauge invariance is maintained as it must.

14.2.5 Gauge-Covariant Wigner Transform

The Wigner transform was introduced by Wigner in 1932 to study quantum corrections to classical statistical mechanics [20]. It enables us to define a quasi-probability distribution in terms of coordinates and momenta quantum mechanically. It has also been useful in formulating quantum mechanics in phase space and elucidating its connection with classical mechanics [7, 15]. Moreover, the transform forms a tool indispensable for deriving the quasiclassical equations of superconductivity.

The original *Wigner transform* may be defined, for example, in terms of the Nambu matrix (14.31) as follows: Let us introduce the "center-of-mass" and "relative" coordinates as

$$\mathbf{r}_{12} \equiv \frac{\mathbf{r}_1 + \mathbf{r}_2}{2}, \qquad \bar{\mathbf{r}}_{12} \equiv \mathbf{r}_1 - \mathbf{r}_2. \tag{14.40}$$

The Wigner transform is defined as the Fourier transform with respect to the relative coordinates,

$$\hat{G}(\varepsilon_n, \mathbf{k}, \mathbf{r}_{12}) = \int \mathrm{d}^3 r\, \mathrm{e}^{-\mathrm{i}\mathbf{k}\cdot\bar{\mathbf{r}}_{12}}\, \hat{G}(\mathbf{r}_1, \mathbf{r}_2; \varepsilon_n),$$

where the $\hat{G}$'s on both sides are different functions distinguished by their arguments. There is no $\mathbf{r}_{12}$ dependence for homogeneous systems. Also for inhomogeneous systems with slow variations, we may expect that the first few terms of the *gradient expansion*, i.e., the expansion of $\hat{G}(\varepsilon_n, \mathbf{k}, \mathbf{r}_{12})$ in terms of gradients in $\mathbf{r}_{12}$, suffice to describe them quantitatively.

However, we encounter a fundamental difficulty when the Wigner transform is applied to charged systems. More specifically, the definition above breaks the gauge invariance with respect to the center-of-mass coordinates. This may be realized by noting that the gauge transformation (14.37) depends on both $\mathbf{r}_1$ and $\mathbf{r}_2$ instead of $(\mathbf{r}_1 + \mathbf{r}_2)/2$, so that the simple Fourier transform with respect to $\mathbf{r}_1 - \mathbf{r}_2$ necessarily breaks the gauge invariance. To remove this difficulty, Stratnovich introduced a modified Wigner transform that may be called the *gauge-invariant Wigner transform* [19]. However, the method is valid only for normal systems with $\underline{G}_{12} = \underline{G}_{21} = \underline{0}$. Here, we apply an extended version for describing superconductors [9].

Let us introduce the line integral:

$$I(\mathbf{r}_1, \mathbf{r}_2) \equiv \frac{e}{\hbar} \int_{\mathbf{r}_2}^{\mathbf{r}_1} \mathbf{A}(\mathbf{s}) \cdot d\mathbf{s}, \tag{14.41}$$

where $\mathbf{s}$ denotes a straight-line path from $\mathbf{r}_2$ to $\mathbf{r}_1$. As may be confirmed easily, factor $e^{iI(\mathbf{r}_1,\mathbf{r}_2)}$ is transformed under the gauge transformation (14.35) as

$$e^{iI(\mathbf{r}_1,\mathbf{r}_2)} = e^{ie\chi(\mathbf{r}_1)/\hbar} e^{iI'(\mathbf{r}_1,\mathbf{r}_2)} e^{-ie\chi(\mathbf{r}_2)/\hbar}. \tag{14.42}$$

Using this factor, we define the matrix:

$$\hat{\Gamma}(\mathbf{r}_1, \mathbf{r}_2) \equiv \begin{bmatrix} \underline{\sigma}_0 e^{iI(\mathbf{r}_1,\mathbf{r}_2)} & \underline{0} \\ \underline{0} & \underline{\sigma}_0 e^{-iI(\mathbf{r}_1,\mathbf{r}_2)} \end{bmatrix}, \tag{14.43}$$

for which the variation under the gauge transformation can be expressed in terms of matrix (14.36),

$$\hat{\Gamma}(\mathbf{r}_1, \mathbf{r}_2) = \hat{\Theta}(\mathbf{r}_1)\hat{\Gamma}'(\mathbf{r}_1, \mathbf{r}_2)\hat{\Theta}^*(\mathbf{r}_2). \tag{14.44}$$

With these preliminaries, we introduce the *gauge-covariant Wigner transform* for (14.31) by

$$\begin{aligned} \hat{G}(\varepsilon_n, \mathbf{k}, \mathbf{r}_{12}) &\equiv \int d^3 \bar{r}\, e^{-i\mathbf{k}\cdot\bar{\mathbf{r}}_{12}}\, \hat{\Gamma}(\mathbf{r}_{12}, \mathbf{r}_1)\hat{G}(\mathbf{r}_1, \mathbf{r}_2; \varepsilon_n)\hat{\Gamma}(\mathbf{r}_2, \mathbf{r}_{12}) \\ &\equiv \begin{bmatrix} \underline{G}(\varepsilon_n, \mathbf{k}, \mathbf{r}_{12}) & \underline{F}(\varepsilon_n, \mathbf{k}, \mathbf{r}_{12}) \\ -\underline{F}^*(\varepsilon_n, -\mathbf{k}, \mathbf{r}_{12}) & -\underline{G}^*(\varepsilon_n, -\mathbf{k}, \mathbf{r}_{12}) \end{bmatrix}, \end{aligned} \tag{14.45}$$

with inverse relation

$$\hat{G}(\mathbf{r}_1, \mathbf{r}_2; \varepsilon_n) = \hat{\Gamma}(\mathbf{r}_1, \mathbf{r}_{12})\frac{1}{V}\sum_{\mathbf{k}} \hat{G}(\varepsilon_n, \mathbf{k}, \mathbf{r}_{12}) e^{i\mathbf{k}\cdot\bar{\mathbf{r}}_{12}}\, \hat{\Gamma}(\mathbf{r}_{12}, \mathbf{r}_2). \tag{14.46}$$

Using (14.37) and (14.44), it follows easily that $\hat{G}(\varepsilon_n, \mathbf{k}, \mathbf{r}_{12})$ changes under the transformation (14.35),

$$\hat{G}(\varepsilon_n, \mathbf{k}, \mathbf{r}_{12}) = \hat{\Theta}(\mathbf{r}_{12})\hat{G}'(\varepsilon_n, \mathbf{k}, \mathbf{r}_{12})\hat{\Theta}^*(\mathbf{r}_{12}). \tag{14.47}$$

Thus, only the center-of-mass coordinate is of relevance to the variation of $\hat{G}(\varepsilon_n, \mathbf{k}, \mathbf{r})$ under the gauge transformation. Note that the diagonal elements in

(14.47) are gauge-invariant as in Stratnovich's transformation [19], whereas the off-diagonal elements characteristic of superconductivity acquire extra phases $e^{\pm 2ie\chi(\mathbf{r}_{12})}$.

Similarly, we transform the mean-field potential (14.32),

$$\hat{\mathscr{U}}_{\rm BdG}(\mathbf{k},\mathbf{r}_{12}) \equiv \int d^3 r\, e^{-i\mathbf{k}\cdot\bar{\mathbf{r}}_{12}}\, \hat{\Gamma}(\mathbf{r}_{12},\mathbf{r}_1)\hat{\mathscr{U}}_{\rm BdG}(\mathbf{r}_1,\mathbf{r}_2)\hat{\Gamma}(\mathbf{r}_2,\mathbf{r}_{12})$$
$$\equiv \begin{bmatrix} \underline{\mathscr{U}}_{\rm HF}(\mathbf{k},\mathbf{r}_{12}) & \underline{\Delta}(\mathbf{k},\mathbf{r}_{12}) \\ -\underline{\Delta}^*(-\mathbf{k},\mathbf{r}_{12}) & -\underline{\mathscr{U}}^*_{\rm HF}(-\mathbf{k},\mathbf{r}_{12}) \end{bmatrix}, \tag{14.48}$$

whose inverse reads

$$\hat{\mathscr{U}}_{\rm BdG}(\mathbf{r}_1,\mathbf{r}_2) = \hat{\Gamma}(\mathbf{r}_1,\mathbf{r}_{12})\frac{1}{V}\sum_{\mathbf{k}} \hat{\mathscr{U}}_{\rm BdG}(\mathbf{k},\mathbf{r}_{12})e^{i\mathbf{k}\cdot\bar{\mathbf{r}}_{12}}\,\hat{\Gamma}(\mathbf{r}_{12},\mathbf{r}_2). \tag{14.49}$$

Note $\underline{\mathscr{U}}_{\rm HF}(\mathbf{k},\mathbf{r}) = \underline{\mathscr{U}}^{\dagger}_{\rm HF}(\mathbf{k},\mathbf{r})$ and $\underline{\Delta}(\mathbf{k},\mathbf{r}) = -\underline{\Delta}^{\rm T}(-\mathbf{k},\mathbf{r})$ due to (8.61) and (8.62).

14.3 Eilenberger Equations

Although equivalent to the BdG equations, the Gor'kov equations for Green's functions with two arguments may be more difficult to resolve. However, they provide a convenient starting point for simplifying the equations. Passing to the Wigner representation, we shall integrate out an irrelevant variable from the Gor'kov equations to obtain the quasiclassical Eilenberger equations [4].

14.3.1 Quasiclassical Green's Function

We introduce the quasiclassical Green's function by (14.59) below and obtain (14.61) and (14.62) it obeys. Our derivation here is from Larkin and Ovchinnikov [12] instead of the ingenious original [4]. For clarity, we only consider the weak-coupling case setting $\underline{\mathscr{U}}_{\rm HF} \to \underline{0}$ in (14.48).

First, let us simplify the kinetic-energy terms of the Gor'kov equation (14.34) in the Wigner representation (14.46). To this end, we rewrite the kinetic-energy operator (14.14) in terms of the center-of-mass and relative coordinates in (14.40),

$$\hat{\mathscr{K}}_1 = \frac{1}{2m}\left[-i\hbar\frac{\partial}{\partial\bar{\mathbf{r}}_{12}} - \frac{1}{2}i\hbar\frac{\partial}{\partial\mathbf{r}_{12}} - e\mathbf{A}(\mathbf{r}_{12}+\bar{\mathbf{r}}_{12}/2)\right]^2 - \mu. \tag{14.50}$$

Next, we approximate the vector potential as $\mathbf{A}(\mathbf{r}_{12}+\bar{\mathbf{r}}_{12}/2)\approx\mathbf{A}(\mathbf{r}_{12})$. As for the phase factor $\hat{\Gamma}(\mathbf{r}_{12},\mathbf{r}_j)$ ($j=1,2$) defined by (14.41) and (14.43), we expand $\mathbf{A}(\mathbf{s})$ in $I(\mathbf{r}_{12},\mathbf{r}_j)$ at $\mathbf{s}=\mathbf{r}_{12}$ and retain only the leading term,

$$I(\mathbf{r}_{12},\mathbf{r}_j)\approx\frac{e}{\hbar}\big(\mathbf{r}_{12}-\mathbf{r}_j\big)\cdot\mathbf{A}(\mathbf{r}_{12})=(-1)^j\frac{e}{2\hbar}\bar{\mathbf{r}}_{12}\cdot\mathbf{A}(\mathbf{r}_{12}). \tag{14.51}$$

After substituting (14.46), we can thereby expand the kinetic-energy terms in (14.34),

$$\begin{aligned}\hat{\mathscr{K}}_1\underline{G}(\mathbf{r}_1,\mathbf{r}_2;\varepsilon_n)&\approx e^{iI(\mathbf{r}_1,\mathbf{r}_{12})+iI(\mathbf{r}_{12},\mathbf{r}_2)}\frac{1}{V}\sum_{\mathbf{k}}\left\{\frac{1}{2m}\left[-i\hbar\frac{\partial}{\partial\bar{\mathbf{r}}_{12}}-\frac{i\hbar}{2}\frac{\partial}{\partial\mathbf{r}_{12}}-e\mathbf{A}(\mathbf{r}_{12})\right.\right.\\&\left.\left.\quad+\hbar\frac{\partial I(\mathbf{r}_1,\mathbf{r}_{12})}{\partial\bar{\mathbf{r}}_{12}}+\hbar\frac{\partial I(\mathbf{r}_{12},\mathbf{r}_2)}{\partial\bar{\mathbf{r}}_{12}}\right]^2-\mu\right\}\underline{G}(\varepsilon_n,\mathbf{k},\mathbf{r}_{12})e^{i\mathbf{k}\cdot\bar{\mathbf{r}}_{12}}\\&=e^{iI(\mathbf{r}_1,\mathbf{r}_{12})+iI(\mathbf{r}_{12},\mathbf{r}_2)}\frac{1}{V}\sum_{\mathbf{k}}\left[\frac{1}{2m}\left(\hbar\mathbf{k}-\frac{i\hbar}{2}\frac{\partial}{\partial\mathbf{r}_{12}}\right)^2-\mu\right]\\&\quad\times\underline{G}(\varepsilon_n,\mathbf{k},\mathbf{r}_{12})e^{i\mathbf{k}\cdot\bar{\mathbf{r}}_{12}},\end{aligned} \tag{14.52}$$

$$\begin{aligned}\hat{\mathscr{K}}_1\underline{F}(\mathbf{r}_1,\mathbf{r}_2;\varepsilon_n)&\approx e^{iI(\mathbf{r}_1,\mathbf{r}_{12})-iI(\mathbf{r}_{12},\mathbf{r}_2)}\frac{1}{V}\sum_{\mathbf{k}}\left\{\frac{1}{2m}\left[\hbar\mathbf{k}-\frac{i\hbar}{2}\frac{\partial}{\partial\mathbf{r}_{12}}-e\mathbf{A}(\mathbf{r}_{12})\right]^2\right.\\&\quad\left.-\mu\right\}\underline{F}(\varepsilon_n,\mathbf{k},\mathbf{r}_{12})e^{i\mathbf{k}\cdot\bar{\mathbf{r}}_{12}},\end{aligned} \tag{14.53}$$

$$\begin{aligned}\hat{\mathscr{K}}_1^*\underline{F}^*(\mathbf{r}_1,\mathbf{r}_2;\varepsilon_n)&\approx e^{-iI(\mathbf{r}_1,\mathbf{r}_{12})+iI(\mathbf{r}_{12},\mathbf{r}_2)}\frac{1}{V}\sum_{\mathbf{k}}\left\{\frac{1}{2m}\left[-\hbar\mathbf{k}+\frac{i\hbar}{2}\frac{\partial}{\partial\mathbf{r}_{12}}-e\mathbf{A}(\mathbf{r}_{12})\right]^2\right.\\&\quad\left.-\mu\right\}\underline{F}^*(\varepsilon_n,-\mathbf{k},\mathbf{r}_{12})e^{i\mathbf{k}\cdot\bar{\mathbf{r}}_{12}},\end{aligned} \tag{14.54}$$

$$\begin{aligned}\hat{\mathscr{K}}_1^*\underline{G}^*(\mathbf{r}_1,\mathbf{r}_2;\varepsilon_n)&\approx e^{-iI(\mathbf{r}_1,\mathbf{r}_{12})-iI(\mathbf{r}_{12},\mathbf{r}_2)}\frac{1}{V}\sum_{\mathbf{k}}\left[\frac{1}{2m}\left(-\hbar\mathbf{k}+\frac{i\hbar}{2}\frac{\partial}{\partial\mathbf{r}_{12}}\right)^2-\mu\right]\\&\quad\times\underline{G}^*(\varepsilon_n,-\mathbf{k},\mathbf{r}_{12})e^{i\mathbf{k}\cdot\bar{\mathbf{r}}_{12}}.\end{aligned} \tag{14.55}$$

As for the interaction term in (14.34), we substitute (14.46) and (14.49), approximate $\mathbf{r}_{13},\mathbf{r}_{32}\approx\mathbf{r}_{12}$, and perform the integration over $\mathbf{r}_3$. Further, we expand the delta function in (14.34) as (6.18). Finally, we introduce the operator

$$\boldsymbol{\partial} \equiv \begin{cases} \nabla & \text{:on } G \text{ or } G^* \\ \nabla - \mathrm{i}\dfrac{2e}{\hbar}\mathbf{A}(\mathbf{r}) & \text{:on } F \\ \nabla + \mathrm{i}\dfrac{2e}{\hbar}\mathbf{A}(\mathbf{r}) & \text{:on } F^* \end{cases} \tag{14.56}$$

to express (14.52)–(14.55) concisely, and neglect second and higher-order terms in $\boldsymbol{\partial}$, which is justified considering that the scale of the spatial variation is much longer than k_{F}^{-1}. Equation (14.34) is thereby transformed into

$$\begin{bmatrix} \left(\mathrm{i}\varepsilon_n - \xi_k + \mathrm{i}\dfrac{\hbar^2\mathbf{k}}{2m}\cdot\boldsymbol{\partial}\right)\underline{\sigma}_0 & \underline{0} \\ \underline{0} & \left(\mathrm{i}\varepsilon_n + \xi_k - \mathrm{i}\dfrac{\hbar^2\mathbf{k}}{2m}\cdot\boldsymbol{\partial}\right)\underline{\sigma}_0 \end{bmatrix} \hat{G}(\varepsilon_n, \mathbf{k}, \mathbf{r}) \\ - \hat{\mathscr{U}}_{\mathrm{BdG}}(\mathbf{k}, \mathbf{r})\hat{G}(\varepsilon_n, \mathbf{k}, \mathbf{r}) = \hat{1}, \tag{14.57}$$

where ξ_k is defined by (9.17), and $\hat{1}$ denotes the 4×4 unit matrix.

Next, we take Hermitian conjugate of (14.57), use symmetries $\hat{\mathscr{U}}^{\dagger}_{\mathrm{BdG}}(\mathbf{k}, \mathbf{r}) = \hat{\mathscr{U}}_{\mathrm{BdG}}(\mathbf{k}, \mathbf{r})$ and $\hat{G}^{\dagger}(\varepsilon_n, \mathbf{k}, \mathbf{r}) = \hat{G}(-\varepsilon_n, \mathbf{k}, \mathbf{r})$ that originate from (8.61), (8.62), (14.29), and (14.30), and replace $\varepsilon_n \to -\varepsilon_n$ to obtain

$$\hat{G}(\varepsilon_n, \mathbf{k}, \mathbf{r}) \begin{bmatrix} \left(\mathrm{i}\varepsilon_n - \xi_k - \mathrm{i}\dfrac{\hbar^2\mathbf{k}}{2m}\cdot\boldsymbol{\partial}\right)\underline{\sigma}_0 & \underline{0} \\ \underline{0} & \left(\mathrm{i}\varepsilon_n + \xi_k + \mathrm{i}\dfrac{\hbar^2\mathbf{k}}{2m}\cdot\boldsymbol{\partial}\right)\underline{\sigma}_0 \end{bmatrix} \\ -\hat{G}(\varepsilon_n, \mathbf{k}, \mathbf{r})\hat{\mathscr{U}}_{\mathrm{BdG}}(\mathbf{k}, \mathbf{r}) = \hat{1}, \tag{14.58}$$

where $\boldsymbol{\partial}$ operates on $\hat{G}(\varepsilon_n, \mathbf{k}, \mathbf{r})$. This equation may also be obtained directly from the differential equation of (14.31) with respect to $\mathbf{r}_2$. We refer to (14.57) and (14.58) as the left and right Gor'kov equations in the Wigner representation.

Now, in terms of (14.45), we introduce the quasiclassical Green's function,

$$\hat{g}(\varepsilon_n, \mathbf{k}_{\mathrm{F}}, \mathbf{r}) \equiv \mathrm{P}\int_{-\infty}^{\infty} \frac{d\xi_k}{\pi}\hat{\sigma}_z \mathrm{i}\hat{G}(\varepsilon_n, \mathbf{k}, \mathbf{r}) \\ \equiv \begin{bmatrix} \underline{g}(\varepsilon_n, \mathbf{k}_{\mathrm{F}}, \mathbf{r}) & -\mathrm{i}\underline{f}(\varepsilon_n, \mathbf{k}_{\mathrm{F}}, \mathbf{r}) \\ -\mathrm{i}\underline{f}^*(\varepsilon_n, -\mathbf{k}_{\mathrm{F}}, \mathbf{r}) & -\underline{g}^*(\varepsilon_n, -\mathbf{k}_{\mathrm{F}}, \mathbf{r}) \end{bmatrix}, \tag{14.59}$$

where P denotes the principal value, $\hat{\sigma}_z$ is given in (12.9), and coefficient $-\mathrm{i}$ in front of $\underline{f}$ is introduced for convenience. It follows from (14.29) and (14.30) that the upper elements $\underline{g}$ and $\underline{f}$ satisfy

$$\underline{g}(\varepsilon_n, \mathbf{k}_{\mathrm{F}}, \mathbf{r}) = -\underline{g}^{\dagger}(-\varepsilon_n, \mathbf{k}_{\mathrm{F}}, \mathbf{r}), \qquad \underline{f}(\varepsilon_n, \mathbf{k}_{\mathrm{F}}, \mathbf{r}) = -\underline{f}^{\mathrm{T}}(-\varepsilon_n, -\mathbf{k}_{\mathrm{F}}, \mathbf{r}). \tag{14.60}$$

To derive the equation for $\hat{g}$, we multiply (14.57) by $\hat{\sigma}_z$ from the left, (14.58) by $\hat{\sigma}_z$ from the right, subtract the latter from the former to eliminate ξ_k, and multiply the resulting equation by $\hat{\sigma}_z$ from the left. We thereby obtain

$$\left[\mathrm{i}\varepsilon_n\hat{\sigma}_z - \hat{\mathscr{U}}_{\mathrm{BdG}}(\mathbf{k},\mathbf{r})\hat{\sigma}_z, \hat{\sigma}_z\hat{G}(\varepsilon_n,\mathbf{k},\mathbf{r})\right] + \mathrm{i}\frac{\hbar^2\mathbf{k}}{m}\cdot\boldsymbol{\partial}\,\hat{\sigma}_z\hat{G}(\varepsilon_n,\mathbf{k},\mathbf{r}) = \hat{0},$$

where $[\hat{A},\hat{B}] \equiv \hat{A}\hat{B} - \hat{B}\hat{A}$. Next, we replace $\mathbf{k}$ in $\hat{\mathscr{U}}_{\mathrm{BdG}}(\mathbf{k},\mathbf{r})$ and $\hbar^2\mathbf{k}/m$ by $\mathbf{k}_{\mathrm{F}}$ as appropriate for the weak-coupling case and subsequently eliminate the remaining k dependence in Green's function by performing the principal-value integral over ξ_k. The resulting equation can be written in terms of $\hat{g}$ of (14.59) as

$$\left[\mathrm{i}\varepsilon_n\hat{\sigma}_z - \hat{\mathscr{U}}_{\mathrm{BdG}}(\mathbf{k}_{\mathrm{F}},\mathbf{r})\hat{\sigma}_z, \hat{g}(\varepsilon_n,\mathbf{k}_{\mathrm{F}},\mathbf{r})\right] + \mathrm{i}\hbar\mathbf{v}_{\mathrm{F}}\cdot\boldsymbol{\partial}\hat{g}(\varepsilon_n,\mathbf{k}_{\mathrm{F}},\mathbf{r}) = \hat{0}, \qquad (14.61)$$

where $\mathbf{v}_{\mathrm{F}} \equiv \hbar\mathbf{k}_{\mathrm{F}}/m$ is the Fermi velocity and $\boldsymbol{\partial}$ is defined by (14.56). Hence, we have derived the main part of the Eilenberger equations.

Because the source term $\hat{1}$ in (14.57) and (14.58) has been canceled in the *left-right subtraction trick*, (14.61) has an arbitrariness about the amplitude of $\hat{g}$, which is removed by Eilenberger's normalization condition,

$$\left[\hat{g}(\varepsilon_n,\mathbf{k}_{\mathrm{F}},\mathbf{r})\right]^2 = \hat{1}. \qquad (14.62)$$

Equation (14.62) may be derived as follows. Operating $\mathrm{i}\hbar\mathbf{v}_{\mathrm{F}}\cdot\boldsymbol{\partial}$ on $\hat{g}^2$ and using (14.61), the resulting equation becomes

$$\begin{aligned}\mathrm{i}\hbar\mathbf{v}_{\mathrm{F}}\cdot\boldsymbol{\partial}\hat{g}^2 &= (\mathrm{i}\hbar\mathbf{v}_{\mathrm{F}}\cdot\boldsymbol{\partial}\hat{g})\hat{g} + \hat{g}(\mathrm{i}\hbar\mathbf{v}_{\mathrm{F}})\cdot\boldsymbol{\partial}\hat{g}\\ &= -\left[\mathrm{i}\varepsilon_n\hat{\sigma}_z - \hat{\mathscr{U}}_{\mathrm{BdG}}\hat{\sigma}_z, \hat{g}\right]\hat{g} - \hat{g}\left[\mathrm{i}\varepsilon_n\hat{\sigma}_z - \hat{\mathscr{U}}_{\mathrm{BdG}}\hat{\sigma}_z, \hat{g}\right]\\ &= -\left[\mathrm{i}\varepsilon_n\hat{\sigma}_z - \hat{\mathscr{U}}_{\mathrm{BdG}}\hat{\sigma}_z, \hat{g}^2\right].\end{aligned}$$

From this differential equation, when $\hat{g}^2 = \hat{1}$ holds at a certain point, the right-hand side vanishes giving $\mathrm{i}\hbar\mathbf{v}_{\mathrm{F}}\cdot\boldsymbol{\partial}\hat{g}^2 = \hat{0}$. Integrating $\mathrm{i}\hbar\mathbf{v}_{\mathrm{F}}\cdot\boldsymbol{\partial}\hat{g}^2 = \hat{0}$ with the initial condition $\hat{g}^2 = \hat{1}$, we conclude that $\hat{g}^2 = \hat{1}$ everywhere in the system.

Condition $\hat{g}^2 = \hat{1}$ does hold for homogeneous s-wave pairing. To see this, we substitute (14.48) with (9.1) and $\underline{\mathscr{U}}_{\mathrm{HF}} \to \underline{0}$ into (14.57) and set $\mathrm{i}(\hbar^2\mathbf{k}/2m)\cdot\boldsymbol{\partial} \to 0$. The resulting Gor'kov equation can be solved easily to obtain the homogeneous Green's function,

$$\hat{G}(\varepsilon_n,\mathbf{k}) = \frac{-1}{\varepsilon_n^2 + \xi_k^2 + |\Delta_k|^2}\begin{bmatrix} \mathrm{i}\varepsilon_n + \xi_k & 0 & 0 & \Delta_k \\ 0 & \mathrm{i}\varepsilon_n + \xi_k & -\Delta_k & 0 \\ 0 & -\Delta_k^* & \mathrm{i}\varepsilon_n - \xi_k & 0 \\ \Delta_k^* & 0 & 0 & \mathrm{i}\varepsilon_n - \xi_k \end{bmatrix}.$$

Note the correspondence with the eigenvalue problem (9.3). Substituting this into (14.59) and performing the integral, we obtain the quasiclassical Green's function for homogenous s-wave pairing,

$$\hat{g}(\varepsilon_n, \mathbf{k}_\mathrm{F}) = \frac{1}{\sqrt{\varepsilon_n^2 + |\Delta_k|^2}} \begin{bmatrix} \varepsilon_n & 0 & 0 & -\mathrm{i}\Delta_k \\ 0 & \varepsilon_n & \mathrm{i}\Delta_k & 0 \\ 0 & -\mathrm{i}\Delta_k^* & -\varepsilon_n & 0 \\ \mathrm{i}\Delta_k^* & 0 & 0 & -\varepsilon_n \end{bmatrix}, \tag{14.63}$$

which obeys (14.62) as is easily verified. Noting that an arbitrary $\hat{g}(\varepsilon_n, \mathbf{k}_\mathrm{F}, \mathbf{r})$ can be produced from (14.63) through a gradual variation in space, we may conclude that (14.62) holds true for s-wave pairing. The same argument may be applied for other pairing symmetries to confirm (14.62).

14.3.2 Pair Potential

Next, we rewrite the self-consistency equation for the pair potential using the quasiclassical Green's function as (14.68) below.

Function $\tilde{\rho}^{(1)}$ in (8.62) is defined by (8.30), which is expressible in terms of the Matsubara Green's function (14.3),

$$\tilde{\rho}^{(1)}_{\alpha_1\alpha_2}(\mathbf{r}_1, \mathbf{r}_2) = -G_{12}(\xi_1, \xi_2; 0) = -\frac{1}{\beta}\sum_{n=-\infty}^{\infty} F_{\alpha_1\alpha_2}(\mathbf{r}_1, \mathbf{r}_2; \varepsilon_n),$$

where we have used (14.4), (14.8), and (14.27). Using it together with (6.18), (14.46), (14.49), we rewrite (8.62) as

$$\underline{\Delta}(\mathbf{k}, \mathbf{r}) = \frac{1}{V}\sum_{\mathbf{k}'} \mathscr{V}_{|\mathbf{k}-\mathbf{k}'|}\frac{1}{\beta}\sum_{n=-\infty}^{\infty} \underline{F}(\varepsilon_n, \mathbf{k}', \mathbf{r}). \tag{14.64}$$

Expanding the pairing interaction as (8.83), we follow the procedure of Sect. 9.2 to replace $\mathscr{V}_\ell(k, k')$ with the effective one,

$$\mathscr{V}_\ell(k, k') \to \mathscr{V}_\ell^{(\mathrm{eff})}(k, k') \equiv \mathscr{V}_\ell^{(\mathrm{eff})}\theta(\varepsilon_\mathrm{c} - |\xi_k|)\theta(\varepsilon_\mathrm{c} - |\xi_{k'}|). \tag{14.65}$$

We also assume that a single ℓ is relevant, use (8.90), approximate $N(\varepsilon_{k'}) \approx N(\varepsilon_\mathrm{F})$, and make a change of variable as $\varepsilon_{k'} \to \xi_{k'}$. Equation (14.64) thereby becomes

$$\frac{\underline{\Delta}(\mathbf{k}, \mathbf{r})}{N(\varepsilon_\mathrm{F})\mathscr{V}_\ell^{(\mathrm{eff})}} = \int_{-\varepsilon_\mathrm{c}}^{\varepsilon_\mathrm{c}} \mathrm{d}\xi_{k'} \int \frac{\mathrm{d}\Omega_{\mathbf{k}'}}{4\pi} 4\pi \sum_{m=-\ell}^{\ell} Y_{\ell m}(\hat{\mathbf{k}})Y_{\ell m}^*(\hat{\mathbf{k}}')\frac{1}{\beta}\sum_{n=\infty}^{\infty} \underline{F}(\varepsilon_n, \mathbf{k}', \mathbf{r}). \tag{14.66}$$

Next, we remove both the coupling constant $N(\varepsilon_{\mathrm{F}})\mathscr{V}_\ell^{(\mathrm{eff})}$ and cutoff energy ε_{c} in favor of the transition temperature T_{c} in zero magnetic field. To this end, we use the T_{c} equation for homogeneous ℓ-wave pairing, which is obtained from (13.35) with $\mathscr{V}_1^{(\mathrm{eff})} \to \mathscr{V}_\ell^{(\mathrm{eff})}$,

$$\frac{-1}{N(\varepsilon_{\mathrm{F}})\mathscr{V}_\ell^{(\mathrm{eff})}} = \int_{-\varepsilon_{\mathrm{c}}}^{\varepsilon_{\mathrm{c}}} \mathrm{d}\xi \frac{1}{2\xi} \tanh \frac{\xi}{2k_{\mathrm{B}}T_{\mathrm{c}}}$$

$$= \int_{-\varepsilon_{\mathrm{c}}}^{\varepsilon_{\mathrm{c}}} \mathrm{d}\xi \left[\frac{1}{2\xi} \tanh \frac{\xi}{2k_{\mathrm{B}}T_{\mathrm{c}}} - \frac{1}{2\xi} \tanh \frac{\xi}{2k_{\mathrm{B}}T}\right] + \int_{-\varepsilon_{\mathrm{c}}}^{\varepsilon_{\mathrm{c}}} \mathrm{d}\xi \frac{1}{\beta} \sum_{n=-\infty}^{\infty} \frac{1}{\xi^2 + \varepsilon_n^2},$$

where we have inserted identity (9.39) with $x = \xi/k_{\mathrm{B}}T$ and $\varepsilon_n = (2n+1)\pi k_{\mathrm{B}}T$ in the integrand. The first term on the right-hand side can be expressed as $\ln(T/T_{\mathrm{c}})$ based on (9.32). As for the second term, we may restrict the summation over n as $-n_{\mathrm{c}} - 1 \le n \le n_{\mathrm{c}}$, taking the limit $\varepsilon_{\mathrm{c}} \to \infty$, and evaluate the integral analytically. The above equation thereby reduces to

$$\frac{-1}{N(\varepsilon_{\mathrm{F}})\mathscr{V}_\ell^{(\mathrm{eff})}} = \ln \frac{T}{T_{\mathrm{c}}} + \frac{1}{\beta} \sum_{n=-n_{\mathrm{c}}-1}^{n_{\mathrm{c}}} \frac{\pi}{|\varepsilon_n|}. \tag{14.67}$$

Substitution of (14.67) into (14.66) yields the self-consistency equation for the pair potential in terms of T_{c} in zero magnetic field,

$$\underline{\Delta}(\mathbf{k}_{\mathrm{F}}, \mathbf{r}) \ln \frac{T_{\mathrm{c}}}{T} = \frac{1}{\beta} \sum_{n=-\infty}^{\infty} \left[\frac{\pi}{|\varepsilon_n|} \underline{\Delta}(\mathbf{k}_{\mathrm{F}}, \mathbf{r}) + \int \mathrm{d}\Omega_{\mathbf{k}'} \sum_{m=-\ell}^{\ell} Y_{\ell m}(\hat{\mathbf{k}}) Y_{\ell m}^*(\hat{\mathbf{k}}') \right.$$
$$\left. \times \int_{-\varepsilon_{\mathrm{c}}}^{\varepsilon_{\mathrm{c}}} \mathrm{d}\xi_{k'} \underline{F}(\varepsilon_n, \mathbf{k}', \mathbf{r}) \right],$$

where we have taken the limit $n_{\mathrm{c}} \to \infty$ on the right-hand side, with no divergence problems, and also replaced an argument of $\underline{\Delta}$ as $\mathbf{k} \to \mathbf{k}_{\mathrm{F}}$. Finally, we make $\varepsilon_{\mathrm{c}} \to \infty$ and express the final integral in terms of the quasiclassical Green's function (14.59). We thereby obtain the self-consistency equation for the pair potential in the quasiclassical formalism as

$$\underline{\Delta}(\mathbf{k}_{\mathrm{F}}, \mathbf{r}) \ln \frac{T_{\mathrm{c}}}{T} = \frac{\pi}{\beta} \sum_{n=-\infty}^{\infty} \left[\frac{\underline{\Delta}(\mathbf{k}_{\mathrm{F}}, \mathbf{r})}{|\varepsilon_n|} - \sum_{m=-\ell}^{\ell} Y_{\ell m}(\hat{\mathbf{k}}) \int \mathrm{d}\Omega_{\mathbf{k}'} Y_{\ell m}^*(\hat{\mathbf{k}}') \underline{f}(\varepsilon_n, \mathbf{k}_{\mathrm{F}}', \mathbf{r})\right], \tag{14.68}$$

where T_{c} denotes the transition temperature of the homogeneous case at zero magnetic field.

14.3.3 Current Density

Next, we express the current density (10.36) using the quasiclassical Green's function, i.e., (14.69) below.

The one-particle density matrix, defined by (8.29), is expressible in terms of the Matsubara Green's function (14.3),

$$\rho^{(1)}(\xi_1, \xi_2) = G_{11}(\xi_1, \xi_2; 0_-) = \frac{1}{\beta} \sum_{n=-\infty}^{\infty} G_{\alpha_1\alpha_2}(\mathbf{r}_1, \mathbf{r}_2; \varepsilon_n) \mathrm{e}^{-\mathrm{i}\varepsilon_n 0_-},$$

where 0_- is an infinitesimal negative constant, and we have used (14.4), (14.8), and (14.27). Writing Green's function above as (14.46), we substitute the resulting expression into (10.36), and simplify the expression of $\mathbf{j}(\mathbf{r})$ in the same way as (14.52). We thereby obtain

$$\begin{aligned} \mathbf{j}(\mathbf{r}_1) &= e \frac{(\hat{\mathbf{p}}_1 - e\mathbf{A}_1) + (-\hat{\mathbf{p}}_2 - e\mathbf{A}_2)}{2m} \frac{1}{\beta} \sum_{n=-\infty}^{\infty} \mathrm{Tr}\, \underline{G}(\mathbf{r}_1, \mathbf{r}_2; \varepsilon_n) \Big|_{\mathbf{r}_2=\mathbf{r}_1} \\ &= \frac{e}{\beta} \sum_{n=-\infty}^{\infty} \frac{1}{V} \sum_{\mathbf{k}} \frac{\hbar \mathbf{k}}{m} \mathrm{Tr}\, \underline{G}(\varepsilon_n, \mathbf{k}, \mathbf{r}_1), \end{aligned}$$

where the trace represents the sum over spin components, and we have set $0_- \to 0$ safely.[5] We transform the sum over $\mathbf{k}$ as (8.90), then approximate $N(\varepsilon_k) \approx N(\varepsilon_F)$ and $\hbar\mathbf{k} \approx \hbar\mathbf{k}_F$, and express the final integral using the quasiclassical Green's function (14.59). We thereby obtain the expression for the current density in the quasiclassical formalism,

$$\mathbf{j}(\mathbf{r}) = -\mathrm{i} \frac{\pi e N(\varepsilon_F)}{\beta} \sum_{n=-\infty}^{\infty} \int \frac{\mathrm{d}\Omega_{\mathbf{k}}}{4\pi} \mathbf{v}_F \mathrm{Tr}\, \underline{g}(\varepsilon_n, \mathbf{k}_F, \mathbf{r}), \tag{14.69}$$

where $\mathbf{v}_F \equiv \hbar\mathbf{k}_F/m$ is the Fermi velocity.

14.3.4 Summary of the Eilenberger Equations

Let us summarize the results of the quasiclassical formalism.

The quasiclassical Green's function is defined by (14.59), which has symmetry (14.60) and satisfies the normalization condition (14.62). The (1,1) submatrix of (14.62) reads $\underline{g}^2 - \underline{f}\,\underline{f}^* = \underline{\sigma}_0$. Hence, $\underline{g} \equiv \underline{g}(\varepsilon_n, \mathbf{k}_F, \mathbf{r})$ is expressible in terms of

[5] Factor 0_- becomes relevant only when we calculate the particle number in terms of $\underline{G}$.

$\underline{f} \equiv \underline{f}(\varepsilon_n, \mathbf{k}_F, \mathbf{r})$ and $\underline{f}^* \equiv \underline{f}^*(\varepsilon_n, -\mathbf{k}_F, \mathbf{r}) = -\underline{f}^\dagger(-\varepsilon_n, \mathbf{k}_F, \mathbf{r})$ as

$$\underline{g}(\varepsilon_n, \mathbf{k}_F, \mathbf{r}) = \mathrm{sgn}(\varepsilon_n)\left[\underline{\sigma}_0 - \underline{f}(\varepsilon_n, \mathbf{k}_F, \mathbf{r})\underline{f}^\dagger(-\varepsilon_n, \mathbf{k}_F, \mathbf{r})\right]^{1/2}, \tag{14.70}$$

where the factor $\mathrm{sgn}(\varepsilon_n) \equiv \theta(\varepsilon_n) - \theta(-\varepsilon_n)$ is introduced to be compatible with (14.63) for homogeneous s-wave pairing. Equation (14.70) tells us that we only need to solve the (1,2)-submatrix element of the Eilenberger equation (14.61). We write it explicitly by including effects of impurity scatterings, which are known to affect the magnetic properties of s-wave superconductors considerably. They give rise to an additional impurity self-energy given by (**Problem** 14.1)

$$\hat{\mathscr{U}}_{\mathrm{imp}}(\varepsilon_n, \mathbf{r}) \equiv -\mathrm{i}\frac{\hbar}{2\tau}\langle \hat{g}(\varepsilon_n, \mathbf{k}_F, \mathbf{r})\rangle_F \hat{\sigma}_z, \tag{14.71}$$

where τ is the *relaxation time*, and $\langle\cdots\rangle_F$ denotes the average over the Fermi surface:

$$\langle A \rangle_F \equiv \int \frac{\mathrm{d}\Omega_{\mathbf{k}}}{4\pi} A(\mathbf{k}_F). \tag{14.72}$$

Let us replace $\hat{\mathscr{U}}_{\mathrm{BdG}} \to \hat{\mathscr{U}}_{\mathrm{BdG}} + \hat{\mathscr{U}}_{\mathrm{imp}}$ in (14.61), substitute (14.48), (14.59), and (14.71) subsequently, and extract the (1,2)-submatrix element noting (14.56) and setting $\underline{\mathscr{U}}_{\mathrm{HF}} \to \underline{0}$. We thereby obtain

$$2\varepsilon_n \underline{f} + \hbar \mathbf{v}_F \cdot \left(\boldsymbol{\nabla} - \mathrm{i}\frac{2e}{\hbar}\mathbf{A}\right)\underline{f} = \underline{\Delta}\underline{g}^* + \underline{g}\underline{\Delta} + \frac{\underline{g}\langle \underline{f}\rangle_F + \langle \underline{f}\rangle_F \underline{g}^* - \langle \underline{g}\rangle_F \underline{f} - \underline{f}\langle \underline{g}^*\rangle_F}{2\tau/\hbar}, \tag{14.73}$$

where $(\underline{f}, \underline{g}^*, \underline{\Delta})$ denote $\underline{f} = \underline{f}(\varepsilon_n, \mathbf{k}_F, \mathbf{r})$, $\underline{g}^* = \underline{g}^*(\varepsilon_n, -\mathbf{k}_F, \mathbf{r})$, and $\underline{\Delta} = \underline{\Delta}(\mathbf{k}_F, \mathbf{r})$, respectively, and $\underline{g} = \underline{g}(\varepsilon_n, \mathbf{k}_F, \mathbf{r})$ is given in terms of $\underline{f}$ by (14.70). This is the main part of the *Eilenberger equations*.

Next, we focus on (14.68) for the pair potential and expand

$$\underline{\Delta}(\mathbf{k}_F, \mathbf{r}) = \sum_{m=-\ell}^{\ell} \underline{\Delta}_{\ell m}(\mathbf{r})\sqrt{4\pi}Y_{\ell m}(\hat{\mathbf{k}}). \tag{14.74}$$

We then multiply the resulting equation by $\sqrt{4\pi}Y^*_{\ell m}(\hat{\mathbf{k}})$, integrate over the solid angle, and use (8.86) to obtain for $\underline{\Delta}_{\ell m}(\mathbf{r})$,

$$\underline{\Delta}_{\ell m}(\mathbf{r}) \ln\frac{T_c}{T} = \frac{\pi}{\beta}\sum_{n=-\infty}^{\infty}\left[\frac{\underline{\Delta}_{\ell m}(\mathbf{r})}{|\varepsilon_n|} - \int \frac{\mathrm{d}\Omega_{\mathbf{k}}}{4\pi}\sqrt{4\pi}Y^*_{\ell m}(\hat{\mathbf{k}})\underline{f}(\varepsilon_n, \mathbf{k}_F, \mathbf{r})\right]. \tag{14.75}$$

Finally, we substitute (14.69) for the current density into (10.30). We thereby obtain Ampère's law that determines the magnetic flux density $\mathbf{B}(\mathbf{r}) = \boldsymbol{\nabla} \times \mathbf{A}(\mathbf{r})$,

$$\nabla \times \mathbf{B}(\mathbf{r}) = -\mathrm{i}\frac{\pi e \mu_0 N(\varepsilon_{\mathrm{F}})}{\beta} \sum_{n=-\infty}^{\infty} \int \frac{\mathrm{d}\Omega_{\mathbf{k}}}{4\pi} \mathbf{v}_{\mathrm{F}} \mathrm{Tr}\, \underline{g}(\varepsilon_n, \mathbf{k}_{\mathrm{F}}, \mathbf{r}). \tag{14.76}$$

Equations (14.73)–(14.76) form a set of self-consistency (i.e. nonlinear) equations for the quasiclassical Green's function $\underline{f}(\varepsilon_n, \mathbf{k}_{\mathrm{F}}, \mathbf{r})$, pair potential $\underline{\Delta}(\mathbf{k}_{\mathrm{F}}, \mathbf{r})$, and magnetic flux density $\mathbf{B}(\mathbf{r})$ in the quasiclassical formalism. It forms an excellent approximation to the Gor'kov equations when $T_{\mathrm{c}} \ll T_{\mathrm{F}} \equiv \varepsilon_{\mathrm{F}}/k_{\mathrm{B}}$ holds.

Comments on the quasiclassical formalism are in order before ending this section. First, there are several different expressions for the quasiclassical free energies [4, 18]; they yield identical results for the bulk free energy but may differ from one another in terms of the surface contribution. Incorporating many-body effects beyond the mean-field theory is discussed in detail in [18]. Boundary conditions for solving the Eilenberger equations are studied, e.g., in [16]. Finally, the Lorentz force for the supercurrent is missing from the above quasiclassical formalism, which can be incorporated appropriately by retaining the next-to-leading order term in (14.51) [3, 9, 10].

14.4 Ginzburg–Landau Equations

We focus on s-wave superconductors near T_{c} and simplify the Eilenberger equations further into the GL equations.

The gap matrix for the homogeneous s-wave pairing is given by (9.1). We assume the same form for both the pair potential and the quasiclassical Green's function:

$$\underline{\Delta}(\mathbf{r}) = \Delta(\mathbf{r})\, \mathrm{i}\underline{\sigma}_y, \qquad \underline{f}(\varepsilon_n, \mathbf{k}_{\mathrm{F}}, \mathbf{r}) = f(\varepsilon_n, \mathbf{k}_{\mathrm{F}}, \mathbf{r})\, \mathrm{i}\underline{\sigma}_y, \tag{14.77}$$

to seek an inhomogeneous solution self-consistently near T_{c} in terms of $\Delta(\mathbf{r})$. It follows from (14.60) that the scalar function f satisfies

$$f(\varepsilon_n, \mathbf{k}_{\mathrm{F}}, \mathbf{r}) = f(-\varepsilon_n, -\mathbf{k}_{\mathrm{F}}, \mathbf{r}). \tag{14.78}$$

Accordingly, (14.70) reduces to a multiple of the unit matrix, $\underline{g} = g\underline{\sigma}_0$, with

$$\begin{aligned} g(\varepsilon_n, \mathbf{k}_{\mathrm{F}}, \mathbf{r}) &= \mathrm{sgn}(\varepsilon_n)\big[1 - f(\varepsilon_n, \mathbf{k}_{\mathrm{F}}, \mathbf{r}) f^*(-\varepsilon_n, \mathbf{k}_{\mathrm{F}}, \mathbf{r})\big]^{1/2} \\ &= g^*(\varepsilon_n, -\mathbf{k}_{\mathrm{F}}, \mathbf{r}). \end{aligned} \tag{14.79}$$

Substitution of these expressions into (14.73) yields a differential equation for $f = f(\varepsilon_n, \mathbf{k}_{\mathrm{F}}, \mathbf{r})$,

$$2\varepsilon_n f + \hbar \mathbf{v}_{\mathrm{F}} \cdot \boldsymbol{\partial} f = 2\Delta g + \hbar \frac{g\langle f \rangle_{\mathrm{F}} - f\langle g \rangle_{\mathrm{F}}}{\tau}, \tag{14.80}$$

where operator $\boldsymbol{\partial}$ is defined by (14.56).

Now, we solve (14.80) perturbatively with respect to $\Delta = \Delta(\mathbf{r})$ by expanding f and g as power series,

$$f = \sum_{\nu=1}^{\infty} f^{(\nu)}, \qquad g = \mathrm{sgn}(\varepsilon_n)\left(1 + \sum_{\nu=2}^{\infty} g^{(\nu)}\right). \tag{14.81}$$

Equality $g^{(1)} = 0$ may be seen from (14.79). Let us substitute these expansions into (14.80) and regard operator $\hbar \mathbf{v}_{\mathrm{F}} \cdot \boldsymbol{\partial}$ as $O(\Delta)$. We thereby obtain an expression for the νth order ($\nu = 1, 2, \cdots$),

$$f^{(\nu)} = \frac{\Delta g^{(\nu-1)}}{\varepsilon_n} - \frac{\hbar \mathbf{v}_{\mathrm{F}} \cdot \boldsymbol{\partial} f^{(\nu-1)}}{2\varepsilon_n} - \sum_{k=0}^{\nu-1} \hbar \frac{f^{(\nu-k)} \langle g^{(k)} \rangle_{\mathrm{F}} - \langle f^{(\nu-k)} \rangle_{\mathrm{F}} g^{(k)}}{2\tau\varepsilon_n}, \tag{14.82}$$

with $f^{(0)} = 0$, $g^{(0)} = \mathrm{sgn}(\varepsilon_n)$, and $g^{(1)} = 0$.

The first-order equation is given as $f^{(1)} = \Delta/|\varepsilon_n| - \hbar(f^{(1)} - \langle f^{(1)} \rangle_{\mathrm{F}})/2\tau|\varepsilon_n|$, which has no $\mathbf{k}_{\mathrm{F}}$ dependence. Hence, we conclude $f^{(1)} = \langle f^{(1)} \rangle_{\mathrm{F}}$ and

$$f^{(1)} = \frac{\Delta}{|\varepsilon_n|}. \tag{14.83}$$

Substituting (14.83) into (14.79) and comparing the resulting expression with the expansion of g in (14.81), we thereby obtain $g^{(2)}$,

$$g^{(2)} = -\mathrm{sgn}(\varepsilon_n) \frac{|\Delta|^2}{2\varepsilon_n^2}. \tag{14.84}$$

Next, setting $\nu = 2$ in (14.82) yields

$$f^{(2)} = -\frac{\hbar \mathbf{v}_{\mathrm{F}} \cdot \boldsymbol{\partial} f^{(1)}}{2\varepsilon_n} - \hbar \frac{f^{(2)} - \langle f^{(2)} \rangle_{\mathrm{F}}}{2\tau|\varepsilon_n|}.$$

As the source term $\propto \mathbf{v}_{\mathrm{F}} \cdot \boldsymbol{\partial} f^{(1)}$ is linear in $\mathbf{v}_{\mathrm{F}}$, we conclude $\langle f^{(2)} \rangle_{\mathrm{F}} = 0$. The corresponding equation for $f^{(2)}$ can be solved easily using (14.83),

$$f^{(2)} = -\frac{\hbar \mathbf{v}_{\mathrm{F}} \cdot \boldsymbol{\partial} f^{(1)}}{2\varepsilon_n(1 + \hbar/2\tau|\varepsilon_n|)} = -\frac{\hbar \mathbf{v}_{\mathrm{F}} \cdot \boldsymbol{\partial} \Delta}{2\varepsilon_n(|\varepsilon_n| + \hbar/2\tau)}. \tag{14.85}$$

To find $g^{(3)}$, we substitute (14.81) into (14.79), extract third-order terms, and use (14.83) and (14.85). We thereby obtain

$$g^{(3)} = -\mathrm{sgn}(\varepsilon_n) \frac{f^{(1)*} f^{(2)} + f^{(2)*} f^{(1)}}{2} = \hbar \mathbf{v}_{\mathrm{F}} \cdot \frac{\Delta^* \boldsymbol{\partial} \Delta - \Delta \boldsymbol{\partial} \Delta^*}{4\varepsilon_n^2(|\varepsilon_n| + \hbar/2\tau)}, \tag{14.86}$$

where $f^* = f^*(-\varepsilon_n, \mathbf{k}_\mathrm{F}, \mathbf{r})$. Finally, we set $\nu = 3$ in (14.82) and subsequently use $f^{(1)} = \langle f^{(1)}\rangle_\mathrm{F}$ and $g^{(2)} = \langle g^{(2)}\rangle_\mathrm{F}$ based on (14.83) and (14.84). This yields

$$\begin{aligned} f^{(3)} &= \frac{\Delta g^{(2)}}{\varepsilon_n} - \frac{\hbar\mathbf{v}_\mathrm{F}\cdot\boldsymbol{\partial} f^{(2)}}{2\varepsilon_n} - \hbar\frac{f^{(3)} - \langle f^{(3)}\rangle_\mathrm{F}}{2\tau|\varepsilon_n|} \\ &= -\frac{|\Delta|^2\Delta}{2|\varepsilon_n|^3} + \frac{(\hbar\mathbf{v}_\mathrm{F}\cdot\boldsymbol{\partial})^2\Delta}{4\varepsilon_n^2(|\varepsilon_n| + \hbar/2\tau)} - \hbar\frac{f^{(3)} - \langle f^{(3)}\rangle_\mathrm{F}}{2\tau|\varepsilon_n|}. \end{aligned}$$

Its Fermi-surface average is easily obtained using (13.14),

$$\langle f^{(3)}\rangle_\mathrm{F} = -\frac{|\Delta|^2\Delta}{2|\varepsilon_n|^3} + \frac{(\hbar v_\mathrm{F})^2\boldsymbol{\partial}^2\Delta}{12\varepsilon_n^2(|\varepsilon_n| + \hbar/2\tau)}. \tag{14.87}$$

We are now ready to derive the GL equations. First, we consider the equation for the pair potential. Let us set $\ell = m = 0$ in (14.75) with $\underline{\Delta}_{00} \to \underline{\Delta}$, substitute (14.77) and $Y_{00}(\hat{\mathbf{k}}) = (4\pi)^{-1/2}$, multiply the equation by $N(\varepsilon_\mathrm{F})$, and substitute the expansion of f in (14.81). We thereby obtain

$$N(\varepsilon_\mathrm{F})\Delta(\mathbf{r})\ln\frac{T}{T_\mathrm{c}} + N(\varepsilon_\mathrm{F})\frac{\pi}{\beta}\sum_{n=-\infty}^{\infty}\left[\frac{\Delta(\mathbf{r})}{|\varepsilon_n|} - \sum_{\nu=1}^{\infty}\langle f^{(\nu)}(\varepsilon_n, \mathbf{k}_\mathrm{F}, \mathbf{r})\rangle_\mathrm{F}\right] = 0, \tag{14.88}$$

where we have used the notation of (14.72). Further, we expand $\ln(T/T_\mathrm{c})$ near T_c as $\ln(T/T_\mathrm{c}) \approx \ln[1 + (T - T_\mathrm{c})/T_\mathrm{c}] \approx (T - T_\mathrm{c})/T_\mathrm{c}$, retain terms of $\nu \le 3$ in the square brackets, and substitute (14.83), (14.85), and (14.87). We thereby obtain an equation for the pair potential as

$$a_2\Delta(\mathbf{r}) + a_4|\Delta(\mathbf{r})|^2\Delta(\mathbf{r}) - b_2\,\boldsymbol{\partial}^2\Delta(\mathbf{r}) = 0, \tag{14.89}$$

where a_2, a_4, and b_2 are defined by

$$a_2 \equiv N(\varepsilon_\mathrm{F})\frac{T - T_\mathrm{c}}{T_\mathrm{c}}, \tag{14.90}$$

$$a_4 \equiv N(\varepsilon_\mathrm{F})\frac{\pi}{2\beta}\sum_{n=-\infty}^{\infty}\frac{1}{|\varepsilon_n|^3} \approx N(\varepsilon_\mathrm{F})\frac{7\zeta(3)}{8(\pi k_\mathrm{B}T_\mathrm{c})^2}, \tag{14.91}$$

$$b_2 \equiv N(\varepsilon_\mathrm{F})\frac{\pi(\hbar v_\mathrm{F})^2}{12\beta}\sum_{n=-\infty}^{\infty}\frac{1}{\varepsilon_n^2(|\varepsilon_n| + \hbar/2\tau)} \approx \frac{(\hbar v_\mathrm{F})^2}{6}a_4\chi. \tag{14.92}$$

The approximation in (14.91) has been obtained with the procedure of (9.41), and χ is defined by

$$\chi \equiv \frac{8}{7\zeta(3)} \sum_{n=0}^{\infty} \frac{1}{(2n+1)^2(2n+1+\hbar/2\pi\tau k_B T_c)}. \tag{14.93}$$

Whereas a_4 and b_2 are positive, a_2 changes sign at $T = T_c$ to be negative for $T < T_c$. Function χ decreases from 1 in the clean limit ($\tau \to \infty$) as τ is decreased. It should be noted that τ is inversely proportional to the impurity concentration, as seen in (14.103).

Next, we focus on Ampère's law. Let us substitute $g = g\underline{\sigma}_0$ into (14.76), expand g as (14.81) and retain terms of $\nu \leq 3$, and use (14.84) and (14.86). The resulting equation can be expressed in terms of coefficient b_2, (14.92),

$$\boldsymbol{\nabla} \times \mathbf{B} = -\mathrm{i}\frac{2e\mu_0 b_2}{\hbar}(\Delta^* \boldsymbol{\partial} \Delta - \Delta \boldsymbol{\partial} \Delta^*). \tag{14.94}$$

Equations (14.89) and (14.94) constitute the *GL equations*. The corresponding free-energy functional is given by

$$F_{\rm sn}[\Delta, \Delta^*, \mathbf{A}] \equiv \int \mathrm{d}^3 r \left[a_2|\Delta|^2 + \frac{a_4}{2}|\Delta|^4 + b_2 \Delta^* \left(-\mathrm{i}\boldsymbol{\nabla} - \frac{2e}{\hbar}\mathbf{A}\right)^2 \Delta + \frac{(\boldsymbol{\nabla} \times \mathbf{A})^2}{2\mu_0} \right]. \tag{14.95}$$

Indeed, one sees easily, applying the transformation of (10.35) to (14.95), that extremal conditions $\delta F_{\rm sn}/\delta \Delta^*(\mathbf{r}) = 0$ and $\delta F_{\rm sn}/\delta \mathbf{A}(\mathbf{r}) = \mathbf{0}$ yield (14.89) and (14.94), respectively. Note that functional (14.95) with coefficients (14.90)–(14.92) forms an inhomogeneous extension of the Landau functional (9.55) with coefficients (9.56).

The homogeneous solution of (14.89) for $T < T_c$ is given by $\Delta_0 \equiv \sqrt{-a_2/a_4}$, which does not depend on the relaxation time τ as seen from (14.90) and (14.91). Hence, we conclude that the thermodynamic properties of homogeneous s-wave superconductors are not affected by impurities [1, 2]. However, it should be pointed out that the statement no longer holds true when gap anisotropy is present. In contrast, the third term in the square brackets of (14.95) represents the kinetic energy due to the spatial variation of Δ. It follows from (14.92), (14.93) and (14.103) that this term strongly depends on the impurity concentration, i.e., the kinetic energy is reduced as relaxation times τ shorten because of increased impurity concentration.

Assuming that spatial variation of the pair potential in (14.94) lies only in its phase as $\Delta(\mathbf{r}) = \Delta_0 \mathrm{e}^{\mathrm{i}\varphi(\mathbf{r})} = \sqrt{-a_2/a_4}\,\mathrm{e}^{\mathrm{i}\varphi(\mathbf{r})}$, we reproduce the London equation (10.39) or (10.40). The corresponding London penetration depth is given by

$$\lambda_{\mathrm{L}} = \sqrt{\frac{\hbar^2 a_4}{8\mu_0 b_2(-a_2)e^2}} = \sqrt{\frac{m}{\mu_0(N/V)2[(T_{\mathrm{c}}-T)/T_{\mathrm{c}}]\chi e^2}}, \tag{14.96}$$

where we have substituted (14.90)–(14.92) and used the relation $2N(\varepsilon_{\mathrm{F}})v_{\mathrm{F}}^2/3 = (N/mV)$. Equation (14.96) in the clean limit $\chi \to 1$ is identical with λ_{L} in (10.40) for $T \lesssim T_{\mathrm{c}}$, as may be confirmed by using (10.22) and (10.47).

The original motivation of Ginzburg and Landau in 1950 [5] was to extend the London theory to incorporate the superconducting order parameter $\Delta(\mathbf{r})$. To this end, they extended the Landau theory for second-order phase transitions described in Sect. 9.5 to inhomogeneous superconductors phenomenologically in a gauge-invariant manner based on the free-energy functional,

$$F_{\mathrm{sn}} = \int \mathrm{d}^3 r \left[\alpha|\Psi|^2 + \frac{\beta}{2}|\Psi|^4 + \Psi^* \frac{\left(-\mathrm{i}\hbar\nabla - e^*\mathbf{A}\right)^2}{2m^*}\Psi + \frac{(\nabla\times\mathbf{A})^2}{2\mu_0} \right]. \tag{14.97}$$

Later in 1959, Gor'kov derived the functional microscopically [6] to show that their order parameter $\Psi(\mathbf{r})$ and constants (α, β) can be expressed in terms of quantities in (14.90)–(14.92) and (14.95); that is,

$$\Psi(\mathbf{r}) = \frac{\sqrt{2m^* b_2}}{\hbar}\Delta(\mathbf{r}), \qquad \alpha = \frac{\hbar^2}{2m^* b_2}a_2, \qquad \beta = \left(\frac{\hbar^2}{2m^* b_2}\right)^2 a_4, \tag{14.98}$$

with $m^* = 2m$ and $e^* = 2e$. Note that m^* and e^* in the considerations of Ginzburg and Landau [5] were set equal to the bare electron mass m and charge e, respectively, as was realized naturally for a period prior to the key concept of pair condensation. They subsequently calculated the surface energy in a magnetic field to show that superconductors can be classified into two groups according to whether the surface energy is positive or negative [5]. We shall discuss this aspect in the next chapter.

Problem

14.1. Scatterings by N_a impurity atoms of the same kind may be described by the Hamiltonian:

$$\mathscr{H}_{\mathrm{imp}} \equiv \int \mathrm{d}\xi_1\, \hat{\psi}^\dagger(\xi_1) \sum_{a=1}^{N_a} U_{\mathrm{imp}}(\mathbf{r}_1 - \mathbf{r}_a)\hat{\psi}(\xi_1), \tag{14.99}$$

where U_{imp} is the impurity potential and $\mathbf{r}_a$ denotes the position of impurity atom a. We incorporate effects described by this Hamiltonian based on the self-consistent Born approximation. In general, the perturbation expansion for superconducting phases can be performed in terms of normal-state Feynman diagrams by replacing the normal Green's function G_{n} with the product $\hat{G}\hat{\sigma}_z$ of the Nambu matrices (14.31) and (12.9) [8]. We thereby obtain the self-energy due to the double scatterings by the same impurity as

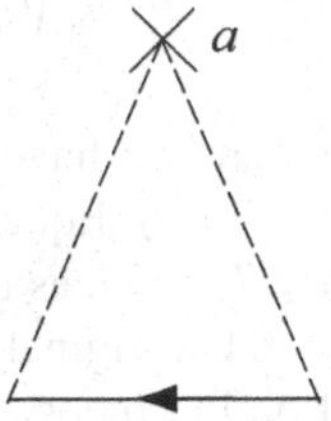

$$\hat{\Sigma}(\mathbf{r}_1,\mathbf{r}_2;\varepsilon_n) = \sum_a U_{\text{imp}}(\mathbf{r}_1-\mathbf{r}_a)\hat{\sigma}_z\hat{G}(\mathbf{r}_1,\mathbf{r}_2;\varepsilon_n)\hat{\sigma}_z U_{\text{imp}}(\mathbf{r}_2-\mathbf{r}_a). \quad (14.100)$$

Solve the following problems.

(a) Let us transform the impurity potential as

$$U_{\text{imp}}(\mathbf{r}_1-\mathbf{r}_a) = \frac{1}{V}\sum_{\mathbf{k}} U_{\mathbf{k}}^{\text{imp}} e^{i\mathbf{k}\cdot(\mathbf{r}_1-\mathbf{r}_a)}. \quad (14.101)$$

Similarly, we expand the self-energy $\hat{\Sigma}$ and the Green's function $\hat{G}$ as (14.46). Further, we assume that $\mathbf{r}_a$ is distributed randomly so that $\sum_a e^{i\mathbf{k}\cdot\mathbf{r}_a} = N_a\delta_{\mathbf{k}\mathbf{0}}$ holds. Show that the Fourier coefficient $\hat{\Sigma}(\varepsilon_n,\mathbf{k},\mathbf{r})$ of the self-energy (14.100) is given by

$$\hat{\Sigma}(\varepsilon_n,\mathbf{k},\mathbf{r}) = \frac{n_a}{V}\sum_{\mathbf{k}'} |U_{\mathbf{k}-\mathbf{k}'}^{\text{imp}}|^2\hat{\sigma}_z\hat{G}(\varepsilon_n,\mathbf{k}',\mathbf{r})\hat{\sigma}_z, \quad (14.102)$$

where $n_a \equiv N_a/V$ denotes the density of impurities.

(b) Let us replace the sum over $\mathbf{k}'$ in (14.102) by an integral as (8.90) and approximate $N(\varepsilon_{k'}) \approx N(\varepsilon_{\text{F}})$. We also consider s-wave impurity scattering where $U_{\mathbf{k}}^{\text{imp}}$ does not depend on $\mathbf{k}$, and introduce the relaxation time τ with

$$\frac{\hbar}{\tau} \equiv 2\pi n_a N(\varepsilon_{\text{F}})|U^{\text{imp}}|^2. \quad (14.103)$$

Hence, τ is inversely proportional to the impurity concentration n_a in the self-consistent Born approximation. Show that the right-hand side of (14.102) can be written in terms of the quasiclassical Green's function (14.59) as (14.71).

References

1. A.A. Abrikosov, L.P. Gor'kov, J. Exp. Theor. Phys. **36**, 319 (1959). (Sov. Phys. JETP **9**, 220 (1959))
2. P.W. Anderson, J. Phys. Chem. Solids **11**, 26 (1959)
3. E. Arahata, Y. Kato, J. Low Temp. Phys. **175**, 346 (2014)
4. G. Eilenberger, Z. Phys. **214**, 195 (1968)
5. V.L. Ginzburg, L.D. Landau, J. Exp. Theor. Phys. **20**, 1064 (1950)
6. L.P. Gor'kov, J. Exp. Theor. Phys. **36**, 1918 (1959). (Sov. Phys. JETP **9**, 1364 (1959)); J. Exp. Theor. Phys. **37**, 1407 (1959) (Sov. Phys. JETP **10**, 998 (1960))
7. H.J. Groenewold, Physica **12**, 405 (1946)
8. T. Kita, J. Phys. Soc. Jpn. **65**, 1355 (1996)
9. T. Kita, Phys. Rev. B **64**, 054503 (2001)
10. T. Kita, Phys. Rev. B **79**, 024521 (2009)
11. T. Kita, J. Phys. Soc. Jpn. **80**, 124704 (2011)
12. A.I. Larkin, Yu.N. Ovchinnikov, J. Exp. Theor. Phys. **55**, 2262 (1968). (Sov. Phys. JETP **28**, 1200 (1969))
13. A.J. Leggett, F. Sols, Found. Phys. **21**, 353 (1991)
14. T. Matsubara, Prog. Theor. Phys. **14**, 351 (1955)
15. J.E. Moyal, Proc. Camb. Philos. Soc. **45**, 99 (1949)
16. Y. Nagato, S. Higashitani, K. Yamada, K. Nagai, J. Low Temp. Phys. **103**, 1 (1996)
17. Y. Nambu, Phys. Rev. **117**, 648 (1960)
18. J.W. Serene, D. Rainer, Phys. Rep. **101**, 221 (1983)
19. R.L. Stratonovich, Dok. Akad. Nauk USSR **1**, 72 (1956). (Sov. Phys. Dokl. **1**, 414 (1956))
20. E.P. Wigner, Phys. Rev. **40**, 749 (1932)

Chapter 15
Abrikosov's Flux-Line Lattice

Abstract Superconductors can be classified into two types according to their response to applied magnetic fields. Whereas *type-I superconductors* exclude the magnetic field completely from the bulk due to the Meissner effect, *type-II superconductors* can retain quantized magnetic fluxes in the bulk over a certain range of magnetic field. In 1957, Abrikosov solved the Ginzburg–Landau equations analytically for a couple of limiting cases to predict that type-II superconductors can form a lattice of quantized flux lines between lower critical field H_{c1} and upper H_{c2}, which was later confirmed by experiments. In this chapter, we elaborate on this flux-line lattice.

15.1 Ginzburg-Landau Equations

Focusing on the region near the transition temperature, Ginzburg and Landau produced a phenomenological version of the free energy as a functional of the superconducting order parameter $\Psi(\mathbf{r})$ and vector potential $\mathbf{A}(\mathbf{r})$ in 1950 [7]. Later, Gor'kov derived it microscopically by extending the BCS theory to inhomogeneous systems. The resulting free energy, measured from the normal-state free energy in zero magnetic field at the same temperature T, is given in terms of the s-wave pair potential $\Delta(\mathbf{r})$, (14.95), i.e.,

$$F_{\mathrm{sn}} = \int \mathrm{d}^3 r \left[a_2|\Delta|^2 + \frac{a_4}{2}|\Delta|^4 + b_2\Delta^* \left(-\mathrm{i}\nabla - \frac{2e}{\hbar}\mathbf{A} \right)^2 \Delta + \frac{(\nabla \times \mathbf{A})^2}{2\mu_0} \right], \tag{15.1}$$

where μ_0 is the vacuum permeability and (a_2, a_4, b_2) are constants given explicitly by (14.90)–(14.92). Whereas a_4 and b_2 are positive, a_2 is proportional to $T - T_c$ and becomes negative for $T < T_c$. Additionally, b_2 is susceptible to variations in the concentration of impurities arising in alloying, becoming smaller as the impurity concentration increases.

As thermodynamic equilibria correspond to the minima of F_{sn}, the pair and vector potentials should obey $\delta F_{\mathrm{sn}}/\delta\Delta^*(\mathbf{r}) = 0$ and $\delta F_{\mathrm{sn}}/\delta\mathbf{A}(\mathbf{r}) = \mathbf{0}$. Following the procedure of (10.35) to calculate them, we obtain a pair of equations

© Springer Japan 2015

T. Kita, *Statistical Mechanics of Superconductivity*, Graduate Texts in Physics,
DOI 10.1007/978-4-431-55405-9_15

$$a_2\Delta + a_4|\Delta|^2\Delta + b_2\left(-\mathrm{i}\nabla - \frac{2e}{\hbar}\mathbf{A}\right)^2\Delta = 0, \tag{15.2}$$

$$\nabla\times\nabla\times\mathbf{A} = \frac{2e\mu_0 b_2}{\hbar}\left[\Delta^*\left(-\mathrm{i}\nabla - \frac{2e}{\hbar}\mathbf{A}\right)\Delta + \Delta\left(\mathrm{i}\nabla - \frac{2e}{\hbar}\mathbf{A}\right)\Delta^*\right]. \tag{15.3}$$

Equation (15.3) represents Ampère's law (10.30). Indeed, the left-hand side can be written in terms of the microscopic magnetic flux density:

$$\mathbf{B}(\mathbf{r}) \equiv \nabla\times\mathbf{A}(\mathbf{r}) \tag{15.4}$$

as $\nabla\times\mathbf{B}(\mathbf{r})$, whereas the right-hand side can be identified as superconducting current density $\mathbf{j}(\mathbf{r})$ multiplied by μ_0.

15.2 Microscopic Flux Density and Magnetization

We elaborate now on how the magnetic flux density (15.4) is determined.

In performing experiments on superconductors in magnetic fields, there are essentially two distinct sources for $\mathbf{B}$. The first is the external current $\mathbf{j}_{\text{ext}}$ that flows far outside the sample and produces the magnetic field $\mathbf{H}$ that obeys Ampère's law:

$$\nabla\times\mathbf{H}(\mathbf{r}) = \mathbf{j}_{\text{ext}}(\mathbf{r}). \tag{15.5}$$

In general, the experimental setup is arranged to produce a uniform field around the sample using, e.g., a Helmholtz coil. The second is the supercurrent $\mathbf{j}$ that flows inside the sample. Flux density $\mathbf{B}(\mathbf{r})$ in the sample is determined by solving (15.3) so that it is connected smoothly to $\mu_0\mathbf{H}$ far outside the sample. The spatial average of $\mathbf{B}(\mathbf{r})$ inside the sample, defined by

$$\bar{\mathbf{B}} \equiv \frac{1}{V}\int \mathbf{B}(\mathbf{r})\,\mathrm{d}^3r, \tag{15.6}$$

is not equal to $\mu_0\mathbf{H}$ generally. Their difference divided by μ_0,

$$\mathbf{M} \equiv \frac{\bar{\mathbf{B}}}{\mu_0} - \mathbf{H} \tag{15.7}$$

defines the *magnetization* due to the supercurrent. Spins may also contribute to $\mathbf{M}$, but in a first approximation their effect in single-element superconductors can be neglected in comparison with that of the supercurrent.

In particular, when $\mathbf{H}$ is applied parallel to the side of a long cylindrical sample, the field produced by the supercurrent is confined inside the sample. We shall consider this situation below, where **H** is derivable from the free energy (15.1) given as a function of $\bar{\mathbf{B}}$ by the thermodynamic relation (**Problem** 15.1):

$$\mathbf{H} = \frac{1}{V}\frac{\partial F_{\rm sn}}{\partial \bar{\mathbf{B}}}. \tag{15.8}$$

A subsequent Legendre transformation:

$$G_{\rm sn}(\mathbf{H}) \equiv F_{\rm sn}(\bar{\mathbf{B}}) - V\bar{\mathbf{B}}\cdot\mathbf{H} \tag{15.9}$$

introduces another free energy $G_{\rm sn}(\mathbf{H})$ as a function of **H** that is controllable experimentally.

Considering (15.6), we express the microscopic flux density inside the sample as

$$\mathbf{B}(\mathbf{r}) = \bar{\mathbf{B}} + \nabla\times\tilde{\mathbf{A}}(\mathbf{r}), \qquad \frac{1}{V}\int[\nabla\times\tilde{\mathbf{A}}(\mathbf{r})]\,{\rm d}^3r = \mathbf{0}. \tag{15.10}$$

Thus, $\tilde{\mathbf{A}}$ represents the spatially varying part of the flux density.

15.3 Dimensionless Equations

We shall rewrite the GL equations (15.2) and (15.3) in dimensionless form, as (15.18) and (15.19). This helps us to simplify the mathematical treatment and also capture the essence of type-II superconductors more clearly.

Let us focus on the region $T \lesssim T_{\rm c}$ where $a_2 < 0$, and perform an appropriate change of variable in (15.1). First, the homogeneous solution of (15.2) is easy to obtain;

$$\Delta_0 = \sqrt{-a_2/a_4}, \tag{15.11}$$

which is realized in zero magnetic field. Substituting this into (15.1), we obtain the zero-field condensation energy in equilibrium per unit volume as $F_{\rm sn}^{\rm (eq)}/V = -(a_4/2)\Delta_0^4 = -a_2^2/2a_4$. This result agrees with (9.53), as confirmed using (14.90), (14.91), and $N(\varepsilon_{\rm F}) = D(\varepsilon_{\rm F})/2V$. Second, we define *thermodynamic critical field* $H_{\rm c}$ by the equality $F_{\rm sn}^{\rm (eq)}/V = -\mu_0 H_{\rm c}^2/2$, which yields

$$H_{\rm c} = \sqrt{a_2^2/\mu_0 a_4}. \tag{15.12}$$

It is the field at which a type-I superconductor changes into the normal state via a first-order phase transition involving latent heat. Third, focusing on the terms of $O(\Delta)$ in (15.2), we note that the b_2 term contains a squared differential operator compared with the a_2 term. Hence, the quantity

$$\xi \equiv \sqrt{b_2/(-a_2)} \tag{15.13}$$

has a unit of length, which is called the *coherence length* and represents the typical scale of variations for amplitude $|\Delta|$. Similarly, we set $\Delta(\mathbf{r}) \to \Delta_0 = \sqrt{-a_2/a_4}$ in (15.3), compare the terms of $O(\mathbf{A})$, and obtain another length scale called the *London penetration depth*:

$$\lambda_{\mathrm{L}} \equiv \sqrt{\frac{\hbar^2 a_4}{8\mu_0 b_2(-a_2)e^2}}, \tag{15.14}$$

which represents a typical scale for variations in the flux density $\mathbf{B}(\mathbf{r})$. The ratio of the above two lengths,

$$\kappa \equiv \frac{\lambda_{\mathrm{L}}}{\xi} = \sqrt{\frac{\hbar^2 a_4}{8\mu_0 b_2^2 e^2}} \tag{15.15}$$

forms the important dimensionless parameter called the *GL parameter*. It follows from (14.90)–(14.93) and (14.103) that κ is temperature independent for $T \lesssim T_{\mathrm{c}}$ but is susceptible to variations in impurity concentration n_a through factor χ in b_2 and is enhanced as n_a increases. Note that n_a may be controlled systematically by alloying. The existence of another characteristic length ξ, apart from λ_{L}, was discovered by Pippard in 1953 [14].

In summary, (15.11), (15.12), and (15.14) are constants that have the units of energy, magnetic field, and length. Using them, we perform a change of variable in (15.1) as

$$\mathbf{r} = \lambda_{\mathrm{L}}\mathbf{r}', \qquad \Delta(\mathbf{r}) = \Delta_0 \Psi'(\mathbf{r}'), \qquad \mathbf{A}(\mathbf{r}) = \sqrt{2}\mu_0 H_{\mathrm{c}} \lambda_{\mathrm{L}} \mathbf{A}'(\mathbf{r}'), \tag{15.16}$$

where factor $\sqrt{2}$ has been introduced for convenience. The corresponding unit of magnetic flux density is given by $\sqrt{2}\mu_0 H_{\mathrm{c}}$. Substituting (15.16) into (15.1), using (15.11)–(15.15), and noting $e < 0$, we rewrite the free energy in the form[1]

$$F_{\mathrm{sn}} = \mu_0 H_{\mathrm{c}}^2 \lambda_{\mathrm{L}}^3 \int d^3 r' \left[-|\Psi'|^2 + \frac{1}{2}|\Psi'|^4 + \Psi'^* \left(-\frac{\mathrm{i}}{\kappa}\nabla' + \mathbf{A}' \right)^2 \Psi' + (\nabla' \times \mathbf{A}')^2 \right]. \tag{15.17}$$

The corresponding dimensionless GL equations are obtained by either calculating $\delta F_{\mathrm{sn}}/\delta \Psi'^*(\mathbf{r}') = 0$ and $\delta F_{\mathrm{sn}}/\delta \mathbf{A}'(\mathbf{r}') = \mathbf{0}$ or applying (15.16) to (15.2) and (15.3). They are given by

[1] Ginzburg and Landau [7] and Abrikosov [2] set $e > 0$ when transforming the free energy functional into a dimensionless form, so that $\mathbf{A}' \to -\mathbf{A}'$ in their expressions. However, the essential results are the same as those described here.

$$\left(-\frac{\mathrm{i}}{\kappa}\nabla' + \mathbf{A}'\right)^2 \Psi' - \Psi' + |\Psi'|^2\Psi' = 0, \tag{15.18}$$

$$\nabla' \times \mathbf{B}' = -\frac{1}{2}\left[\Psi'^*\left(-\frac{\mathrm{i}}{\kappa}\nabla' + \mathbf{A}'\right)\Psi' + \Psi'\left(\frac{\mathrm{i}}{\kappa}\nabla' + \mathbf{A}'\right)\Psi'^*\right], \tag{15.19}$$

where $\mathbf{B}' \equiv \nabla' \times \mathbf{A}'$ denotes the microscopic magnetic flux density.

15.4 Upper Critical Field and Distinction Between Type-I and II

First, we derive an expression for the *upper critical field* H_{c2} at which the superconducting phase transition occurs.

Assuming a continuous transition in a magnetic field, we may linearize (15.18) in terms of Ψ',

$$\left(-\frac{\mathrm{i}}{\kappa}\nabla' + \mathbf{A}'\right)^2 \Psi' = \Psi'. \tag{15.20}$$

As the supercurrent is negligible at the transition point, we can also set $\mathbf{B}'(\mathbf{r}') = \bar{\mathbf{B}}'$. Let us choose the z' axis along $\bar{\mathbf{B}}'$, assume a uniform solution along z', and consider a region of unit length along the z' axis from this point on. We also adopt the Landau gauge:

$$\mathbf{A}'(\mathbf{r}') = (0, \bar{B}'x', 0) \tag{15.21}$$

for describing $\bar{\mathbf{B}}' = \nabla' \times \mathbf{A}'$ and introduce creation and annihilation operators

$$\left.\begin{matrix}\hat{a}^\dagger \\ \hat{a}\end{matrix}\right\} \equiv \frac{1}{\sqrt{2\kappa\bar{B}'}}\left[\mp\frac{\partial}{\partial x'} - \mathrm{i}\left(\frac{\partial}{\partial y'} + \mathrm{i}\kappa\bar{B}'x'\right)\right] \tag{15.22}$$

satisfying $\hat{a}\hat{a}^\dagger - \hat{a}^\dagger\hat{a} = 1$. Equation (15.20) thereby becomes

$$\left(\hat{a}^\dagger\hat{a} + \frac{1}{2}\right)\frac{2\bar{B}'}{\kappa}\Psi' = \Psi', \tag{15.23}$$

which has the same form as the Schrödinger equation for the one-dimensional harmonic oscillator [15]. Hence, we conclude that the eigenvalues of $\hat{a}^\dagger\hat{a}$ are non-negative integers. The maximum field $\bar{B}' = B'_{c2}$ at which (15.23) has a solution can be obtained by replacing $\hat{a}^\dagger\hat{a}$ above with the smallest eigenvalue 0 as

$$B'_{c2} = \kappa, \qquad H_{c2} = \sqrt{2}\kappa H_c, \tag{15.24}$$

where the second identity in the ordinary units has been derived based on (15.16). This H_{c2} is called the *upper critical field*.

For the magnetic field to penetrate into the bulk, condition $H_{c2} \geq H_c$ must hold, from which we obtain for type-II superconductors the criterion

$$\kappa \geq \frac{1}{\sqrt{2}}. \tag{15.25}$$

15.5 Flux-Line Lattice Near H_{c2}

Next, we consider the region $H \lesssim H_{c2}$ in calculating the pair potential $\Delta'(\mathbf{r}')$ and microscopic magnetic flux density $\mathbf{B}'(\mathbf{r}')$.

Supercurrent $\mathbf{j}'(\mathbf{r}')$ also becomes finite in this region to produce a finite contribution to the average flux density (15.6). The corresponding vector potential may be chosen to reproduce (15.10) as

$$\mathbf{A}' = \bar{B}'x'\hat{\mathbf{y}}' + \tilde{\mathbf{A}}', \tag{15.26}$$

where $\hat{\mathbf{y}}'$ denotes the unit vector along the y' direction. We fix the gauge of $\tilde{\mathbf{A}}'$ so that

$$\nabla' \cdot \tilde{\mathbf{A}}' = 0 \tag{15.27}$$

is satisfied. Let us substitute (15.26) into the third term in the square brackets of (15.17). We then find that its operator part is expressible in terms of (15.22),

$$\left(-\frac{\mathrm{i}}{\kappa}\nabla' + \mathbf{A}'\right)^2 \approx \hat{h}_0 + \mathrm{i}\sqrt{\frac{2\bar{B}'}{\kappa}}\left[-(\tilde{A}'_x + \mathrm{i}\tilde{A}'_y)\hat{a} + (\tilde{A}'_x - \mathrm{i}\tilde{A}'_y)\hat{a}^\dagger\right], \tag{15.28}$$

where $\hat{h}_0$ is defined by

$$\hat{h}_0 \equiv \left(-\frac{\mathrm{i}}{\kappa}\nabla' + \bar{B}'x'\hat{\mathbf{y}}'\right)^2 = \left(\hat{a}^\dagger\hat{a} + \frac{1}{2}\right)\frac{2\bar{B}'}{\kappa}, \tag{15.29}$$

and we have neglected a term of $O(\tilde{\mathbf{A}}'^2)$ as appropriate near H_{c2}.

15.5.1 *Constructing Basis Functions*

To find the equilibrium structure of $\Delta'(\mathbf{r}')$, we use operator (15.29) to construct a complete set of basis functions in which to expand $\Delta'(\mathbf{r}')$. Those who are not interested in the derivation may proceed to the next section.

Hamiltonian (15.29) is identical to that for an electron in a uniform magnetic field in two dimensions [13]. Its eigenvalues distinguish the *Landau levels* given by

$$\varepsilon'_N = \left(N + \frac{1}{2}\right)\frac{2\bar{B}'}{\kappa} \qquad (N = 0, 1, 2, \cdots), \tag{15.30}$$

with a macroscopic degree of degeneracy for each level. Every eigenfunction $\phi_0(\mathbf{r}')$ for $N = 0$ obeys the first-order differential equation $\hat{a}\,\phi_0(\mathbf{r}') = 0$. Noting that operator (15.22) does not depend on y', we set $\phi_0(\mathbf{r}') \propto e^{iq'_y y'}\varphi_0(x')$ characterized by wavenumber q'_y and substitute it into $\hat{a}\,\phi_0(\mathbf{r}') = 0$. The resulting equation for $\varphi_0(x')$ can be solved easily, and we find the eigenfunctions take the form,

$$\phi_0(\mathbf{r}') \propto \exp\left[iq'_y y' - \frac{(x' + q'_y l'^2_c)^2}{2l'^2_c}\right], \tag{15.31}$$

where

$$l'_c \equiv \frac{1}{\sqrt{\kappa\bar{B}'}}, \qquad l_c = \frac{\lambda_L}{\sqrt{\kappa(\bar{B}/\sqrt{2}\mu_0 H_c)}} = \sqrt{\frac{\hbar}{2|e|\bar{B}}}, \tag{15.32}$$

denotes the magnetic length for a bound pair with charge $2e$, as may be realized from the second expression given in the ordinary units. To find the number of possible q'_y's in (15.31), we consider a square region of sides L' ($\gg l'_c$) and impose the periodic boundary condition along the y' direction [13]. We thereby obtain an expression for q'_y as $q'_y = 2\pi n_y/L'$ in terms of integers n_y. Let us substitute this into inequality $0 \le -q'_y l'^2_c \le L'$, which denotes that the central coordinate of the wave function (15.31) along the x' axis lies within the system. We thereby obtain the number of allowed values for n_y as $L'^2/2\pi l'^2_c$. Alternatively, there is a single state per area of $2\pi l'^2_c$.

If we choose $\Delta'(\mathbf{r}') \propto \phi_0(\mathbf{r}')$ in terms of (15.31), we can describe a state where a region of width $\sim 2\sqrt{2}l'_c$ around $x' = -q'_y l'^2_c$ is superconducting. However, it is clearly more favorable to realize superconductivity over the entire system. Hence, Abrikosov considered a linear combination of (15.31) to find that a quantized flux-line lattice should be stable [2].

The basic issue here is how to construct extended wave functions from (15.31). To this end, we use the magnetic translation operator in the Landau gauge [5, 10]:

$$\hat{T}_{\mathbf{R}'} \equiv e^{-\mathbf{R}'\cdot(\boldsymbol{\nabla}' + i\kappa\bar{B}'y'\hat{\mathbf{x}}')} = e^{-i\kappa\bar{B}'R'_x(y'-R'_y/2)}e^{-\mathbf{R}'\cdot\boldsymbol{\nabla}'}, \tag{15.33}$$

where $\hat{\mathbf{x}}'$ denotes the unit vector along the x' axis. For the second equality, we have used the identity $e^{\hat{C}+\hat{D}} = e^{-\frac{1}{2}[\hat{C},\hat{D}]}e^{\hat{C}}e^{\hat{D}}$ that holds for a pair of operators $\hat{C}$ and $\hat{D}$ satisfying $[\hat{C},[\hat{C},\hat{D}]] = [\hat{D},[\hat{C},\hat{D}]] = 0$, where $[\hat{C},\hat{D}] \equiv \hat{C}\hat{D} - \hat{D}\hat{C}$. It follows from the commutation relation $[(\boldsymbol{\nabla}' + i\kappa\bar{B}'y'\hat{\mathbf{x}}')_i, (\boldsymbol{\nabla}' + i\kappa\bar{B}'x'\hat{\mathbf{y}}')_j] = 0$ $(i, j = x', y')$ that $\hat{T}_{\mathbf{R}'}$ and $\hat{h}_0$, (15.29), commute

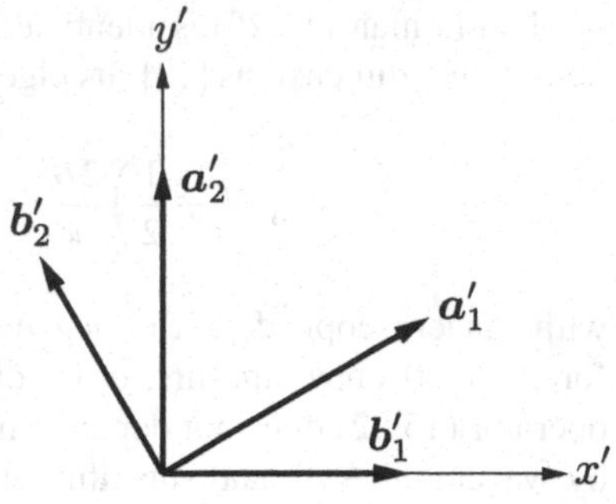

Fig. 15.1 A pair of primitive translation vectors $(\mathbf{a}_1', \mathbf{a}_2')$ of the flux-line lattice, and the corresponding pair of reciprocal lattice vectors $(\mathbf{b}_1', \mathbf{b}_2')$. The area of the unit cell is given by $|\mathbf{a}_1' \times \mathbf{a}_2'| = 2\pi l_c'^2$

$$\left[\hat{T}_{\mathbf{R}'}, \hat{h}_0\right] = 0. \tag{15.34}$$

Using this relation, we transform the degenerate bases of $\hat{h}_0$ into eigenstates of $\hat{T}_{\mathbf{R}'}$, which will be shown to form a complete orthonormal set that is suitable for describing the flux-line lattice [10]. First, let us introduce a set of primitive translation vectors $\mathbf{a}_1'$ and $\mathbf{a}_2'$ as (see Fig. 15.1)

$$\mathbf{a}_1' = (a_{1x}', a_{1y}'), \qquad \mathbf{a}_2' = (0, a_2'), \qquad a_{1x}' a_2' = 2\pi l_c'^2, \tag{15.35}$$

where the areas spanned by $\mathbf{a}_1'$ and $\mathbf{a}_2'$ are set equal to $2\pi l_c'^2$, i.e., the basic area for a single quantum state. The corresponding reciprocal lattice vectors are given by

$$\mathbf{b}_1' = 2\pi \frac{\mathbf{a}_2' \times \hat{\mathbf{z}}'}{|\mathbf{a}_1' \times \mathbf{a}_2'|}, \qquad \mathbf{b}_2' = 2\pi \frac{\hat{\mathbf{z}}' \times \mathbf{a}_1'}{|\mathbf{a}_1' \times \mathbf{a}_2'|}, \tag{15.36}$$

satisfying $\mathbf{a}_j' \cdot \mathbf{b}_j' = 2\pi\delta_{ij}$ $(i, j = x, y)$. Next, we write the translation vector $\mathbf{R}'$ of $\hat{T}_{\mathbf{R}'}$ in terms of integers n_1 and n_2,

$$\mathbf{R}' = n_1 \mathbf{a}_1' + n_1 \mathbf{a}_2'. \tag{15.37}$$

Further, we use an even number $\mathcal{N}_f \gg 1$ to impose periodic boundary conditions $\hat{T}_{\mathcal{N}_f \mathbf{a}_j'} = 1$ $(j = 1, 2)$. We can thereby distinguish the degenerate eigenstates belonging to ε_N' in (15.30) by the magnetic Bloch vector:

$$\mathbf{q}' = \frac{m_1}{\mathcal{N}_f}\mathbf{b}_1' + \frac{m_2}{\mathcal{N}_f}\mathbf{b}_2' \qquad \left(m_j = -\frac{\mathcal{N}_f}{2} + 1, -\frac{\mathcal{N}_f}{2} + 2, \cdots, \frac{\mathcal{N}_f}{2}\right), \tag{15.38}$$

which specifies the eigenstate of $\hat{T}_{\mathbf{R}'}$. To be specific, the eigenfunction associated with $N = 0$ and $\mathbf{q}' = \mathbf{0}$ is given by [10]

$$\phi_{00}(\mathbf{r}') = \frac{1}{\mathcal{N}_\mathrm{f}\sqrt{\sqrt{\pi}\, l'_\mathrm{c} a'_2}} \sum_{n=-\mathcal{N}_\mathrm{f}/2+1}^{\mathcal{N}_\mathrm{f}/2} \exp\left[-\mathrm{i}\frac{na'_{1x}}{l'^2_\mathrm{c}}\left(y' - \frac{na'_{1y}}{2}\right) - \frac{(x'-na'_{1x})^2}{2l'^2_\mathrm{c}}\right]. \tag{15.39}$$

Further, the eigenfunction of $N = 0$ and $\mathbf{q}' \neq \mathbf{0}$ can be obtained by a magnetic translation from $\phi_{00}(\mathbf{r})$,

$$\phi_{0\mathbf{q}'}(\mathbf{r}') = \hat{T}_{l'^2_\mathrm{c}\mathbf{q}'\times\hat{\mathbf{z}}'}\phi_{00}(\mathbf{r}'). \tag{15.40}$$

Eigenfunctions of the higher Landau levels obey the recursion relations [15]:

$$\phi_{N\mathbf{q}'}(\mathbf{r}') = \frac{1}{\sqrt{N}}\hat{a}^\dagger \phi_{N-1\mathbf{q}'}(\mathbf{r}'). \tag{15.41}$$

It follows from (15.38) that the number of distinct wave vectors is equal to the number $\mathcal{N}_\mathrm{f}^{\,2}$ of possible states in the area spanned by $\mathcal{N}_\mathrm{f}\mathbf{a}'_1$ and $\mathcal{N}_\mathrm{f}\mathbf{a}'_2$. Also, the eigenfunctions can be shown to satisfy orthonormality relations $\langle\phi_{0\mathbf{q}'_1}|\phi_{0\mathbf{q}'_2}\rangle = \delta_{\mathbf{q}'_1\mathbf{q}'_2}$. Hence, the degenerate states belonging to the lowest Landau level are now transformed into an orthonormal set of magnetic Bloch states that satisfy [10]

$$\hat{T}_{\mathbf{R}'}\phi_{0\mathbf{q}'}(\mathbf{r}') = \mathrm{e}^{-\mathrm{i}\mathbf{q}'\cdot\mathbf{R}' - \mathrm{i}\pi n_1 n_2}\phi_{0\mathbf{q}'}(\mathbf{r}'). \tag{15.42}$$

To find the zeros of $\phi_{00}(\mathbf{r}')$, we use the following symmetry relations:

$$\phi_{00}(\mathbf{r}' - \mathbf{R}') = \mathrm{e}^{\mathrm{i}\kappa\bar{B}' R'_x (y' - R'_y/2) - \mathrm{i}\pi n_1 n_2}\phi_{00}(\mathbf{r}'), \tag{15.43}$$

$$\phi_{00}(-\mathbf{r}') = \phi_{00}(\mathbf{r}'). \tag{15.44}$$

The first equality is obtained from (15.42) by setting $\mathbf{q}' = \mathbf{0}$ and rearranging its left-hand side using the second expression of (15.33). The second equality obtains by replacing $\mathbf{r}' \to -\mathbf{r}'$ in (15.39) and subsequently changing $n \to -n$ in the summation. Let us set $\mathbf{r}' = \mathbf{R}'/2$ in (15.43) and use (15.44) on the left-hand side. We thereby obtain $\phi_{00}(\mathbf{R}'/2) = \mathrm{e}^{-\mathrm{i}\pi n_1 n_2}\phi_{00}(\mathbf{R}'/2)$, concluding that

$$\phi_{00}(n_1\mathbf{a}'_1/2 + n_2\mathbf{a}'_2/2) = 0 \quad \text{when } n_1 n_2 \text{ is odd.} \tag{15.45}$$

Therefore, the zeros of $\phi_{00}(\mathbf{r}')$ are distributed periodically, their total number being equal to $\mathcal{N}_\mathrm{f}^{\,2}$. One may also confirm numerically that the wave function changes its phase by -2π for a counterclockwise rotation around every zero point. Figure 15.2 plots the amplitude of the sum in (15.39) for $a'_2/a'_1 = 1$ and $\theta_\mathrm{A} \equiv \cos^{-1}(\mathbf{a}'_1 \cdot \mathbf{a}'_2/a'_1 a'_2) = \pi/3$. We can clearly observe a periodic hexagonal arrangement of zeros.

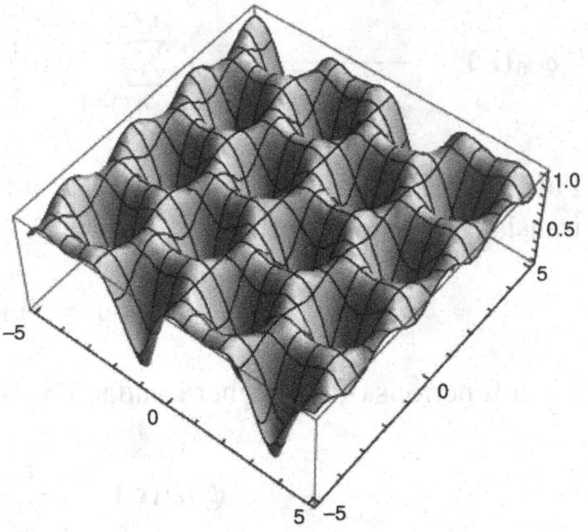

Fig. 15.2 Plot of the amplitude of the sum in (15.39) for $a_2'/a_1' = 1$ and $\theta_A \equiv \cos^{-1}(\mathbf{a}_1' \cdot \mathbf{a}_2'/a_1'a_2') = \pi/3$ over $|x'|, |y'| \leq 5l_c'$ in units of $l_c' = 1$

15.5.2 *Minimization of the Free Energy Functional*

Basis functions (15.39) and (15.40) with translational symmetry (15.42) are suitable for describing superconductivity that extends over the whole system. Moreover, states with different $\mathbf{q}'$'s are connected by magnetic translations, as given in (15.40), so that they are essentially the same. Hence, we choose a single magnetic Bloch state, e.g., that of $\mathbf{q}' = \mathbf{0}$, to expand the pair potential $\Delta'(\mathbf{r}')$ for $H' \lesssim H_{c2}'$,

$$\Delta'(\mathbf{r}') = c_0\sqrt{V'}\phi_{00}(\mathbf{r}'). \tag{15.46}$$

Choosing a single $\mathbf{q}'$ represents spontaneously broken translational symmetry of the flux-line lattice. As $\phi_{00}(\mathbf{r}') \propto 1/\sqrt{V'}$ from the normalization condition, we have extracted the factor $\sqrt{V'}$ to make c_0 of order 1.

Correspondingly, we assume periodicity $\tilde{\mathbf{A}}'(\mathbf{r}' + \mathbf{R}') = \tilde{\mathbf{A}}'(\mathbf{r}')$ for the second term of (15.26) and expand it as

$$\tilde{\mathbf{A}}'(\mathbf{r}') = \sum_{\mathbf{K}'} \tilde{\mathbf{A}}'(\mathbf{K}')\, e^{i\mathbf{K}'\cdot\mathbf{r}'}, \tag{15.47}$$

where $\mathbf{K}'$ is a reciprocal lattice vector expressible in terms of (15.36) and integers m_j $(j = 1, 2)$ as

$$\mathbf{K}' = m_1\mathbf{b}_1' + m_2\mathbf{b}_2'. \tag{15.48}$$

It follows from $\tilde{\mathbf{A}}'(\mathbf{r}')$ being real and (15.27) that

$$\tilde{\mathbf{A}}'^{*}(\mathbf{K}') = \tilde{\mathbf{A}}'(-\mathbf{K}'), \qquad \mathbf{K}' \cdot \tilde{\mathbf{A}}'(\mathbf{K}') = 0 \tag{15.49}$$

holds.

We now use the Ritz method [6] to obtain an approximate solution of (15.18) and (15.19) by minimizing the free energy in terms of the variational parameters, i.e., expansion parameters $(c_0, \tilde{\mathbf{A}}'_{\mathbf{K}'})$, angle $\theta_{\mathrm{A}} \equiv \cos^{-1}(\mathbf{a}'_1 \cdot \mathbf{a}'_2/a'_1 a'_2)$ between the two primitive vectors, and their ratio a'_2/a'_1. Specifically, we substitute (15.26), (15.28), (15.46), and (15.47) into (15.17), use (15.41), and perform the integration. We thereby transform F_{sn} into

$$F_{\mathrm{sn}} \approx \mu_0 H_{\mathrm{c}}^2 V \left\{ \left(\frac{\bar{B}'}{\kappa} - 1 \right) |c_0|^2 + \frac{|c_0|^4}{2} I^{(4)}_{00,00} + \bar{B}'^2 + \sum_{\mathbf{K}'} K'^2 |\tilde{\mathbf{A}}'(\mathbf{K}')|^2 \right.$$
$$\left. + \mathrm{i}\sqrt{\frac{2\bar{B}'}{\kappa}} \sum_{\mathbf{K}'} I_{01}(-\mathbf{K}') \left[\tilde{A}'_x(\mathbf{K}') - \mathrm{i}\tilde{A}'_y(\mathbf{K}') \right] |c_0|^2 \right\}, \tag{15.50}$$

where $I^{(4)}_{00,00}$ and $I_{N_1 N_2}(\mathbf{K}')$ are defined by

$$I^{(4)}_{00,00} \equiv V' \int |\phi_{00}(\mathbf{r}')|^4 \, \mathrm{d}^3 r' = \frac{\sqrt{2\pi} l'_{\mathrm{c}}}{a'_2} \sum_{n_1 n_2} \exp\left[-\frac{(n_1^2 + n_2^2) a'^2_{1x}}{2 l'^2_{\mathrm{c}}} + \mathrm{i} \frac{n_1 n_2 a'_{1x} a'_{1y}}{l'^2_{\mathrm{c}}} \right], \tag{15.51}$$

$$I_{N_1 N_2}(\mathbf{K}') \equiv \int \phi^*_{N_1 \mathbf{0}}(\mathbf{r}') \phi_{N_2 \mathbf{0}}(\mathbf{r}') \, \mathrm{e}^{-\mathrm{i}\mathbf{K}' \cdot \mathbf{r}'} \mathrm{d}^3 r'$$
$$= \frac{1}{\sqrt{N_2}} \left[\sqrt{N_1} I_{N_1 - 1, N_2 - 1}(\mathbf{K}') - \frac{\mathrm{i}(K'_x + \mathrm{i}K'_y)}{\sqrt{2\kappa \bar{B}'}} I_{N_1, N_2 - 1}(\mathbf{K}') \right]. \tag{15.52}$$

The second expression of (15.52) has been obtained by substituting (15.41) with $N \to N_2$ and $\mathbf{q}' \to \mathbf{0}$, performing integration by parts in terms of $\hat{a}^\dagger$ in (15.22), and writing $\hat{a}\,\phi_{N_1 \mathbf{0}} = \sqrt{N_1} \phi_{N_1 - 1 \mathbf{0}}$. Using (15.49) and (15.52), we can express the $I_{01}(-\mathbf{K}')$ term in (15.50) as

$$I_{01}(-\mathbf{K}') \left[\tilde{A}'_x(\mathbf{K}') - \mathrm{i}\tilde{A}'_y(\mathbf{K}') \right] = I_{00}(-\mathbf{K}') \frac{\left[\mathbf{K}' \times \tilde{\mathbf{A}}'(\mathbf{K}') \right] \cdot \hat{\mathbf{z}}'}{\sqrt{2\kappa \bar{B}'}}. \tag{15.53}$$

Let us substitute (15.53) into (15.50) and calculate the extremal conditions $\partial F_{\mathrm{sn}}/\partial |c_0|^2 = 0$ and $\partial F_{\mathrm{sn}}/\tilde{\mathbf{A}}'(-\mathbf{K}') = \mathbf{0}$. They are given by

$$\frac{\bar{B}'}{\kappa} - 1 + I^{(4)}_{00,00} |c_0|^2 + \frac{\mathrm{i}}{\kappa} \sum_{\mathbf{K}'} I_{00}(-\mathbf{K}') \left[\mathbf{K}' \times \tilde{\mathbf{A}}'(\mathbf{K}') \right] \cdot \hat{\mathbf{z}}' = 0, \tag{15.54}$$

$$\tilde{\mathbf{A}}'(\mathbf{K}') = \frac{\mathrm{i}}{2\kappa K'^2} I_{00}(\mathbf{K}') \, \hat{\mathbf{z}}' \times \mathbf{K}' |c_0|^2. \tag{15.55}$$

Further, we put (15.55) into (15.54) and use the identity (**Problem** 15.2):

$$I^{(4)}_{00,00} = \sum_{\mathbf{K}'} I_{00}(\mathbf{K}')I_{00}(-\mathbf{K}'), \tag{15.56}$$

to obtain $|c_0|^2$ as

$$|c_0|^2 = \frac{2\kappa(\kappa - \bar{B}')}{(2\kappa^2 - 1)I^{(4)}_{00,00}}. \tag{15.57}$$

By substituting (15.53) and (15.54) into (15.50) and then using (15.55), (15.56), and (15.57), the equilibrium free energy is expressed concisely as

$$\begin{aligned} F_{\rm sn} &= \mu_0 H_{\rm c}^2 V \left[-\frac{|c_0|^4}{2} I^{(4)}_{00,00} + \bar{B}'^2 + \sum_{\mathbf{K}'} K'^2 |\tilde{\mathbf{A}}'(\mathbf{K}')|^2 \right] \\ &= \mu_0 H_{\rm c}^2 V \left[-\frac{(\kappa - \bar{B}')^2}{(2\kappa^2 - 1)I^{(4)}_{00,00}} + \bar{B}'^2 \right]. \end{aligned} \tag{15.58}$$

Further, it should be minimized with respect to $I^{(4)}_{00,00}$, which contains information about the lattice structure. A numerical evaluation of (15.51) [12] verifies that $I^{(4)}_{00,00}$ takes its minimum for the hexagonal lattice of Fig. 15.2,

$$I^{(4)}_{00,00} = 1.16. \tag{15.59}$$

That the hexagonal lattice is stable follows naturally by recalling that it is the densest-packing structure in two dimensions. Nevertheless, as the temperature is lowered from $T_{\rm c}$, the Fermi surface anisotropy changes the stability of the structure and $H_{\rm c2}$ to such an extent that the hexagonal lattice is seldom observed in certain materials [3, 11].

Next, we calculate the magnetic field H based on (15.8) by differentiating (15.50) in terms of $\bar{B} = \sqrt{2}\mu_0 H_{\rm c}\bar{B}'$. As $\partial F_{\rm sn}/\partial|c_0|^2 = 0$ and $\partial F_{\rm sn}/\tilde{\mathbf{A}}'(-\mathbf{K}') = \mathbf{0}$ hold in equilibrium, we only need to consider the explicit $\bar{B}'$ dependences. Noting that $I^{(4)}_{00,00}$ and $I_{N_1 N_2}(\mathbf{K}')$ are constants whereas $K'^2 \propto l_{\rm c}'^{-2} \propto \bar{B}'$, the differentiation yields

$$H = \frac{H_{\rm c}}{\sqrt{2}} \left[2\bar{B}' + \frac{|c_0|^2}{\kappa} + O(|c_0|^4) \right] \approx \frac{\bar{B}}{\mu_0} + \frac{H_{\rm c2} - H}{(2\kappa^2 - 1)I^{(4)}_{00,00}}.$$

Comparing this expression with (15.7), we obtain the magnetization for $H \lesssim H_{\rm c2}$,

$$M = -\frac{H_{\rm c2} - H}{(2\kappa^2 - 1)I^{(4)}_{00,00}}. \tag{15.60}$$

Its gradient in terms of $H_{c2} - H$ increases as κ is decreased towards $1/\sqrt{2}$, and we finally have a discontinuous transition at $\kappa = 1/\sqrt{2}$ between the normal and Meissner states.

The above consideration can be extended to lower fields by retaining terms of $O(\tilde{\mathbf{A}}'^2)$ in (15.28) and including contributions from higher Landau levels ($N > 0$) in (15.46) [10]. The solid line in Fig. 15.3 plots the normalized magnetization $M' \equiv M/\sqrt{2}H_c$ as a function of $H' \equiv H/\sqrt{2}H_c$ for $\kappa = 5$, whereas the dotted line represents the approximate curve of (15.60) near H_{c2} ($H'_{c2} = \kappa$). As the magnetic field is reduced from H'_{c2}, $-M'$ starts to deviate upwards from (15.60), indicating an accelerating demagnetization. The upward curvature continues to increase through $H'_c = 1/\sqrt{2} = 0.707$ and eventually stops at a certain field $H'_{c1} < H'_c$, where complete diamagnetism ($\bar{B}' = 0$) is achieved because of the Meissner effect.

15.6 Lower Critical Field H_{c1}

Equation (10.43) infers that the magnetic flux in superconductors should be quantized in units of Φ_0 in consequence of the single-valuedness of the macroscopic wave function. In every type-II superconductor, there exists a lower critical field H_{c1} at which a single flux starts to penetrate into the bulk. To study H_{c1}, we here solve the GL equations in cylindrical coordinates; see also Sect. 16.3 on this topic for a detailed study at all temperatures.

Let us express the wave function and vector potential for a single flux quantum using $\mathbf{r}' = (r' \cos\varphi', r' \sin\varphi', z')$,

$$\Psi'(\mathbf{r}') = f(r')\,\mathrm{e}^{-\mathrm{i}\varphi'}, \qquad \mathbf{A}'(\mathbf{r}') = A'(r')\hat{\boldsymbol{\varphi}}', \tag{15.61}$$

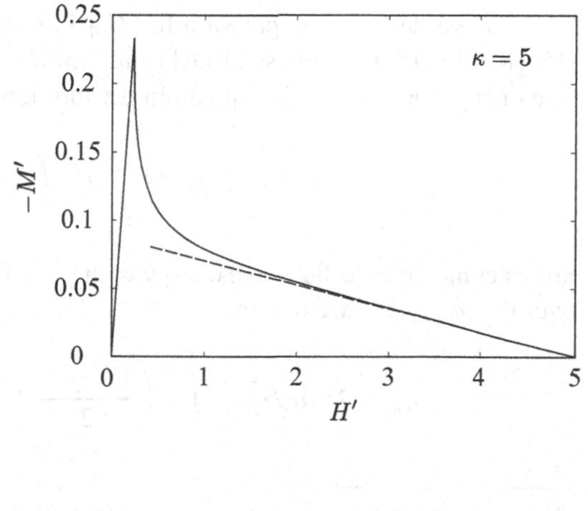

Fig. 15.3 Magnetization curve for $\kappa = 5$

where $f'(r')$ and $A'(r')$ are real functions and $\hat{\boldsymbol{\varphi}}' \equiv \hat{\mathbf{z}}' \times \hat{\mathbf{r}}'$ denotes the unit vector along the φ' direction.[2] This $\Psi'(\mathbf{r}')$ changes its phase by -2π for a counterclockwise rotation around the z axis. Hence, $f(0) = 0$ should hold because of the single-valuedness of the wave function. The corresponding microscopic flux density $\mathbf{B}'(\mathbf{r}') = \nabla' \times \mathbf{A}'(\mathbf{r}')$ is given in cylindrical coordinates by [4]

$$\mathbf{B}'(\mathbf{r}') = \frac{\hat{\mathbf{z}}'}{r'} \frac{\mathrm{d}}{\mathrm{d}r'}[r'A'(r')] = -\frac{\hat{\mathbf{z}}'}{r'} \frac{\mathrm{d}}{\mathrm{d}r'}[r'Q'(r')], \tag{15.62}$$

where we have introduced a new function,

$$Q'(r') \equiv -A'(r') + \frac{1}{\kappa r'}, \tag{15.63}$$

to express the derivative of $r'A'(r') = -r'Q'(r') + 1/\kappa$ in terms of $r'Q'(r')$. Substituting (15.61) into (15.18) and (15.19) and performing the differentiations in the cylindrical coordinates [4], for example $\kappa^{-1}\nabla'\varphi' = (\kappa r')^{-1}\hat{\boldsymbol{\varphi}}'$, we thereby obtain the GL equations for a single flux quantum,

$$-\frac{1}{\kappa^2 r'} \frac{\mathrm{d}}{\mathrm{d}r'}\left(r' \frac{\mathrm{d}f'}{\mathrm{d}r'}\right) + Q'^2 f' = f'(1 - f'^2), \tag{15.64}$$

$$\frac{\mathrm{d}}{\mathrm{d}r'}\left[\frac{1}{r'} \frac{\mathrm{d}}{\mathrm{d}r'}(r'Q')\right] = Q' f'^2. \tag{15.65}$$

Equations (15.64) and (15.65) should be solved with boundary conditions:

$$f'(0) = 0, \quad f'(\infty) = 1, \quad r'Q'(r')\bigg|_{r'=0} = \frac{1}{\kappa}, \quad Q'(\infty) = 0. \tag{15.66}$$

Next, we derive an expression for H_{c1} in terms of f' and B'. Let us substitute (15.18) into (15.17) and use (15.61) and (15.62). We thereby obtain the equilibrium free energy for a single flux quantum per unit length along the z' axis,

$$F_{\mathrm{sn}}^{(\mathrm{eq})} = \mu_0 H_{\mathrm{c}}^2 \lambda_{\mathrm{L}}^3 \int \mathrm{d}^2 r' \left(-\frac{f'^4}{2} + B'^2\right).$$

Subtracting the zero-field free energy with $f' = 1$ and $B' = 0$ from it and integrating over $0 \le \varphi' \le 2\pi$, we obtain

$$\varepsilon_1 \equiv 2\pi\mu_0 H_{\mathrm{c}}^2 \lambda_{\mathrm{L}}^3 \int_0^\infty \left(\frac{1 - f'^4}{2} + B'^2\right) r' \mathrm{d}r', \tag{15.67}$$

[2] The minus sign in the phase of $\Psi'(\mathbf{r}')$ originates from the negative electronic charge $e < 0$. From classical mechanics, when the field is applied along the positive z direction, an electron rotates in a clockwise manner around the z axis, and the current flows in the opposite direction.

which represents the energy of formation for a singly quantized vortex. An assembly of such vortices with area density σ yields the flux density:

$$\bar{B} \equiv \sigma \frac{h}{2|e|} = \sigma \frac{\pi\hbar}{|e|}.$$

Now, the lower critical field H_{c1} is obtained from the condition that the free energy (15.9) as a function of H is equal to the zero-field value, i.e.,

$$0 = \frac{G_{sn}(H_{c1}) - G_{sn}(0)}{V} = \frac{\sigma\varepsilon_1}{\lambda_L} - \bar{B}H_{c1} = \sigma\left[\frac{\varepsilon_1}{\lambda_L} - \frac{\pi\hbar}{|e|}H_{c1}\right].$$

Hence, we obtain

$$H_{c1} = \frac{|e|\varepsilon_1}{\pi\hbar\lambda_L} = H_c\frac{\kappa}{\sqrt{2}}\int_0^\infty \left(\frac{1-f'^4}{2} + B'^2\right) r'\mathrm{d}r', \tag{15.68}$$

where we have used (15.11)–(15.15) in expressing the prefactor.

Solving (15.64) and (15.65) for a given κ and substituting the results into (15.68), we obtain $H_{c1} = H_{c1}(\kappa)$. While this procedure needs numerical studies in general, we can find analytic solutions for some limiting cases.

First, $\kappa = 1/\sqrt{2}$ is special in that (15.64) and (15.65) have first integrals [8]:

$$\frac{\mathrm{d}f'}{\mathrm{d}r'} = \frac{Q'f'}{\sqrt{2}}, \qquad B' = -\frac{1}{r'}\frac{\mathrm{d}}{\mathrm{d}r'}(r'Q') = \frac{1-f'^2}{\sqrt{2}}. \tag{15.69}$$

Indeed, substitution of the second equality into (15.65) yields the first equality, and that of the first equality into the left-hand side of (15.64) for $\kappa = 1/\sqrt{2}$ using (15.69) confirms the second equality. We use this B' in (15.68) for $\kappa = 1/\sqrt{2}$ and write the resulting expression as

$$H_{c1} = \frac{H_c}{2}\int_0^\infty (1-f'^2)r'\mathrm{d}r' = \frac{H_c}{\sqrt{2}}\int_0^\infty B'r'\mathrm{d}r' = \frac{H_c}{\sqrt{2}}\left[-r'Q'\right]_{r'=0}^{\infty} = H_c. \tag{15.70}$$

Combining this result with (15.24), we conclude $H_{c1} = H_{c2} = H_c$ at $\kappa = 1/\sqrt{2}$. Thus, we have confirmed that $\kappa = 1/\sqrt{2}$ is the critical value that separates superconductors into type-I and type-II.

Second, we focus on the region $r' \ll 1$ near the vortex core, where (15.63) can be approximated as $Q'(r') \approx 1/\kappa r'$. Substituting this into (15.64) yields

$$\frac{1}{\kappa^2}\left[-\frac{1}{r'}\frac{\mathrm{d}}{\mathrm{d}r'}\left(r'\frac{\mathrm{d}f'}{\mathrm{d}r'}\right) + \frac{f'}{r'^2}\right] \approx f'(1-f'^2). \tag{15.71}$$

Further, near the vortex core, we set the right-hand side to zero to obtain

$$f'(r') \approx c_0 r' \qquad (r' \approx 0). \tag{15.72}$$

Thus, $f'(r')$ grows linearly for $r' \approx 0$ with some gradient c_0. We expect $c_0 \sim \kappa$, as κ^{-1} is the coherence length in the dimensionless units that gives the scale of the variation in $f'(r')$; see (15.15) on this point. Let us substitute (15.72) and $Q'(r') \approx 1/\kappa r'$ into (15.65), integrate both sides over $0 \leq r' \leq r'_1$, use (15.62), and set $r'_1 \to r'$. We thereby obtain

$$B'(r') \approx B'(0) - \frac{c_0^2}{2\kappa} r'^2. \tag{15.73}$$

Thus, the magnetic flux density decreases quadratically from the vortex center.

Third, we can approximate $f' \approx 1$ for $r' \gtrsim \kappa^{-1}$. Substitution of this into (15.65) finds that the equation for Q' is the modified Bessel's equation of first order [1, 4]. Thus, the solution is given by

$$Q'(r) \approx cK_1(r'), \tag{15.74}$$

where c is a constant. Using this expression in (15.62), we find the flux density,

$$B'(r') \approx cK_0(r'). \tag{15.75}$$

The asymptotic expression of $K_0(r')$ for $r' \to \infty$ [1, 4] implies that the flux density for $r' \gg 1$ decreases exponentially, $(\pi/2r')^{1/2} \mathrm{e}^{-r'}$.

Fourth, we can simplify the problem for the extreme type-II limit of $\kappa \gg 1$. In this case, we expect that $f'(r')$ grows rapidly over $0 \leq r' \lesssim \kappa^{-1}$ and asymptotes to 1. Hence, f' for $\kappa^{-1} \ll r' \ll 1$ should obey (15.71). Let us parameterize the solution in this region as $f'(r') \approx 1 - ar'^{\nu}$ ($|a| \ll 1$) in terms of unknown constants a and ν, substitute it into (15.71), and use the leading terms to determine a and ν. We thereby obtain $a = 1/2\kappa^2$ and $\nu = -2$, and hence

$$f'(r') \approx 1 - \frac{1}{2\kappa^2 r'^2}. \tag{15.76}$$

As for the magnetic field, we can use $K_1(r') \approx 1/r'$ for $r' \ll 1$ to equate the leading terms of (15.63) and (15.74) for $\kappa^{-1} \ll r' \ll 1$. This yields $c = 1/\kappa$ for the coefficient of (15.74). Substituting this into (15.75), we thereby obtain the flux density for $r' \gtrsim \kappa^{-1}$,

$$B'(r') = \frac{K_0(r')}{\kappa} \approx -\frac{\ln r'}{\kappa}, \tag{15.77}$$

where the second expression holds for $r' \ll 1$. We now estimate (15.68) for H_{c1} using these results. To this end, we note that the main contribution to the integral for $\kappa \gg 1$ originates from the region $\kappa^{-1} \ll r' \lesssim 1$, whereas that from $r' \gtrsim 1$ is exponentially small. Hence, we may replace the lower and upper limits of (15.68) with κ^{-1} and 1, respectively, substitute (15.76) and (15.77), and perform the integration. The contribution from the flux density can be neglected, and we obtain

$$H_{c1} \approx H_c \frac{\kappa}{\sqrt{2}} \int_{\kappa^{-1}}^{1} \frac{1 - f'^4}{2} r' \mathrm{d}r' = H_c \frac{\ln \kappa}{\sqrt{2}\kappa}.$$

By solving (15.64) and (15.65) numerically, a better expression for $\kappa \gg 1$ has been obtained [9],

$$H_{c1} = H_c \frac{\ln \kappa + 0.497}{\sqrt{2}\kappa}. \tag{15.78}$$

As may be expected from (15.70) and (15.78), the ratio H_{c1}/H_c decreases from 1 at $\kappa = 1/\sqrt{2}$ monotonically as κ is increased.

We note that the London equation (10.46) with a singularity $n = -1$ at $\mathbf{r}_0 = \mathbf{0}$ yields (**Problem** 15.3)

$$\mathbf{B}(\mathbf{r}) = \frac{\hat{\mathbf{z}}\Phi_0}{2\pi\lambda_L^2} K_0(r/\lambda_L), \tag{15.79}$$

which agrees with (15.77) for the extreme type-II superconductors, as may be confirmed by multiplying (15.77) by $\sqrt{2}\mu_0 H_c$ and using (15.11)–(15.15). Thus, the London equation can be regarded as the extreme type-II limit of the GL equations near T_c, but also has the advantage of being applicable at lower temperatures.

Although we have not considered it here, the possibility of doubly or triply quantized vortices may be excluded in the bulk based on the observation that distributing singly quantized vortices homogeneously on a macroscopic scale realizes the most homogeneous distribution of magnetic flux density.

Problems

15.1. Show that vector $\mathbf{H}$ defined by (15.8) satisfies (15.5).

15.2. Prove (15.56) between (15.51) and (15.52) for $N_1 = N_2 = 0$.

15.3. Show that (10.46) for $n = -1$ and $\mathbf{r}_0 = \mathbf{0}$ gives (15.79) as the solution.

References

1. M. Abramowitz, I.A. Stegun (eds.), *Handbook of Mathematical Functions: With Formulas, Graphs, and Mathematical Tables* (Dover, New York, 1965)
2. A.A. Abrikosov, J. Exp. Theor. Phys. **32**, 1442 (1957). (Sov. Phys. JETP **5**, 1174 (1957)) Corrections are summarized as follows. First, the square lattice given by Eqs. (20)–(22) is metastable, and the equilibrium is given by the hexagonal lattice with $\beta = 1.16$ [12]. See Eq. (15.59) in the text. Second, numerical coefficients of Eqs. (36) and (37) should be corrected as 0.08 → 0.497 and 0.18 → 0.282 [9].

3. H.M. Adachi, M. Ishikawa, T. Hirano, M. Ichioka, K. Machida, J. Phys. Soc. Jpn. **80**, 113702 (2011)
4. G.B. Arfken, H.J. Weber, *Mathematical Methods for Physicists* (Academic, New York, 2012)
5. E. Brown, Phys. Rev. **133**, A1038 (1964)
6. I.M. Gelfand, S.V. Fomin, *Calculus of Variations* (Prentice-Hall, Englewood Cliffs, 1963)
7. V.L. Ginzburg, L.D. Landau, J. Exp. Theor. Phys. **20**, 1064 (1950)
8. J.L. Harden, V. Arp, Cryogenics **3**, 105 (1963)
9. C.-R. Hu, Phys. Rev. B **6**, 1756 (1972)
10. T. Kita, J. Phys. Soc. Jpn. **67**, 2067 (1998)
11. T. Kita, M. Arai, Phys. Rev. B **70**, 224522 (2004)
12. W.H. Kleiner, L.M. Roth, S.H. Autler, Phys. Rev. **133**, A1226 (1964)
13. L.D. Landau, E.M. Lifshitz, *Quantum Mechanics: Non-relativistic Theory*, 3rd edn. (Butterworth-Heinemann, Oxford, 1991)
14. A.B. Pippard, Proc. R. Soc. Lond. A **216**, 547 (1953)
15. J.J. Sakurai, *Modern Quantum Mechanics*, rev. edn. (Addison-Wesley, Reading, 1994)

Chapter 16
Surfaces and Vortex Cores

Abstract We discuss topics concerning inhomogeneous superconducting states of s-wave pairing that are realized near boundaries and vortex cores. First, we consider a normal-superconducting interface to show that an electron approaching the interface from the normal side experiences a peculiar reflection called *Andreev reflection*, which backscatters a hole; the energy flow is substantially blocked through the interface because of this reflection. Next, we study quasiparticles around a vortex core to find that there exist localized quasiparticle states called *Caroli–de Gennes–Matricon mode* below the bulk energy gap; they recover a T-linear term in the specific heat and are also responsible for the electric resistivity when vortices are forced to move. Finally, we use the quasiclassical Eilenberger equations to study in detail an isolated s-wave vortex and its local density of states, Figs. 16.3–16.5.

16.1 Andreev Reflection

We consider a normal-superconducting interface as given in Fig. 16.1 to study how a normal electron or hole incident on the boundary is scattered and transmitted.

A starting point is the BdG equations. For weak-coupling s-wave pairing, they read

$$\begin{bmatrix} \hat{\mathscr{H}} & \Delta(\mathbf{r}) \\ \Delta^*(\mathbf{r}) & -\hat{\mathscr{H}}^* \end{bmatrix} \begin{bmatrix} u(\mathbf{r}) \\ v(\mathbf{r}) \end{bmatrix} = E \begin{bmatrix} u(\mathbf{r}) \\ v(\mathbf{r}) \end{bmatrix}, \tag{16.1}$$

with $\hat{\mathscr{H}} = \hat{\mathbf{p}}^2/2m - \mu$. This is a direct extension of the $(1,4)$ elements of (9.3) to inhomogeneous systems. Equations for the $(2,3)$ elements are obtained from (16.1) by changing the signs of Δ and v simultaneously. Hence, they are identical to (16.1) so that it suffices to consider only (16.1). As for the pair potential, we here adopt a model form:

$$\Delta(\mathbf{r}) = \Delta_0 \theta(x), \tag{16.2}$$

© Springer Japan 2015

T. Kita, *Statistical Mechanics of Superconductivity*, Graduate Texts in Physics,
DOI 10.1007/978-4-431-55405-9_16

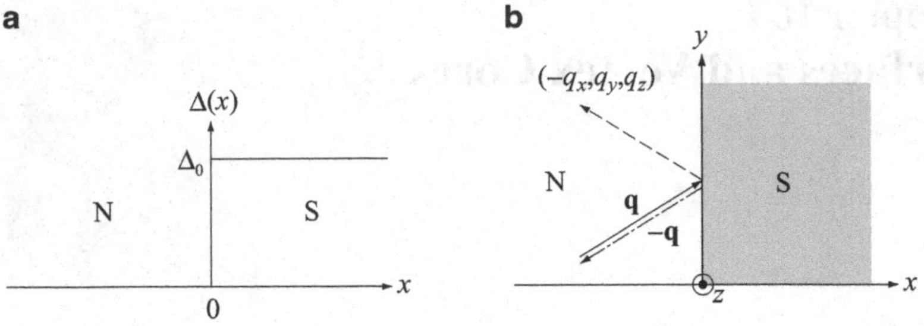

Fig. 16.1 (**a**) Model pair potential around a normal-superconducting (NS) interface. (**b**) Andreev reflection (*chain line*) at an NS interface

which enables an analytic treatment. In principle, the pair potential should be determined self-consistently and is expected to change smoothly near the interface. However, we may capture the essence of the scattering by studying this simplified model.

Let us express the eigenvector in (16.1) as

$$\begin{bmatrix} u(\mathbf{r}) \\ v(\mathbf{r}) \end{bmatrix} = \mathrm{e}^{i\mathbf{k}_\mathrm{F}\cdot\mathbf{r}} \begin{bmatrix} \tilde{u}(\mathbf{r}) \\ \tilde{v}(\mathbf{r}) \end{bmatrix}, \tag{16.3}$$

where $\mathbf{k}_\mathrm{F}$ denotes a Fermi wave vector. Functions $(\tilde{u}, \tilde{v})$ thereby introduced are expected to vary slowly compared with k_F^{-1}. Next, we substitute (16.3) into (16.1) and approximate the kinetic energy of the upper element as

$$\begin{aligned} \hat{\mathscr{K}} u(\mathbf{r}) &= \left(\frac{\hat{\mathbf{p}}^2}{2m} - \mu\right) \mathrm{e}^{i\mathbf{k}_\mathrm{F}\cdot\mathbf{r}} \tilde{u}(\mathbf{r}) = \mathrm{e}^{i\mathbf{k}_\mathrm{F}\cdot\mathbf{r}} \left[\frac{(-i\hbar\nabla + \hbar\mathbf{k}_\mathrm{F})^2}{2m} - \mu\right] \tilde{u}(\mathbf{r}) \\ &\approx \mathrm{e}^{i\mathbf{k}_\mathrm{F}\cdot\mathbf{r}} \mathbf{v}_\mathrm{F} \cdot (-i\hbar\nabla)\tilde{u}(\mathbf{r}), \end{aligned}$$

with $\mathbf{v}_\mathrm{F} \equiv \hbar\mathbf{k}_\mathrm{F}/m$. Here we have used $\mu = \hbar^2 k_\mathrm{F}^2/2m$ and also neglected the $\nabla^2\tilde{u}$ term. Equation (16.1) is thereby simplified to

$$\begin{bmatrix} -i\hbar\mathbf{v}_\mathrm{F}\cdot\nabla & \Delta(\mathbf{r}) \\ \Delta^*(\mathbf{r}) & i\hbar\mathbf{v}_\mathrm{F}\cdot\nabla \end{bmatrix} \begin{bmatrix} \tilde{u}(\mathbf{r}) \\ \tilde{v}(\mathbf{r}) \end{bmatrix} = E \begin{bmatrix} \tilde{u}(\mathbf{r}) \\ \tilde{v}(\mathbf{r}) \end{bmatrix}. \tag{16.4}$$

This is the Andreev approximation to the BdG equations [2], which is essentially identical to the quasiclassical approximation adopted in Sect. 14.3.

Equation (16.4) on the normal side reduces to

$$\begin{bmatrix} -i\hbar\mathbf{v}_\mathrm{F}\cdot\nabla & 0 \\ 0 & i\hbar\mathbf{v}_\mathrm{F}\cdot\nabla \end{bmatrix} \begin{bmatrix} \tilde{u}(\mathbf{r}) \\ \tilde{v}(\mathbf{r}) \end{bmatrix} = E \begin{bmatrix} \tilde{u}(\mathbf{r}) \\ \tilde{v}(\mathbf{r}) \end{bmatrix}.$$

Its eigenfunctions have the general form:

$$\begin{bmatrix} \tilde{u}(\mathbf{r}) \\ \tilde{v}(\mathbf{r}) \end{bmatrix} = \begin{bmatrix} A_1 e^{i\mathbf{q}\cdot\mathbf{r}} \\ A_2 e^{-i\mathbf{q}\cdot\mathbf{r}} \end{bmatrix}, \tag{16.5}$$

with eigenvalues:

$$E_{\mathbf{q}} = \hbar \mathbf{v}_F \cdot \mathbf{q}. \tag{16.6}$$

Because $\mathbf{q}$ is a correction to $\mathbf{k}_F$ with $q \ll k_F$, we should choose the direction of $\mathbf{q}$ either parallel to $\mathbf{k}_F$ ($E_{\mathbf{q}} > 0$) or antiparallel to $\mathbf{k}_F$ ($E_{\mathbf{q}} < 0$) by definition. Keeping this particle-hole symmetry in mind, we shall focus our attention on the case $\mathbf{q} \parallel \mathbf{k}_F$ ($E_{\mathbf{q}} > 0$) below.

Equation (16.4) far inside the superconductor is expressible generally in terms of a plane wave as

$$\begin{bmatrix} \tilde{u}(\mathbf{r}) \\ \tilde{v}(\mathbf{r}) \end{bmatrix} = \begin{bmatrix} B_1 \\ B_2 \end{bmatrix} e^{i\mathbf{q}'\cdot\mathbf{r}}, \tag{16.7}$$

as we have seen in Sect. 8.3.4. Substituting into (16.4) with $\Delta(\mathbf{r}) = \Delta_0$, we obtain

$$\begin{bmatrix} \xi_{\mathbf{q}'} & \Delta_0 \\ \Delta_0^* & -\xi_{\mathbf{q}'} \end{bmatrix} \begin{bmatrix} B_1 \\ B_2 \end{bmatrix} = E_{\mathbf{q}'} \begin{bmatrix} B_1 \\ B_2 \end{bmatrix}, \tag{16.8}$$

with $\xi_{\mathbf{q}'} \equiv \hbar \mathbf{v}_F \cdot \mathbf{q}'$. This eigenvalue problem is identical to (9.4) for homogeneous systems. Hence, we can immediately write down the positive eigenvalue and its eigenfunction as

$$E_{\mathbf{q}'} = \sqrt{\xi_{\mathbf{q}'}^2 + |\Delta_0|^2}, \tag{16.9}$$

$$\begin{bmatrix} B_1 \\ B_2 \end{bmatrix} = C \begin{bmatrix} \sqrt{(E_{\mathbf{q}'} + \xi_{\mathbf{q}'})/2E_{\mathbf{q}'}} \\ \Delta_0^* / \sqrt{2E_{\mathbf{q}'}(E_{\mathbf{q}'} + \xi_{\mathbf{q}'})} \end{bmatrix}, \tag{16.10}$$

where C is a constant.

We now match (16.5) and (16.7) continuously at $x = 0$ by setting $E_{\mathbf{q}} = E_{\mathbf{q}'} = E > 0$ based on the energy conservation through the barrier.[1] Noting (16.10), we can express the relevant condition as

$$A_1 = C\sqrt{\frac{E + \xi_{\mathbf{q}'}}{2E}}, \qquad A_2 = C\frac{\Delta_0^*}{\sqrt{2E(E + \xi_{\mathbf{q}'})}}. \tag{16.11}$$

[1] This matching condition also makes derivatives $\partial u(\mathbf{r})/\partial x$ and $\partial v(\mathbf{r})/\partial x$ continuous at $x = 0$ to the lowest order in q/k_F. Homogeneity along the y and z directions implies $q'_y = q_y$ and $q'_z = q_z$ between (16.5) and (16.7).

We shall focus on instances $\xi_{\mathbf{q}'} > 0$ below, deferring $\xi_{\mathbf{q}'} < 0$ to **Problem** 16.1. Then, it follows from (16.9) and (16.11) that we can express $|C|^2/|A_1|^2$ and $|C|^2/|A_2|^2$ as

$$\frac{|C|^2}{|A_1|^2} = \frac{2E}{E + \sqrt{E^2 - |\Delta_0|^2}}, \qquad \frac{|C|^2}{|A_2|^2} = \frac{2E}{E - \sqrt{E^2 - |\Delta_0|^2}}. \tag{16.12}$$

Hence, inequality $|A_1| > |A_2|$ holds for $\xi_{\mathbf{q}'} > 0$, implying that the upper element of (16.5) should be regarded as the incident wave in the transmission process from the normal side.

Next, we derive a formula for the quasiparticle current to obtain the transmission coefficient. To this end, we generalize the wave function as

$$\begin{bmatrix} u(\mathbf{r}) \\ v(\mathbf{r}) \end{bmatrix} \longrightarrow \begin{bmatrix} u(\mathbf{r},t) \\ v(\mathbf{r},t) \end{bmatrix} = \mathrm{e}^{-\mathrm{i}Et/\hbar} \begin{bmatrix} u(\mathbf{r}) \\ v(\mathbf{r}) \end{bmatrix}. \tag{16.13}$$

Equation (16.1) is then expressible as

$$\mathrm{i}\hbar \frac{\partial u}{\partial t} - \left(\frac{\hat{\mathbf{p}}^2}{2m} - \mu \right) u - \Delta v = 0, \tag{16.14}$$

$$\mathrm{i}\hbar \frac{\partial v}{\partial t} + \left(\frac{\hat{\mathbf{p}}^2}{2m} - \mu \right) v - \Delta^* u = 0, \tag{16.15}$$

with $u = u(\mathbf{r}, t)$ and $v = v(\mathbf{r}, t)$. Let us multiply (16.14) and (16.15) by u^* and v^*, respectively, add them to form a single equation, and subtract its complex conjugate. We thereby obtain the continuity equation for $\rho_{\mathrm{qp}} \equiv |u|^2 + |v|^2$ in the form $\partial \rho_{\mathrm{qp}}/\partial t + \nabla \cdot \mathbf{j}_{\mathrm{qp}} = 0$ with

$$\mathbf{j}_{\mathrm{qp}} \equiv \frac{-\mathrm{i}\hbar}{2m} \left(u^* \nabla u - u \nabla u^* - v^* \nabla v + v \nabla v^* \right). \tag{16.16}$$

Because the extra t dependence introduced in (16.13) cancels out in (16.16), we now return to the original wave function (16.1) without time dependence.

Substituting (16.3) and (16.5) into (16.16), we find the quasiparticle current on the normal side,

$$\mathbf{j}^{\mathrm{n}}_{\mathrm{qp}} = \frac{\hbar}{m} \left[|A_1|^2 (\mathbf{k}_{\mathrm{F}} + \mathbf{q}) - |A_2|^2 (\mathbf{k}_{\mathrm{F}} - \mathbf{q}) \right] \approx \frac{\hbar \mathbf{k}_{\mathrm{F}}}{m} |A_1|^2 - \frac{\hbar \mathbf{k}_{\mathrm{F}}}{m} |A_2|^2. \tag{16.17}$$

The current on the superconducting side is found similarly using (16.3), (16.7), (16.9), and (16.10) with $\xi_{\mathbf{q}'} > 0$ as

$$\mathbf{j}^{\mathrm{s}}_{\mathrm{qp}} = \frac{\hbar (\mathbf{k}_{\mathrm{F}} + \mathbf{q}')}{m} \left(|B_1|^2 - |B_2|^2 \right) \approx \frac{\hbar \mathbf{k}_{\mathrm{F}}}{m} |C|^2 \frac{\xi_{\mathbf{q}'}}{E} = \frac{\hbar \mathbf{k}_{\mathrm{F}}}{m} |C|^2 \frac{\sqrt{E^2 - |\Delta_0|^2}}{E}. \tag{16.18}$$

We are now ready to calculate the transmission coefficient for an incident wave from the normal side. Combined with $|A_1| > |A_2|$ from (16.12), we can identify the z component of (16.17) with coefficient $|A_1|^2$ and $k_{Fz} > 0$ as the incident current from the normal side. The transmission coefficient, which we denote by $\mathscr{T}$, is defined as the ratio of the transmitted current density $j^{s}_{qp,z}$ relative to the incident current density. Hence, we obtain $\mathscr{T}$ using (16.12), (16.17), and (16.18),

$$\mathscr{T} = \frac{|C|^2}{|A_1|^2}\frac{\sqrt{E^2 - |\Delta_0|^2}}{E}\theta(E - |\Delta_0|) = \frac{2\sqrt{E^2 - |\Delta_0|^2}}{E + \sqrt{E^2 - |\Delta_0|^2}}\theta(E - |\Delta_0|), \tag{16.19}$$

where the step function has been introduced to express that there is no transmitted wave for $E < |\Delta_0|$. Thus, the flow of quasiparticles across the SN boundary is completely blocked for $E < |\Delta_0|$. This causes a reduction in energy flow from the normal side across the barrier at low temperatures, which explains the steep increase in the low-temperature thermal resistance in the intermediate state with a periodic SN arrangement [2].

It is interesting to note that wave function (16.5) is given as a linear combination of an incident particle with wave vector $\mathbf{q}$ and reflected hole with wave vector $-\mathbf{q}$, as depicted in Fig. 16.1b. This reflection, called the *Andreev reflection* [2], is in marked contrast to the normal-state reflection (broken line in Fig. 16.1b) in that all three components of $\mathbf{q}$ are reversed simultaneously upon reflection. It should also be emphasized that this law is relevant to the envelope functions $(\tilde{u}, \tilde{v})$ with $q \ll k_F$; see, e.g., [4] for a more detailed treatment of the transmission and reflection at the SN boundary without the Andreev approximation.

16.2 Vortex-Core States

As a second topic, we consider low-energy excitations localized around an isolated s-wave vortex.

In the presence of the magnetic field, the BdG equations in the Andreev approximation are given by

$$\hat{\mathscr{H}}_{BdG}\tilde{\mathbf{u}} = E\tilde{\mathbf{u}}, \tag{16.20}$$

where

$$\hat{\mathscr{H}}_{BdG} \equiv \begin{bmatrix} \mathbf{v}_F \cdot (-i\hbar\nabla - e\mathbf{A}) & \Delta \\ \Delta^* & \mathbf{v}_F \cdot (i\hbar\nabla - e\mathbf{A}) \end{bmatrix}, \qquad \tilde{\mathbf{u}} \equiv \begin{bmatrix} \tilde{u} \\ \tilde{v} \end{bmatrix} \tag{16.21}$$

with $e < 0$. This is modified from (16.4) to incorporate a vector potential $\mathbf{A}$. We seek a solution for an isolated s-wave vortex that is homogeneous along the z direction and isotropic in the xy plane. The corresponding pair potential and vector potential are expressible in two-dimensional polar coordinates $\mathbf{r} = (r\cos\varphi, r\sin\varphi)$ as

$$\Delta(\mathbf{r}) = \Delta(r)\mathrm{e}^{-\mathrm{i}\varphi}, \qquad \mathbf{A}(\mathbf{r}) = A(r)\hat{\boldsymbol{\varphi}}, \tag{16.22}$$

where we choose the gauge so that $\Delta(r)$ is real and positive and $\hat{\boldsymbol{\varphi}} \equiv \hat{\mathbf{z}} \times \hat{\mathbf{r}}$ denotes the unit vector along the φ direction. We note that (16.22) is identical with (15.61) for studying an isolated vortex within the GL formalism. As discussed around (15.72), amplitude $\Delta(r)$ grows linearly from $r = 0$ over the coherence length ξ.

Next, we introduce the projection of $\mathbf{v}_\mathrm{F}$ onto the xy plane and the corresponding unit vector,

$$\mathbf{v}_{\mathrm{F}\perp} \equiv \mathbf{v}_\mathrm{F} - (\hat{\mathbf{z}} \cdot \mathbf{v}_\mathrm{F})\hat{\mathbf{z}}, \qquad \hat{\mathbf{v}}_\perp \equiv \frac{\mathbf{v}_{\mathrm{F}\perp}}{|\mathbf{v}_{\mathrm{F}\perp}|}. \tag{16.23}$$

Using $\hat{\mathbf{v}}_\perp$, we can express $\mathbf{r}$ alternatively as

$$\mathbf{r} = s\hat{\mathbf{v}}_\perp + b\hat{\mathbf{z}} \times \hat{\mathbf{v}}_\perp, \tag{16.24}$$

which corresponds to a rotation of the coordinate system by

$$\varphi_\mathbf{v} \equiv \arctan \frac{\hat{v}_{\perp y}}{\hat{v}_{\perp x}}, \tag{16.25}$$

as seen in Fig. 16.2. Coordinate b is called the *impact parameter*. In this new coordinate system, the differential operators in (16.21) are expressible as

$$\mathbf{v}_\mathrm{F} \cdot [\mp\mathrm{i}\hbar\nabla - eA(r)\hat{\boldsymbol{\varphi}}] = v_{\mathrm{F}\perp}\left[\mp\mathrm{i}\hbar\frac{\partial}{\partial s} + \frac{b}{r}eA(r)\right], \tag{16.26}$$

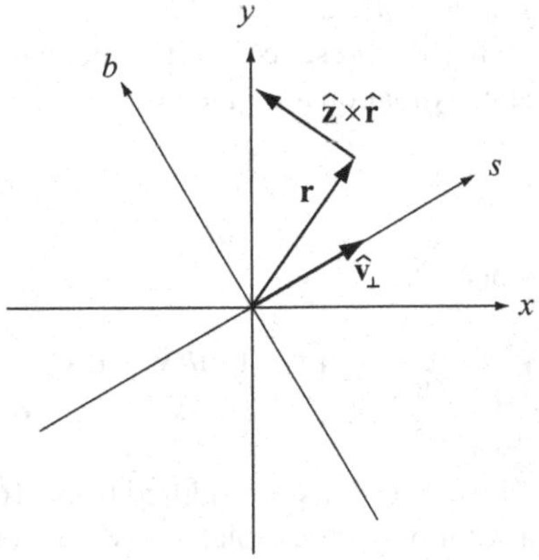

Fig. 16.2 Two coordinate systems of a two-dimensional space

where we have used $\hat{\mathbf{v}}_\perp \cdot \hat{\boldsymbol{\varphi}} = \hat{\mathbf{v}}_\perp \cdot (\hat{\mathbf{z}} \times \mathbf{r})/r = -b/r$. Next, we introduce the rotation matrix:

$$\underline{R} \equiv \begin{bmatrix} e^{i\varphi_v/2} & 0 \\ 0 & e^{-i\varphi_v/2} \end{bmatrix} = e^{i(\varphi_v/2)\underline{\sigma}_z}, \tag{16.27}$$

where $\underline{\sigma}_z$ is given in (8.42).

With these preliminaries, we multiply (16.20) by $\underline{R}$ from the left and insert the identity matrix $\underline{R}^{-1}\underline{R}$ between $\hat{\mathscr{H}}_{\mathrm{BdG}}$ and $\tilde{\mathbf{u}}$. It is thereby transformed into an eigenvalue problem for Hamiltonian $\hat{\mathscr{H}}'_{\mathrm{BdG}} \equiv \underline{R}\hat{\mathscr{H}}_{\mathrm{BdG}}\underline{R}^{-1}$ and eigenvector $\tilde{\mathbf{u}}' \equiv \underline{R}\tilde{\mathbf{u}}$. Using (16.21), (16.22), (16.26), and (16.27), we can express $\hat{\mathscr{H}}'_{\mathrm{BdG}}$ concisely as

$$\hat{\mathscr{H}}'_{\mathrm{BdG}}(b) = v_{\mathrm{F}\perp}\left[-i\hbar\underline{\sigma}_z\frac{\partial}{\partial s} + \frac{b}{r}eA(r)\underline{\sigma}_0\right] + \Delta(r)\left(\underline{\sigma}_x\cos\tilde{\varphi} + \underline{\sigma}_y\sin\tilde{\varphi}\right), \tag{16.28}$$

with

$$r = \sqrt{s^2+b^2}, \qquad \tilde{\varphi} \equiv \varphi - \varphi_v = \arctan\frac{b}{s}. \tag{16.29}$$

We solve $\hat{\mathscr{H}}'_{\mathrm{BdG}}(b)\tilde{\mathbf{u}}' = E\tilde{\mathbf{u}}'$ perturbatively from $b = 0$. First, Hamiltonian (16.28) for $b = 0$ reduces to

$$\hat{\mathscr{H}}'_{\mathrm{BdG}}(0) = -i\hbar v_{\mathrm{F}\perp}\frac{\partial}{\partial s}\underline{\sigma}_z + \Delta(s)\mathrm{sgn}(s)\underline{\sigma}_x, \tag{16.30}$$

with $\mathrm{sgn}(s) \equiv \pm 1$ for $s \gtrless 0$, which has eigenvalue 0. Indeed, the differential equation $\hat{\mathscr{H}}'_{\mathrm{BdG}}(0)\tilde{\mathbf{u}}'^{(0)} = \mathbf{0}$ has the (unnormalized) eigenfunction

$$\tilde{\mathbf{u}}'^{(0)}(s) = \begin{bmatrix} 1 \\ -i \end{bmatrix} e^{-K(s)}, \qquad K(s) \equiv \frac{1}{\hbar v_{\mathrm{F}\perp}}\int_0^s \Delta(s')\mathrm{sgn}(s')\mathrm{d}s'. \tag{16.31}$$

Next, we regard $\hat{\mathscr{H}}'^{(1)}_{\mathrm{BdG}} \equiv \hat{\mathscr{H}}'_{\mathrm{BdG}}(b) - \hat{\mathscr{H}}'_{\mathrm{BdG}}(0)$ and E as a perturbation. The first-order equation is given by

$$\hat{\mathscr{H}}'^{(1)}_{\mathrm{BdG}}\tilde{\mathbf{u}}'^{(0)} + \hat{\mathscr{H}}'_{\mathrm{BdG}}(0)\tilde{\mathbf{u}}'^{(1)} = E\tilde{\mathbf{u}}'^{(0)}. \tag{16.32}$$

Hamiltonian $\hat{\mathscr{H}}'^{(1)}_{\mathrm{BdG}}$ above is found by extracting terms of $O(b)$ from (16.28) with $\sin\tilde{\varphi} = b/\sqrt{s^2+b^2} \approx b/|s|$,

$$\hat{\mathscr{H}}'^{(1)}_{\mathrm{BdG}} \approx \frac{b}{|s|}\left[v_{\mathrm{F}\perp}eA(s)\underline{\sigma}_0 + \Delta(s)\underline{\sigma}_y\right]. \tag{16.33}$$

Now, we multiply (16.32) by $\tilde{\mathbf{u}}'^{(0)\dagger}$ from the left, use $\tilde{\mathbf{u}}'^{(0)\dagger}\hat{\mathscr{H}}'_{\rm BdG}(0) = \mathbf{0}^{\dagger}$, and integrate the resulting equation over $0 \le s \le \infty$. We thereby obtain

$$E_b = -\frac{b\displaystyle\int_0^{\infty}\frac{\Delta(s) - v_{\rm F\perp}eA(s)}{s}\mathrm{e}^{-2K(s)}\mathrm{d}s}{\displaystyle\int_0^{\infty}\mathrm{e}^{-2K(s)}\mathrm{d}s}. \tag{16.34}$$

This is the *Caroli-de Gennes-Matricon mode* for an isolated vortex [5], which plays a central role in describing low-energy and dissipative properties of s-wave type-II superconductors in magnetic fields.

To see how E_b grows from $E_{b=0} = 0$, let us adopt a model pair potential $\Delta(r) = \Delta_0 \tanh(r/\xi_\perp)$ with $\xi_\perp \equiv \hbar v_{\rm F\perp}/\Delta_0$. For this pair potential, function $K(s)$ in (16.31) for $s > 0$ becomes $K(s) = \ln\bigl[\cosh(s/\xi_\perp)\bigr]$. Substituting this into (16.34) and neglecting $A(r)$, we find $E_b = -0.85\frac{b}{\xi_\perp}\Delta_0$ around $b = 0$. Thus, the low-energy excitations are localized in the core region; see Fig. 16.5 below for the Caroli-de Gennes-Matricon mode, and also [6] for a fully self-consistent solution of the BdG equations for an isolated s-wave vortex.

16.3 Quasiclassical Study of an Isolated Vortex

As already noted, the Eilenberger equations form a useful and convenient basis for studying inhomogeneous superconductors theoretically. We here apply them to an isolated vortex of a clean s-wave superconductor to clarify numerically the details of the spatial variations of the pair potential, magnetic flux density, and local density of states near the vortex core. The equations to be solved are summarized in (16.64)–(16.67), and results are presented in Fig. 16.3 for the pair potential and magnetic flux density, and Fig. 16.5 for the local density of states.

16.3.1 Eilenberger Equations in Magnetic Fields

We consider a clean s-wave type-II superconductor in a weak magnetic field. It follows from (14.80), (14.88), and (14.76) that the corresponding Eilenberger equations are given by

$$2\varepsilon_n f + \hbar\mathbf{v}_{\rm F}\cdot\left(\boldsymbol{\nabla} - \mathrm{i}\frac{2e}{\hbar}\mathbf{A}\right) f = 2\Delta g, \tag{16.35}$$

$$\Delta(\mathbf{r}) = g_0\frac{2\pi}{\beta}\sum_{n=0}^{n_c}\langle f(\varepsilon_n, \mathbf{k}_{\rm F}, \mathbf{r})\rangle_{\rm F}, \tag{16.36}$$

$$\nabla \times \nabla \times \mathbf{A}(\mathbf{r}) = -\mathrm{i}\frac{2\pi e \mu_0 N(\varepsilon_\mathrm{F})}{\beta} \sum_{n=-\infty}^{\infty} \langle \mathbf{v}_\mathrm{F} g(\varepsilon_n, \mathbf{k}_\mathrm{F}, \mathbf{r}) \rangle_\mathrm{F}. \tag{16.37}$$

Here $\varepsilon_n \equiv (2n+1)\pi k_\mathrm{B} T$ is the Matsubara energy, $f \equiv f(\varepsilon_n, \mathbf{k}_\mathrm{F}, \mathbf{r}) = f(-\varepsilon_n, -\mathbf{k}_\mathrm{F}, \mathbf{r})$ is the quasiclassical Green's function, $\mathbf{v}_\mathrm{F}$ denotes the Fermi velocity, $e < 0$ is the electronic charge, $\mathbf{A}$ and Δ are the vector and pair potentials, respectively, $\langle \cdots \rangle_\mathrm{F}$ is the Fermi-surface average, and g is given in terms of f by

$$\begin{aligned} g(\varepsilon_n, \mathbf{k}_\mathrm{F}, \mathbf{r}) &\equiv \mathrm{sgn}(\varepsilon_n)[1 - f(\varepsilon_n, \mathbf{k}_\mathrm{F}, \mathbf{r}) f^*(\varepsilon_n, -\mathbf{k}_\mathrm{F}, \mathbf{r})]^{1/2} \\ &= g^*(\varepsilon_n, -\mathbf{k}_\mathrm{F}, \mathbf{r}) = -g^*(-\varepsilon_n, \mathbf{k}_\mathrm{F}, \mathbf{r}). \end{aligned} \tag{16.38}$$

Quantity g_0 in (16.36) denotes the dimensionless coupling constant, which from (14.67) for $\ell = 0$ is expressible in the form

$$\frac{1}{g_0} = \ln\frac{T}{T_\mathrm{c}} + \frac{2\pi}{\beta}\sum_{n=0}^{n_\mathrm{c}} \frac{1}{\varepsilon_n}, \tag{16.39}$$

where $n_\mathrm{c} > 0$ denotes a cutoff determined at each temperature in terms of a fixed cutoff energy $\varepsilon_\mathrm{c} \sim 40 k_\mathrm{B} T_\mathrm{c}$ through

$$(2n_\mathrm{c} + 1)\pi k_\mathrm{B} T = \varepsilon_\mathrm{c}. \tag{16.40}$$

Coupling constant (16.39) with a finite sum forms a convenient basis for numerical studies of the Eilenberger equations. This is confirmed for homogeneous systems for which the solution of (16.35) is given as

$$f = \frac{\Delta}{\sqrt{\varepsilon_n^2 + |\Delta|^2}}, \qquad g = \frac{\varepsilon_n}{\sqrt{\varepsilon_n^2 + |\Delta|^2}}. \tag{16.41}$$

Substituting this f into (16.36) and solving numerically, we can reproduce the temperature dependence of the energy gap $\Delta(T)$ given in Fig. 9.2 excellently.

Let us measure the energy, length, and vector potential in units of

$$\Delta_0, \qquad \xi_0 \equiv \frac{\hbar v_\mathrm{F}}{\Delta_0}, \qquad A_0 \equiv \frac{\hbar}{2|e|\xi_0}, \tag{16.42}$$

where Δ_0 is the energy gap for the homogeneous systems at $T = 0$. We next make a change of variables,

$$\begin{aligned} &\mathbf{r} = \xi_0 \mathbf{r}', \qquad \varepsilon_n = \Delta_0 \varepsilon_n', \qquad \Delta(\mathbf{r}) = \Delta_0 \Delta'(\mathbf{r}'), \qquad \mathbf{A}(\mathbf{r}) = A_0 \mathbf{A}'(\mathbf{r}'), \\ &k_\mathrm{B} T = \Delta_0 T', \qquad f(\varepsilon_n, \mathbf{k}_\mathrm{F}, \mathbf{r}) = f'(\varepsilon_n', \mathbf{k}_\mathrm{F}, \mathbf{r}'), \qquad g(\varepsilon_n, \mathbf{k}_\mathrm{F}, \mathbf{r}) = g'(\varepsilon_n', \mathbf{k}_\mathrm{F}, \mathbf{r}'). \end{aligned} \tag{16.43}$$

The s-wave transition temperature in these units is given by $T_c' = e^{\gamma}/\pi \approx 0.567$ as seen from (9.36). We also introduce the London penetration depth λ_{L0} at $T = 0$ and dimensionless parameter κ_0 by

$$\lambda_{L0} \equiv \sqrt{\frac{dA_0}{\mu_0|e|N(\varepsilon_F)\Delta_0 v_F}}, \qquad \kappa_0 \equiv \frac{\lambda_{L0}}{\xi_0}, \tag{16.44}$$

with $d = 2, 3$ denoting the dimension of the isotropic Fermi surface under consideration. Equations (16.35)–(16.37) are then expressible in terms of primed quantities and κ_0 as

$$2\varepsilon_n' f' + \hat{\mathbf{v}} \cdot (\boldsymbol{\nabla}' + \mathrm{i}\mathbf{A}') f' = 2\Delta' g', \tag{16.45}$$

$$\Delta'(\mathbf{r}') = 2\pi g_0 T' \sum_{n=0}^{n_c} \langle f(\varepsilon_n', \mathbf{k}_F, \mathbf{r}') \rangle_F, \tag{16.46}$$

$$\boldsymbol{\nabla}' \times \boldsymbol{\nabla}' \times \mathbf{A}'(\mathbf{r}') = \mathrm{i}\frac{2d\pi T'}{\kappa_0^2} \sum_{n=0}^{n_c} \big\langle \hat{\mathbf{v}} \big[g(\varepsilon_n', \mathbf{k}_F, \mathbf{r}') - g^*(\varepsilon_n', \mathbf{k}_F, \mathbf{r}') \big] \big\rangle_F, \tag{16.47}$$

where $\hat{\mathbf{v}}$ denotes the unit vector along $\mathbf{v}_F$. Equation (16.47) has been obtained by mapping the sum over $n < 0$ in (16.37) onto $n \geq 0$, then using the symmetry of (16.38), and finally omitting the sum over $n \geq n_c$ to give an excellent first approximation. Thus, we only need to consider those Matsubara energies satisfying $0 < \varepsilon_n \leq \varepsilon_c$. Hereafter, we remove primes from these equations.

It is worth noting that λ_{L0} in (16.44) for $d = 3$ is identical with that defined in the London equation (10.40) at $T = 0$. In addition, parameter κ_0 in (16.44) is connected with the GL parameter κ defined by (15.15) with (14.90)–(14.93) as $\kappa_0 = 0.5e^{-\gamma}\sqrt{7\zeta(3)/6}\,\kappa \approx 0.33\kappa$ in the clean limit of $d = 3$.

16.3.2 Transformation to a Riccati-Type Equation

Two alternative methods have been developed to solve (16.45) numerically by removing unphysical solutions that explode exponentially as we proceed with the numerical integration. The first one is called the *explosion method*, which takes the commutator of two exploding solutions to obtain a physical solution [9, 10, 16]. The other performs a transformation to a Riccati-type equation [9, 12, 15]. We here adopt the latter approach and rearrange (16.45) to (16.51) below.

In (16.45), let us express f and g for $\varepsilon_n > 0$ alternatively as

$$f = \frac{2a}{1 + a\bar{a}}, \qquad g = \frac{1 - a\bar{a}}{1 + a\bar{a}}, \tag{16.48}$$

with $\bar{a}(\varepsilon_n, \mathbf{k}_F, \mathbf{r}) \equiv a^*(\varepsilon_n, -\mathbf{k}_F, \mathbf{r})$, so that $g^2 + f\bar{f} = 1$ is satisfied automatically. Substituting (16.48) into (16.45), we obtain

$$2\varepsilon_n a + \hat{\mathbf{v}} \cdot (\boldsymbol{\nabla} + \mathrm{i}\mathbf{A})a - \frac{a\hat{\mathbf{v}} \cdot \boldsymbol{\nabla}(a\bar{a})}{1 + a\bar{a}} = \Delta(1 - a\bar{a}). \tag{16.49}$$

The corresponding equation for $\bar{a}$ is given by

$$2\varepsilon_n \bar{a} - \hat{\mathbf{v}} \cdot (\boldsymbol{\nabla} - \mathrm{i}\mathbf{A})\bar{a} + \frac{\bar{a}\hat{\mathbf{v}} \cdot \boldsymbol{\nabla}(a\bar{a})}{1 + a\bar{a}} = \Delta^*(1 - a\bar{a}). \tag{16.50}$$

Let us multiply each of (16.49) and (16.50) by $\bar{a}$ and a, respectively, from the left and subtract the latter from the former. We thereby obtain

$$\frac{\hat{\mathbf{v}} \cdot \boldsymbol{\nabla}(a\bar{a})}{1 + a\bar{a}} = \Delta\bar{a} - \Delta^* a.$$

Substitution of this expression back into (16.49) yields

$$\hat{\mathbf{v}} \cdot (\boldsymbol{\nabla} + \mathrm{i}\mathbf{A})a = -2\varepsilon_n a + \Delta - \Delta^* a^2. \tag{16.51}$$

Taking $\varepsilon_n > 0$, the solution of (16.51) for homogeneous systems is obtained as

$$a = \frac{\Delta}{\varepsilon_n + \sqrt{\varepsilon_n^2 + |\Delta|^2}}, \tag{16.52}$$

where we have chosen one of the two formal solutions by requiring that $a \to 0$ for $\Delta \to 0$ based on (16.41) and (16.48).

16.3.3 Equations for an Isolated Vortex

We seek a solution of the Eilenberger equation for an isolated s-wave vortex that is homogeneous along the direction of the magnetic field, which is taken parallel to the z axis. To minimize the numerical computation without losing the essential features of an isolated vortex, we consider a superconductor with a cylindrical Fermi surface ($d = 2$) whose side is arranged to lie along the magnetic field. Accordingly, we consider its variation in the xy plane. First, we write $\mathbf{r}$ and $\hat{\mathbf{v}}$ as

$$\mathbf{r} = (r\cos\varphi, r\sin\varphi), \qquad \hat{\mathbf{v}} = (\cos\varphi_{\mathbf{v}}, \sin\varphi_{\mathbf{v}}), \tag{16.53}$$

with $\hat{\mathbf{v}} \parallel \mathbf{k}_F$ for the cylindrical symmetry. See Fig. 16.2 with replacement $\hat{\mathbf{v}}_\perp \to \hat{\mathbf{v}}$ for the present model. Using $\hat{\mathbf{v}}$, we can express $\mathbf{r}$ alternatively as

$$\mathbf{r} = s\hat{\mathbf{v}} + b\hat{\mathbf{z}} \times \hat{\mathbf{v}}, \qquad \begin{cases} s = r\cos\tilde{\varphi} \\ b = r\sin\tilde{\varphi} \end{cases}, \qquad \tilde{\varphi} \equiv \varphi - \varphi_{\mathbf{v}}. \tag{16.54}$$

Second, we write the pair and vector potentials for an isolated vortex with cylindrical symmetry as

$$\Delta(\mathbf{r}) = \Delta(r)\mathrm{e}^{-\mathrm{i}\varphi}, \qquad \mathbf{A}(\mathbf{r}) = A(r)\hat{\mathbf{z}}\times\hat{\mathbf{r}}, \tag{16.55}$$

where $\Delta^*(r) = \Delta(r)$ and $r \equiv \sqrt{s^2+b^2}$; see also (15.61). Now, the differential operator of (16.51) is expressible concisely in the new coordinates as

$$\hat{\mathbf{v}}\cdot(\boldsymbol{\nabla}+\mathrm{i}\mathbf{A}) = \frac{\partial}{\partial s} - \mathrm{i}\frac{b}{r}A(r). \tag{16.56}$$

Moreover, $\boldsymbol{\nabla}\times\boldsymbol{\nabla}\times\mathbf{A}$ is transformed into [3]

$$\boldsymbol{\nabla}\times\boldsymbol{\nabla}\times\mathbf{A} = -\hat{\mathbf{z}}\times\hat{\mathbf{r}}\frac{\mathrm{d}B(r)}{\mathrm{d}r}, \qquad B(r) \equiv \frac{1}{r}\frac{\mathrm{d}}{\mathrm{d}r}rA(r). \tag{16.57}$$

We also rewrite a as

$$a(\varepsilon_n, \mathbf{k}_\mathrm{F}, \mathbf{r}) = \tilde{a}(\varepsilon_n, s, b)\mathrm{e}^{-\mathrm{i}\varphi_\mathrm{v}}. \tag{16.58}$$

Substituting (16.55)–(16.58) into (16.51), we obtain

$$\left(\frac{\partial}{\partial s} - \mathrm{i}\frac{b}{r}A\right)\tilde{a} = -2\varepsilon_n\tilde{a} + \Delta\mathrm{e}^{-\mathrm{i}\tilde{\varphi}} - \Delta\mathrm{e}^{\mathrm{i}\tilde{\varphi}}\tilde{a}^2, \tag{16.59}$$

with $A = A(r)$ and $\Delta = \Delta(r)$, which depends only on the relative angle $\tilde{\varphi} \equiv \varphi - \varphi_\mathrm{v}$ between $\mathbf{r}$ and $\hat{\mathbf{v}}$. Taking the complex conjugate of (16.59), we obtain the symmetry relation:

$$\tilde{a}^*(\varepsilon_n, s, b) = \tilde{a}(\varepsilon_n, s, -b). \tag{16.60}$$

The two results imply that we only need to solve (16.59) in the (s, b) plane for a single direction of $\hat{\mathbf{v}}$ and $b \geq 0$, e.g., $\hat{\mathbf{v}} = \hat{\mathbf{x}}$ and $b \geq 0$. Next, we insert (16.58) into (16.48) and use (16.60) and $\mathrm{e}^{-\mathrm{i}\varphi_{-\mathrm{v}}} = \mathrm{e}^{-\mathrm{i}(\varphi_\mathrm{v}+\pi)} = -\mathrm{e}^{-\mathrm{i}\varphi_\mathrm{v}}$. We can thereby express f and g as

$$f(\varepsilon_n, \mathbf{k}_\mathrm{F}, \mathbf{r}) = \tilde{f}(\varepsilon_n, s, b)\mathrm{e}^{-\mathrm{i}\varphi_\mathrm{v}}, \qquad g(\varepsilon_n, \mathbf{k}_\mathrm{F}, \mathbf{r}) = \tilde{g}(\varepsilon_n, s, b), \tag{16.61}$$

with

$$\tilde{f}(\varepsilon_n, s, b) \equiv \frac{2\tilde{a}(\varepsilon_n, s, b)}{1-\tilde{a}(\varepsilon_n, s, b)\tilde{a}(\varepsilon_n, -s, b)} = \tilde{f}^*(\varepsilon_n, s, -b), \tag{16.62}$$

$$\tilde{g}(\varepsilon_n, s, b) = \frac{1+\tilde{a}(\varepsilon_n, s, b)\tilde{a}(\varepsilon_n, -s, b)}{1-\tilde{a}(\varepsilon_n, s, b)\tilde{a}(\varepsilon_n, -s, b)} = \tilde{g}^*(\varepsilon_n, s, -b). \tag{16.63}$$

Now, we are ready to summarize the equations to be solved numerically. The first is given by (16.59), i.e.,

$$\left(\frac{\partial}{\partial s} - \mathrm{i}\frac{b}{r}A\right)\tilde{a} = -2\varepsilon_n\tilde{a} + \Delta \mathrm{e}^{-\mathrm{i}\tilde{\varphi}} - \Delta \mathrm{e}^{\mathrm{i}\tilde{\varphi}}\tilde{a}^2, \tag{16.64}$$

with $r = \sqrt{s^2+b^2}$, $\tilde{\varphi} = \arctan(b/s)$, $A = A(r)$, $\tilde{a} = \tilde{a}(\varepsilon_n, s, b)$, $\varepsilon_n = (2n+1)\pi T$, and $\Delta = \Delta(r)$. Next, let us substitute (16.55) and (16.61) into (16.46) to obtain a simplified equation for the pair potential as

$$\Delta(r) = 2g_0\pi T \sum_{n=0}^{n_c}\int_0^{2\pi}\frac{\mathrm{d}\tilde{\varphi}}{2\pi}\tilde{f}(\varepsilon_n, r\cos\tilde{\varphi}, r\sin\tilde{\varphi})\mathrm{e}^{\mathrm{i}\tilde{\varphi}}, \tag{16.65}$$

where we have made a change of integration variables as $\varphi_{\mathrm{v}} \to \tilde{\varphi} \equiv \varphi - \varphi_{\mathrm{v}}$ in averaging over the cylindrical Fermi surface. Finally, we substitute (16.57) and (16.61) into (16.47) for $d = 2$, take its scalar product with $\hat{\mathbf{r}} \times \hat{\mathbf{z}}$, and write $(\hat{\mathbf{r}} \times \hat{\mathbf{z}}) \cdot \hat{\mathbf{v}} = \sin\tilde{\varphi}$ based on (16.54). We then obtain a differential equation for $B(r)$ as

$$\frac{\mathrm{d}B(r)}{\mathrm{d}r} = -j(r), \qquad j(r) \equiv \frac{8\pi T}{\kappa_0^2}\sum_{n=0}^{n_c}\int_0^{2\pi}\frac{\mathrm{d}\tilde{\varphi}}{2\pi}\mathrm{Im}\tilde{g}(\varepsilon_n, r\cos\tilde{\varphi}, r\sin\tilde{\varphi})\sin\tilde{\varphi}, \tag{16.66}$$

where Im denotes the imaginary part. After integrating (16.66), the vector potential is obtained based on (16.57) by

$$A(r) = \frac{1}{r}\int_0^r r'B(r')\mathrm{d}r'. \tag{16.67}$$

Equations (16.64)–(16.67) form a set of self-consistency equations for $\tilde{a}(\varepsilon_n, s, b)$, $\Delta(r)$, and $A(r)$. Using the symmetry of (16.62), we can reduce the integral of (16.65) to that of the real part of $\tilde{f}(\varepsilon_n, s, b)\mathrm{e}^{\mathrm{i}\tilde{\varphi}}$ over $0 \le \tilde{\varphi} \le \pi$. In addition, one can show based on the symmetry of (16.63) that integrals over $0 \le \tilde{\varphi} < \pi$ and $\pi \le \tilde{\varphi} < 2\pi$ in (16.66) yield an identical contribution. Hence, we need to perform the self-consistent calculations only over $b \ge 0$.

16.3.4 Numerical Procedures

The numerical procedures to solve (16.64)–(16.67) at a given temperature T are summarized as follows. First, we solve the ordinary differential equation (16.64) with some trial pair and vector potentials such as

$$\Delta(r) = \Delta_T \tanh r, \qquad A(r) = \frac{1-(1+r/\kappa_0)\mathrm{e}^{-r/\kappa_0}}{r},$$

where Δ_T is the energy gap of homogeneous systems that can be obtained by the procedure described below (16.41). To this end, we choose a rectangular region of $-r_c \le s \le r_c$ and $0 \le b \le r_c$ with $r_c \gtrsim 5$, fix ε_n and b, and integrate (16.64) numerically along the straight-line path from $(s, b) = (-r_c, b)$ up to $(s, b) = (r_c, b)$ [14] by imposing the initial condition:

$$\tilde{a}(\varepsilon_n, -r_c, b) = \tilde{a}^{(0)}(\varepsilon_n, -r_c, b) - \mathrm{i}\frac{b}{2R_c^2}[1 - R_c A(R_c)]\frac{\tilde{a}^{(0)}(\varepsilon_n, -r_c, b)}{\sqrt{\varepsilon_n^2 + \Delta_T^2}}, \tag{16.68}$$

with $R_c \equiv \sqrt{r_c^2 + b^2}$. Here the first term is defined as

$$\tilde{a}^{(0)}(\varepsilon_n, -r_c, b) \equiv \frac{\Delta_T}{\varepsilon_n + \sqrt{\varepsilon_n^2 + \Delta_T^2}} \frac{-r_c - ib}{R_c}, \tag{16.69}$$

which is obtained by setting the left-hand side of (16.64) equal to zero with $\Delta(R_c) \to \Delta_T$. The second term originates from the first-order perturbation with respect to the left-hand side of (16.64). Next, we use $\tilde{a}(\varepsilon_n, s, b)$ obtained in this manner on discrete points in $-r_c \le s \le r_c$ and $0 \le b \le r_c$ to perform integrations of (16.65) and (16.66) numerically with interpolations [14] for updating $\Delta(r)$ and $j(r)$ over $0 \le r \le r_c$. Current $j(r)$ thereby obtained is used subsequently to integrate (16.66) for $B(r)$ with the initial condition:

$$B(r_c) = cK_0(r_c/\lambda_L) \tag{16.70}$$

given in terms of the modified Bessel function $K_0(x)$ [1, 3] as established by the London theory; see (15.75) on this point. As shown in **Problem** 16.2 below, constants λ_L and c above can be determined in terms of $j(r)$ by solving

$$\left[\lambda_L + \frac{r_c K_0(r_c/\lambda_L)}{2K_1(r_c/\lambda_L)}\right]\lambda_L r_c j(r_c) = 1 - \int_0^{r_c} dr_1 r_1 \int_{r_1}^{r_c} dr_2 j(r_2), \tag{16.71}$$

$$c = \frac{\lambda_L j(r_c)}{K_1(r_c/\lambda_L)}. \tag{16.72}$$

This consideration also indicates that we can write $1 - R_c A(R_c) = c\lambda_L R_c K_1(R_c/\lambda_L)$ in (16.68). Finally, the vector potential is calculated using (16.67). Potentials $\Delta(r)$ and $A(r)$ thereby obtained are used in the iteration of (16.64)–(16.67). When updating $\Delta(r)$ and $A(r)$ at this stage, it is better to mix old and new potentials with certain relative weights, e.g., 0.5 vs. 0.5, to avoid numerical oscillations. The procedure should be repeated until numerical convergence in $\Delta(r)$ and $A(r)$ is reached.

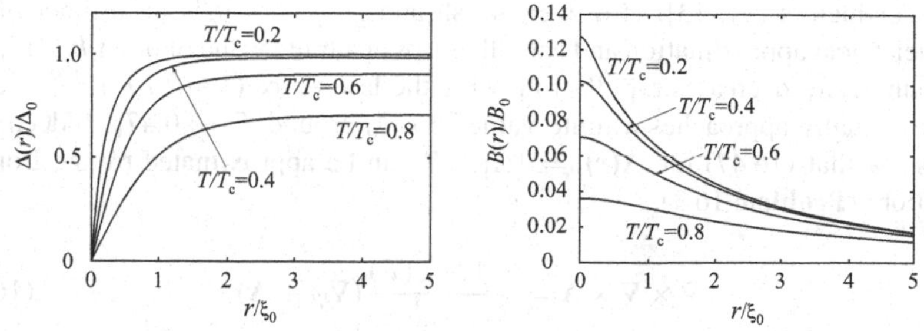

Fig. 16.3 Spatial variations of the pair potential $\Delta(r)$ and magnetic flux density $B(r)$ around an isolated vortex centered at $r = 0$ for $\kappa_0 = 5$ expressed in units given in (16.42) and $B_0 \equiv \hbar/2|e|\xi_0^2$

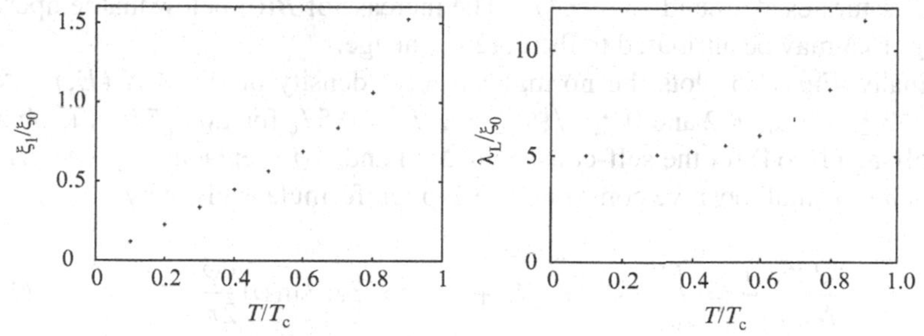

Fig. 16.4 Temperature dependences of ξ_1 and λ_L for $\kappa_0 = 5.0$ in units given in (16.42) and $B_0 \equiv \hbar/2|e|\xi_0^2$

16.3.5 Results

Figure 16.3 plots the spatial variations of the pair potential $\Delta(r)$ and magnetic flux density $B(r)$ near the vortex core calculated for $\kappa_0 = 5.0$. The pair potential is seen to increase linearly from the core center, as

$$\Delta(r) \sim \frac{r}{\xi_1}\Delta_\mathrm{T} \qquad (r \gtrsim 0) \tag{16.73}$$

towards the homogeneous value Δ_T, and the core size, which may be estimated by parameter $\xi_1 = \xi_1(T)$, is seen to diminish substantially as the temperature is lowered. Contrastingly, $B(r)$ decreases quadratically from the core center, and its behavior for $r \gtrsim 3\xi_1$, where $\Delta(r) \approx \Delta_T$ holds, is well described by the formula $B(r) \approx cK_0(r/\lambda_\mathrm{L})$; see also (15.73) and (15.75) on these points. The value of $B(0)$ is seen to increase as the temperature is lowered partly because of the decrease in the penetration depth $\lambda_\mathrm{L}(T)$.

Figure 16.4 plots ξ_1 and λ_L as a function of temperature to show these features more quantitatively. Most remarkably, the core size ξ_1 is seen to approach zero as $T \to 0$, which is known as the *Kramer-Pesch effect* for clean type-II superconductors [11, 13]. However, its shrinkage $\xi_1 \to 0$ is an artifact of the quasiclassical approximation and actually stops at a finite value of order k_F^{-1} [7, 11]. In contrast, λ_L decreases rapidly following the law $\lambda_L \propto (1 - T/T_c)^{-1/2}$ near T_c and eventually approaches a finite value $\sim \kappa_0 \xi_0$ around $T \sim 0.4T_c$. Indeed, one can show that (16.47) for $\Delta(\mathbf{r}) = |\Delta| e^{i\varphi(\mathbf{r})}$ can be approximated by the London equation[2] (**Problem** 16.3),

$$\nabla \times \nabla \times \mathbf{A} = -\frac{1 - Y(T)}{\kappa_0^2}(\nabla\varphi + \mathbf{A}), \tag{16.74}$$

where $Y(T)$ is the Yosida function (10.17); the equation implies $\lambda_L(T) = \kappa_0[1 - Y(T)]^{-1/2}$, as seen from (10.39) and (10.40). Hence, the zero-temperature value of $\lambda_L(T)$ saturates at around $T \sim 0.4T_c$. The increase of $B(0)$ below that temperature in Fig. 16.3 may be attributed to the core shrinkage.

Finally, Fig. 16.5 plots the normalized local density of states $N_s(E, r)/N(\varepsilon_F)$ over $-2 \le E/\Delta_0 \le 2$ and $0 \le r/\xi_0 \le 5$ at $T = 0.5T_c$ for $\kappa_0 = 5.0$. It is obtained by solving (16.64) for the self-consistent $\Delta(r)$ and $A(r)$ replacing $\varepsilon_n \to -iE + \delta$, where δ is a small positive constant. The explicit formula is given by

$$\frac{N_s(E, r)}{N(\varepsilon_F)} = \int_0^{2\pi} \mathrm{Re}\, \bar{g}(-iE + \delta, r\cos\bar{\varphi}, r\sin\bar{\varphi}) \frac{d\bar{\varphi}}{2\pi}, \tag{16.75}$$

where $\bar{g}$ is given by (16.63) and Re denotes the real part. Indeed, one confirms that $\mathrm{Re}\, g(\varepsilon_n \to -iE + \delta)$ in (16.41) with $\delta \to 0_+$ yields the homogeneous superconducting density of states given by (9.48) or (12.29). We have set here $\delta = 0.1\Delta_0$ to avoid any numerical divergence. We observe that $N_s(E, r)$ for $r \sim 5\xi_0$ is almost identical with the bulk density of states (9.48) smeared by δ. Near the core center, however, there is another structure with a sharp zero-energy peak at $r = 0$, which represents the *Caroli-de Gennes-Matricon mode* given by (16.34). This local density of states around a vortex core was observed experimentally with the scanning-tunneling microscope by Hess et al. [8].

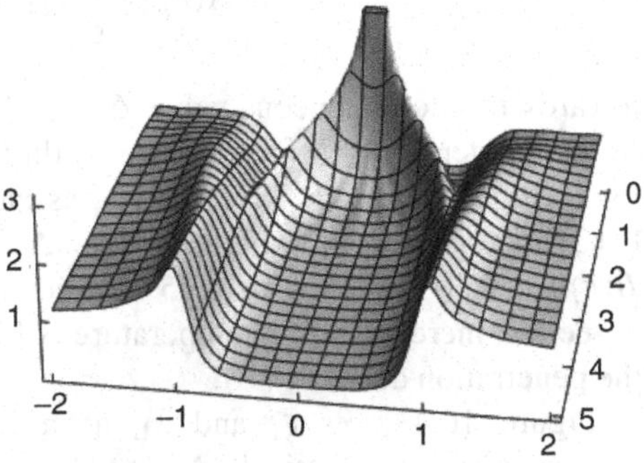

Fig. 16.5 Normalized local density of states $N_s(E, r)/N(\varepsilon_F)$ plotted over $-2 \le E/\Delta_0 \le 2$ and $0 \le r/\xi_0 \le 5$ at $T = 0.5T_c$ for $\kappa_0 = 5.0$ and $\delta = 0.1\Delta_0$

[2]Quantity $\varphi(\mathbf{r})$ here denotes the phase of $\Delta(\mathbf{r})$; it is not the polar angle in two dimensions.

Problems

16.1. With $\xi_{\mathbf{q}'} < 0$ in (16.11), obtain the corresponding equation to (16.19).

16.2. Derive (16.71) and (16.72) using the condition:

$$\int_0^\infty \mathrm{d}r_1 r_1 B(r_1) = 1, \tag{16.76}$$

which results from (10.43) for the flux quantization with $n = -1$ in units of (16.42).

16.3. Derive (16.74) from (16.47) by regarding the gradient term in (16.45) as a perturbation and setting $\Delta(\mathbf{r}) = |\Delta|\mathrm{e}^{\mathrm{i}\varphi(\mathbf{r})}$.

References

1. M. Abramowitz, I.A. Stegun (eds.), *Handbook of Mathematical Functions: With Formulas, Graphs, and Mathematical Tables* (Dover, New York, 1965)
2. A.F. Andreev, J. Exp. Theor. Phys. **46**, 1823 (1964). (Sov. Phys. JETP **19**, 1228 (1964))
3. G.B. Arfken, H.J. Weber, *Mathematical Methods for Physicists* (Academic, New York, 2012)
4. G.E. Blonder, M. Tinkham, T.M. Klapwijk, Phys. Rev. B **25**, 4515 (1982)
5. C. Caroli, P.G. de Gennes, J. Matricon, Phys. Lett. **9**, 307 (1964)
6. F. Gygi, M. Schlüter, Phys. Rev. B **43**, 7609 (1991)
7. N. Hayashi, T. Isoshima, M. Ichioka, K. Machida, Phys. Rev. Lett. **80**, 2921 (1998)
8. H.F. Hess, R.B. Robinson, R.C. Dynes, J.M. Valles Jr., J.V. Waszczak, Phys. Rev. Lett. **62**, 214 (1989)
9. M. Ichioka, N. Hayashi, N. Enomoto, K. Machida, Phys. Rev. B **53**, 15316 (1996)
10. U. Klein, J. Low Temp. Phys. **69**, 1 (1987)
11. L. Kramer, W. Pesch, Z. Phys. **269**, 59 (1974)
12. Y. Nagato, K. Nagai, J. Hara, J. Low Temp. Phys. **93**, 33 (1993)
13. W. Pesch, L. Kramer, J. Low Temp. Phys. **15**, 367 (1974)
14. W.H. Press, S.A. Teukolsky, W.T. Vetterling, B.P. Flannery, *Numerical Recipes: The Art of Scientific Computing*, 3rd edn. (Cambridge University Press, Cambridge, 2007)
15. N. Schopohl, K. Maki, Phys. Rev. B **52**, 490 (1995)
16. E.V. Thuneberg, J. Kurkijärvi, D. Rainer, Phys. Rev. B **29**, 3913 (1984)

Chapter 17
Solutions to Problems

Problems of Chapter 1

1.1

(a) Both differentiations in (1.5) gives $2x$ so that the equality holds.
(b) Performing the integration of (1.3) for the gradient $\partial z/\partial x = 2xy + 1$, we obtain

$$z(x, y) = x^2 y + x + g(y),$$

where we have incorporated a constant $-(x_0^2 y_0 + x_0)$ into the definition of the unknown function $g(y)$. Let us differentiate the above equation with respect to y and set $\partial z/\partial y = x^2 + 2y$. We thereby obtain $g'(y) = 2y$, which gives $g(y) = y^2 + C$ with C denoting a constant. Substitution of this result into the above $z(x, y)$ yields $z(x, y) = x^2 y + x + y^2 + C$.

1.2 Integrating (1.9) and (1.10) over a cycle of the heat engine in contact with a single heat bath of temperature T yields

$$0 = \oint \mathrm{d}'Q + \oint \mathrm{d}'W, \qquad \oint \mathrm{d}'Q \le \oint T\mathrm{d}S = T\oint \mathrm{d}S = 0,$$

where we have used the fact that the internal energy and entropy are potentials. Combining the results, we obtain an inequality for the total work $-\Delta W$ performed on the exterior,

$$-\Delta W \equiv -\oint \mathrm{d}'W = \oint \mathrm{d}'Q \le 0.$$

Hence, the statement is proved.

© Springer Japan 2015

T. Kita, *Statistical Mechanics of Superconductivity*, Graduate Texts in Physics,
DOI 10.1007/978-4-431-55405-9_17

1.3

(a) It follows from (1.8) that

$$\left(\frac{\partial P}{\partial T}\right)_V = \frac{nR}{V - nb}, \qquad T\left(\frac{\partial P}{\partial T}\right)_V - P = a\frac{n^2}{V^2}$$

holds. Substitution of them into (1.33) and (1.34) gives

$$\mathrm{d}S = \frac{C_V}{T}\mathrm{d}T + \frac{nR}{V - nb}\mathrm{d}V, \qquad \mathrm{d}U = C_V\mathrm{d}T + a\frac{n^2}{V^2}\mathrm{d}V.$$

(b) Maxwell's relation for $\mathrm{d}U$ above reads

$$\frac{\partial C_V}{\partial V} = \frac{\partial}{\partial T}\left(a\frac{n^2}{V^2}\right) = 0,$$

which implies that C_V is independent of V. The same conclusion also results from Maxwell's relation for $\mathrm{d}S$.

(c) Noting that C_V is a constant, we integrate $\mathrm{d}S$ and $\mathrm{d}U$ in (a) above to obtain

$$S = C_V \ln T + nR\ln(V - nb) + S_0, \qquad U = C_V T - a\frac{n^2}{V} + U_0,$$

where S_0 and U_0 are constants of integration.

(d) Entropy does not change in reversible adiabatic processes. Keeping this fact in mind, we rearrange the expression for S in (c) into the form $S = C_V \ln T(V - nb)^{nR/C_V} + S_0$. We thereby conclude $T(V - nb)^{nR/C_V} = \text{const}$ in reversible adiabatic processes.

(e) Equalities $\mathrm{d}'Q = \mathrm{d}'W = 0$ hold in adiabatic free expansions. Hence, it follows from the first law of thermodynamics that the internal energy does not change. Combining this fact with the expression for U obtained in (c), we obtain

$$C_V T_1 - a\frac{n^2}{V_1} = C_V T_2 - a\frac{n^2}{V_2} \longleftrightarrow \Delta T \equiv T_2 - T_1 = \frac{an^2}{C_V}\left(\frac{1}{V_2} - \frac{1}{V_1}\right) < 0.$$

Problems of Chapter 2

2.1

(a) Can be shown elementarily.

(b) Using the equality of (a), one may calculate the expectation as

$$\langle k\rangle = \sum_{k=1}^{n} kP_k^n = np\sum_{k=1}^{n} P_{k-1}^{n-1} = np\sum_{k'=0}^{n-1} P_{k'}^{n-1} = np.$$

One can also show $k^2 P_k^n = [k(k-1)+k]P_k^n = n(n-1)p^2 P_{k-2}^{n-2} + npP_{k-1}^{n-1}$, which is used to rearrange $\langle k^2 \rangle$ as

$$\langle k^2 \rangle = \sum_{k=1}^{n} k^2 P_k^n = n(n-1)p^2 \sum_{k=2}^{n} P_{k-2}^{n-2} + np \sum_{k=1}^{n} P_{k-1}^{n-1} = n(n-1)p^2 + np.$$

Hence, we obtain $\sigma_k \equiv \sqrt{\langle k^2 \rangle - \langle k \rangle^2} = \sqrt{np(1-p)}$.

2.2 A necessary condition for $g(x) \equiv x + y + \lambda(x^2 + y^2 - 1)$ being extremal is given by $\partial g(x,y)/\partial x = \partial g(x,y)/\partial y = 0$, i.e., $0 = 1 + 2\lambda x = 1 + 2\lambda y$. Coupled with $x^2 + y^2 = 1$, the equalities yield $x = y = \pm\frac{1}{\sqrt{2}}$. Substituting both into $x + y$, we find the point to be $(x,y) = \left(\frac{1}{\sqrt{2}}, \frac{1}{\sqrt{2}}\right)$.

Problems of Chapter 3

3.1 Substitute (3.24) into the left-hand side of (3.26) and use (3.19), (3.23), (3.18), and (3.12) successively. Denoting integration over ξ by $\bar{\xi}$, the demonstration of this proof follows:

$$\begin{aligned}
\hat{\psi}(\xi_1)|\Phi_\nu\rangle &= \frac{1}{\sqrt{N!}}\hat{\psi}(\xi_1)\hat{\psi}^\dagger(\bar{\xi}_1')\hat{\psi}^\dagger(\bar{\xi}_2')\cdots\hat{\psi}^\dagger(\bar{\xi}_N')|0\rangle\Phi_\nu(\bar{\xi}_1',\bar{\xi}_2',\cdots,\bar{\xi}_N') \\
&= \frac{1}{\sqrt{N!}}\Big[\hat{\psi}^\dagger(\bar{\xi}_2')\cdots\hat{\psi}^\dagger(\bar{\xi}_N')|0\rangle\Phi_\nu(\xi_1,\bar{\xi}_2',\cdots,\bar{\xi}_N') \\
&\quad +\sigma\hat{\psi}^\dagger(\bar{\xi}_1')\hat{\psi}^\dagger(\bar{\xi}_3')\cdots\hat{\psi}^\dagger(\bar{\xi}_N')|0\rangle\Phi_\nu(\bar{\xi}_1',\xi_1,\bar{\xi}_3',\cdots,\bar{\xi}_N') + \cdots \\
&\quad +\sigma^{N-1}\hat{\psi}^\dagger(\bar{\xi}_1')\cdots\hat{\psi}^\dagger(\bar{\xi}_{N-1}')|0\rangle\Phi_\nu(\bar{\xi}_1',\cdots,\bar{\xi}_{N-1}',\xi_1) \\
&\quad +\sigma^N\hat{\psi}^\dagger(\bar{\xi}_1')\hat{\psi}^\dagger(\bar{\xi}_2')\cdots\hat{\psi}^\dagger(\bar{\xi}_N')\hat{\psi}(\xi_1)|0\rangle\Phi_\nu(\bar{\xi}_1',\cdots,\bar{\xi}_{N-1}',\bar{\xi}_N')\Big] \\
&= \frac{1}{\sqrt{N!}}\Big[\hat{\psi}^\dagger(\bar{\xi}_2')\cdots\hat{\psi}^\dagger(\bar{\xi}_N')|0\rangle\Phi_\nu(\xi_1,\bar{\xi}_2',\cdots,\bar{\xi}_N') \\
&\quad +\sigma^2\hat{\psi}^\dagger(\bar{\xi}_1')\hat{\psi}^\dagger(\bar{\xi}_3')\cdots\hat{\psi}^\dagger(\bar{\xi}_N')|0\rangle\Phi_\nu(\xi_1,\bar{\xi}_1',\bar{\xi}_3',\cdots,\bar{\xi}_N') + \cdots \\
&\quad +\sigma^{2(N-1)}\hat{\psi}^\dagger(\bar{\xi}_1')\cdots\hat{\psi}^\dagger(\bar{\xi}_{N-1}')|0\rangle\Phi_\nu(\xi_1,\bar{\xi}_1',\cdots,\bar{\xi}_{N-1}')\Big] \\
&= \frac{N}{\sqrt{N!}}\hat{\psi}^\dagger(\bar{\xi}_2')\cdots\hat{\psi}^\dagger(\bar{\xi}_N')|0\rangle\Phi_\nu(\xi_1,\bar{\xi}_2',\cdots,\bar{\xi}_N') \\
&= \sqrt{N}\,|\bar{\xi}_2',\cdots,\bar{\xi}_N'\rangle\Phi_\nu(\xi_1,\bar{\xi}_2',\cdots,\bar{\xi}_N').
\end{aligned}$$

3.2 The first equation is proved by substituting (3.24) and (3.25) and using (3.22), (3.12), and (3.33) successively as follows

$$\begin{aligned}\langle\Phi_{\nu'}|\Phi_\nu\rangle &= \Phi^*_{\nu'}(\bar{\xi}'_1,\cdots,\bar{\xi}'_N)\langle\bar{\xi}'_1,\cdots,\bar{\xi}'_N|\bar{\xi}_1,\cdots,\bar{\xi}_N\rangle\Phi_\nu(\bar{\xi}_1,\cdots,\bar{\xi}_N)\\ &= \Phi^*_{\nu'}(\bar{\xi}'_1,\cdots,\bar{\xi}'_N)\frac{1}{N!}\sum_{\hat{P}}\sigma^P\delta(\bar{\xi}'_1,\bar{\xi}_{p_1})\cdots\delta(\bar{\xi}'_N,\bar{\xi}_{p_N})\Phi_\nu(\bar{\xi}_1,\cdots,\bar{\xi}_N)\\ &= \frac{1}{N!}\sum_{\hat{P}}\sigma^P\Phi^*_{\nu'}(\bar{\xi}_{p_1},\cdots,\bar{\xi}_{p_N})\Phi_\nu(\bar{\xi}_1,\cdots,\bar{\xi}_N)\\ &= \frac{1}{N!}\sum_{\hat{P}}\Phi^*_{\nu'}(\bar{\xi}_1,\cdots,\bar{\xi}_N)\Phi_\nu(\bar{\xi}_1,\cdots,\bar{\xi}_N) = \frac{1}{N!}\sum_{\hat{P}}\delta_{\nu'\nu} = \delta_{\nu'\nu}.\end{aligned}$$

Similarly, the second relation is proved by substituting (3.24) and (3.25) and applying (3.34), (3.21), and (3.30) successively as follows

$$\begin{aligned}\sum_\nu|\Phi_\nu\rangle\langle\Phi_\nu| &= |\bar{\xi}_1,\cdots,\bar{\xi}_N\rangle\langle\bar{\xi}'_1,\cdots,\bar{\xi}'_N|\sum_\nu\Phi_\nu(\bar{\xi}_1,\cdots,\bar{\xi}_N)\Phi^*_\nu(\bar{\xi}'_1,\cdots,\bar{\xi}'_N)\\ &= |\bar{\xi}_1,\cdots,\bar{\xi}_N\rangle\langle\bar{\xi}'_1,\cdots,\bar{\xi}'_N|\frac{1}{N!}\sum_{\hat{P}}\sigma^P\delta(\bar{\xi}'_1,\bar{\xi}_{p_1})\cdots\delta(\bar{\xi}'_N,\bar{\xi}_{p_N})\\ &= |\bar{\xi}_1,\cdots,\bar{\xi}_N\rangle\frac{1}{N!}\sum_{\hat{P}}\sigma^P\langle\bar{\xi}_{p_1},\cdots,\bar{\xi}_{p_N}| = |\bar{\xi}_1,\cdots,\bar{\xi}_N\rangle\langle\bar{\xi}_1,\cdots,\bar{\xi}_N|=1.\end{aligned}$$

3.3 The proof proceeds by substituting (3.26) into the right-hand side of (3.37) and applying (3.22) and (3.12) successively to obtain

$$\begin{aligned}&\langle\Phi_{\nu'}|\hat{\psi}^\dagger(\bar{\xi}_1)\hat{h}^{(1)}_1\hat{\psi}(\bar{\xi}_1)|\Phi_\nu\rangle\\ &\quad= N\frac{1}{(N-1)!}\sum_{\hat{P}}\sigma^P\Phi^*_{\nu'}(\bar{\xi}_1,\bar{\xi}_{p_2},\cdots,\bar{\xi}_{p_N})\hat{h}^{(1)}_1\Phi_\nu(\bar{\xi}_1,\bar{\xi}_2,\cdots,\bar{\xi}_N)\\ &\quad= N\Phi^*_{\nu'}(\bar{\xi}_1,\bar{\xi}_2,\cdots,\bar{\xi}_N)\hat{h}^{(1)}_1\Phi_\nu(\bar{\xi}_1,\bar{\xi}_2,\cdots,\bar{\xi}_N)\\ &\quad= N\Phi^*_{\nu'}(\bar{\xi}_2,\bar{\xi}_1,\bar{\xi}_3,\cdots,\bar{\xi}_N)\hat{h}^{(1)}_1\Phi_\nu(\bar{\xi}_2,\bar{\xi}_1,\bar{\xi}_3,\cdots,\bar{\xi}_N)\\ &\quad= \Phi^*_{\nu'}(\bar{\xi}_1,\bar{\xi}_2,\bar{\xi}_3,\cdots,\bar{\xi}_N)\sum_{j=1}^N\hat{h}^{(1)}_j\Phi_\nu(\bar{\xi}_1,\bar{\xi}_2,\bar{\xi}_3,\cdots,\bar{\xi}_N).\end{aligned}$$

3.4 Substitute $\sigma = -1$ and (3.56) into (3.54), insert the resulting wave function and (3.19) into (3.44), and use (3.50) and (3.51) successively. Specifically, the transformation proceeds as follows:

$$
\begin{aligned}
|\Phi_\nu\rangle &= |\bar{\xi}_1, \cdots, \bar{\xi}_N\rangle \Phi_\nu(\bar{\xi}_1, \cdots, \bar{\xi}_N) \\
&= \frac{1}{N!} \sum_{\hat{P}} (-1)^P \hat{\psi}^\dagger(\bar{\xi}_1)\varphi_{q_{P1}}(\bar{\xi}_1) \cdots \hat{\psi}^\dagger(\bar{\xi}_N)\varphi_{q_{PN}}(\bar{\xi}_N)|0\rangle \\
&= \frac{1}{N!} \sum_{\hat{P}} (-1)^P \hat{c}^\dagger_{q_{P1}} \cdots \hat{c}^\dagger_{q_{PN}} |0\rangle = \hat{c}^\dagger_{q_1} \cdots \hat{c}^\dagger_{q_N} |0\rangle .
\end{aligned}
$$

3.5 The first identity is proved by expanding the commutator as

$$
\begin{aligned}
\left[\hat{c}, (\hat{c}^\dagger)^n\right]_+ &= \left[\hat{c}, \hat{c}^\dagger\right]_+ (\hat{c}^\dagger)^{n-1} + \hat{c}^\dagger \left[\hat{c}, \hat{c}^\dagger\right]_+ (\hat{c}^\dagger)^{n-2} \\
&\quad + (\hat{c}^\dagger)^2 \left[\hat{c}, \hat{c}^\dagger\right]_+ (\hat{c}^\dagger)^{n-3} + \cdots + (\hat{c}^\dagger)^{n-1} \left[\hat{c}, \hat{c}^\dagger\right]_+
\end{aligned}
$$

and substituting $\left[\hat{c}, \hat{c}^\dagger\right]_+ = 1$ on the right-hand side. The second identity is obtained by expanding $g(\hat{c}^\dagger)$ in the commutator as

$$
g(\hat{c}^\dagger) = \sum_{n=0}^{\infty} \frac{g^{(n)}(0)}{n!} (\hat{c}^\dagger)^n
$$

and using the first equality.

Problems of Chapter 4

4.1 The wave number for $d = 1$ is given by $k_n \equiv 2\pi n/L$ $(n = 0, \pm 1, \pm 2, \cdots)$, which lies in $-\infty \le k_n \le \infty$ with a common spacing $\Delta k_n \equiv k_{n+1} - k_n = 2\pi/L$. Thus, the density of states is calculated as

$$
\begin{aligned}
D(\epsilon) &= (2s+1) \sum_n \delta(\epsilon - \varepsilon_n) \\
&= (2s+1) \frac{L}{2\pi} \sum_n \Delta k_n \delta(\epsilon - \varepsilon_n) && \text{sum} \to \text{integral} \\
&\approx \frac{(2s+1)L}{2\pi} \int_{-\infty}^{\infty} dk_n \delta(\epsilon - \varepsilon_n) && k_n \equiv -k_n' \text{ for } k_n < 0 \\
&= \frac{(2s+1)L}{\pi} \int_0^{\infty} dk_n \delta(\epsilon - \varepsilon_n) && k_n = \left(\frac{2m}{\hbar^2}\right)^{1/2} \varepsilon_n^{1/2} \\
&= \frac{(2s+1)L}{\pi} \left(\frac{2m}{\hbar^2}\right)^{1/2} \int_0^{\infty} \frac{d\varepsilon_n}{2\varepsilon_n^{1/2}} \delta(\epsilon - \varepsilon_n)
\end{aligned}
$$

$$= \frac{(2s+1)L}{2\pi}\left(\frac{2m}{\hbar^2}\right)^{1/2}\frac{1}{\epsilon^{1/2}}\theta(\epsilon).$$

As for $d = 2$, we express the wave vector as $\mathbf{k} = (k\cos\varphi, k\sin\varphi)$ with $\mathrm{d}^2k = k\mathrm{d}k\mathrm{d}\varphi$ $(0 \le k \le \infty, 0 \le \varphi \le 2\pi)$ and make a change of variable as $k = (2m/\hbar^2)^{1/2}\varepsilon_k^{1/2}$. The density of states is thereby transformed as

$$\begin{aligned} D(\epsilon) &= (2s+1)\sum_{\mathbf{k}}\delta(\epsilon-\varepsilon_k) \\ &\approx (2s+1)\left(\frac{L}{2\pi}\right)^2\int \mathrm{d}^2k\,\delta(\epsilon-\varepsilon_k) \\ &= (2s+1)\frac{L^2}{(2\pi)^2}\int_0^{2\pi}\mathrm{d}\varphi\int_0^{\infty}\mathrm{d}kk\delta(\epsilon-\varepsilon_k) \qquad k=\left(\frac{2m}{\hbar^2}\right)^{1/2}\varepsilon_k^{1/2} \\ &= \frac{(2s+1)L^2}{2\pi}\frac{2m}{\hbar^2}\int_0^{\infty}\frac{\mathrm{d}\varepsilon_k}{2\varepsilon_k^{1/2}}\varepsilon_k^{1/2}\delta(\epsilon-\varepsilon_k) \\ &= \frac{(2s+1)L^2}{4\pi}\frac{2m}{\hbar^2}\theta(\epsilon). \end{aligned}$$

4.2

(a) The chemical potential is determined from (4.12) with $\sigma = -1$, where the distribution function at $T = 0$ reduces to the step function, (4.32). Substituting the density of states for $d = 2$ obtained in Problem 4.1 and setting $\mu(0) = \varepsilon_F$, we can transform the equation as

$$N = \int_0^{\varepsilon_F} D(\epsilon)\mathrm{d}\epsilon = \frac{L^2 m}{\pi\hbar^2}\varepsilon_F.$$

We thereby obtain the Fermi energy ε_F and Fermi wave vector $k_F \equiv (2m\varepsilon_F/\hbar^2)^{1/2}$ as

$$\varepsilon_F = \frac{\pi\hbar^2 N}{mL^2}, \qquad k_F = \left(\frac{2\pi N}{L^2}\right)^{1/2}.$$

(b) Equation (4.12) for the chemical potential corresponds to the case of $g(\tilde{\epsilon}) \to D(\epsilon)$ in (4.41). As $D(\epsilon)$ is constant in two dimensions, there is no power-law temperature dependence in μ according to the Sommerfeld expansion, i.e., $\mu \approx \epsilon_F$ holds at low temperatures. A weak exponential T dependence in μ still exists at low temperatures, which can be reproduced by removing the approximation $\mu/k_BT \to \infty$ in (4.38).

4.3

(a) Let us introduce variables $\varepsilon_\eta \equiv n_\eta \hbar\omega$ ($\eta = x, y, z$), the spacing between adjacent pair of levels being constant and given by $\Delta\varepsilon_\eta = \hbar\omega$. Using them, we rewrite the density of states as

$$
\begin{aligned}
D(\epsilon) &= \sum_{n_x=0}^{\infty}\sum_{n_y=0}^{\infty}\sum_{n_z=0}^{\infty} \delta(\epsilon - \varepsilon_x - \varepsilon_y - \varepsilon_z - 3\hbar\omega/2) \qquad (\tilde{\epsilon} \equiv \epsilon - 3\hbar\omega/2) \\
&= \frac{1}{(\hbar\omega)^3}\sum_{n_x=0}^{\infty}\sum_{n_y=0}^{\infty}\sum_{n_z=0}^{\infty} \Delta\varepsilon_x \Delta\varepsilon_y \Delta\varepsilon_z \delta(\tilde{\epsilon} - \varepsilon_x - \varepsilon_y - \varepsilon_z) \\
&\approx \frac{1}{(\hbar\omega)^3}\int_0^\infty d\varepsilon_x \int_0^\infty d\varepsilon_y \int_0^\infty d\varepsilon_z\, \delta(\tilde{\epsilon} - \varepsilon_x - \varepsilon_y - \varepsilon_z) \\
&= \frac{\theta(\tilde{\epsilon})}{(\hbar\omega)^3}\int_0^{\tilde{\epsilon}} d\varepsilon_x \int_0^{\tilde{\epsilon}-\varepsilon_x} d\varepsilon_y = \frac{\tilde{\epsilon}^2}{2(\hbar\omega)^3}\theta(\tilde{\epsilon}).
\end{aligned}
$$

(b) This case corresponds to $A = 1/2(\hbar\omega)^3$ and $\eta = 3$ in (4.52). Substituting both into (4.53) yields the expression of T_0 stated in the problem.

(c) Repeat the calculation of (4.48) using the present density of states to obtain N_0/N as

$$
\begin{aligned}
\frac{N_0}{N} &= 1 - \int_{\varepsilon_0}^\infty \frac{D(\epsilon)/N}{e^{(\epsilon-\varepsilon_0)/k_B T} - 1} d\epsilon = 1 - \frac{(k_B T)^3}{2(\hbar\omega)^3 N}\int_0^\infty \frac{x^2}{e^x - 1} dx \\
&= 1 - \frac{\zeta(3)\Gamma(3)(k_B T)^3}{2(\hbar\omega)^3 N} = 1 - \left(\frac{T}{T_0}\right)^3.
\end{aligned}
$$

(d) Repeat the calculation of (4.49) using the present density of states and setting $\varepsilon_0 = 3\hbar\omega/2$. We thereby obtain internal energy U as

$$
\begin{aligned}
U &= N_0\varepsilon_0 + \int_{\varepsilon_0}^\infty \frac{D(\epsilon)\epsilon}{e^{(\epsilon-\varepsilon_0)/k_B T} - 1} d\epsilon \\
&= N_0\varepsilon_0 + \frac{1}{2(\hbar\omega)^3}\int_{\varepsilon_0}^\infty \frac{(\epsilon-\varepsilon_0)^2(\varepsilon_0 + \epsilon - \varepsilon_0)}{e^{(\epsilon-\varepsilon_0)/k_B T} - 1} d\epsilon \\
&= \left[N_0 + \int_{\varepsilon_0}^\infty \frac{D(\epsilon)}{e^{(\epsilon-\varepsilon_0)/k_B T} - 1} d\epsilon\right]\varepsilon_0 + \frac{(k_B T)^4}{2(\hbar\omega)^3}\int_0^\infty \frac{x^3}{e^x - 1} dx \\
&= N\varepsilon_0 + \frac{\zeta(4)\Gamma(4)}{2(\hbar\omega)^3}(k_B T)^4 \\
&= N\left[\varepsilon_0 + \frac{3\zeta(4)}{\zeta(3)} k_B T_0 \left(\frac{T}{T_0}\right)^4\right].
\end{aligned}
$$

The heat capacity is subsequently obtained based on $C = \partial U/\partial T$ as

$$C = Nk_{\mathrm{B}}\frac{12\zeta(4)}{\zeta(3)}\left(\frac{T}{T_0}\right)^3.$$

Problems of Chapter 5

5.1 First, in taking an average, enumerate distinct decompositions by marking each pair of operators with a common symbol on top. Hence,

$$\begin{aligned}
&\langle\hat{\psi}^\dagger(\xi_1')\hat{\psi}^\dagger(\xi_2')\hat{\psi}^\dagger(\xi_3')\hat{\psi}(\xi_3)\hat{\psi}(\xi_2)\hat{\psi}(\xi_1)\rangle \\
&\quad= \langle\dot{\hat{\psi}}^\dagger(\xi_1')\ddot{\hat{\psi}}^\dagger(\xi_2')\acute{\hat{\psi}}^\dagger(\xi_3')\acute{\hat{\psi}}(\xi_3)\ddot{\hat{\psi}}(\xi_2)\dot{\hat{\psi}}(\xi_1)\rangle \\
&\qquad+\langle\dot{\hat{\psi}}^\dagger(\xi_1')\ddot{\hat{\psi}}^\dagger(\xi_2')\acute{\hat{\psi}}^\dagger(\xi_3')\ddot{\hat{\psi}}(\xi_3)\acute{\hat{\psi}}(\xi_2)\dot{\hat{\psi}}(\xi_1)\rangle \\
&\qquad+\langle\dot{\hat{\psi}}^\dagger(\xi_1')\ddot{\hat{\psi}}^\dagger(\xi_2')\acute{\hat{\psi}}^\dagger(\xi_3')\acute{\hat{\psi}}(\xi_3)\dot{\hat{\psi}}(\xi_2)\ddot{\hat{\psi}}(\xi_1)\rangle \\
&\qquad+\langle\dot{\hat{\psi}}^\dagger(\xi_1')\ddot{\hat{\psi}}^\dagger(\xi_2')\acute{\hat{\psi}}^\dagger(\xi_3')\ddot{\hat{\psi}}(\xi_3)\dot{\hat{\psi}}(\xi_2)\acute{\hat{\psi}}(\xi_1)\rangle \\
&\qquad+\langle\dot{\hat{\psi}}^\dagger(\xi_1')\ddot{\hat{\psi}}^\dagger(\xi_2')\acute{\hat{\psi}}^\dagger(\xi_3')\dot{\hat{\psi}}(\xi_3)\ddot{\hat{\psi}}(\xi_2)\acute{\hat{\psi}}(\xi_1)\rangle \\
&\qquad+\langle\dot{\hat{\psi}}^\dagger(\xi_1')\ddot{\hat{\psi}}^\dagger(\xi_2')\acute{\hat{\psi}}^\dagger(\xi_3')\dot{\hat{\psi}}(\xi_3)\acute{\hat{\psi}}(\xi_2)\ddot{\hat{\psi}}(\xi_1)\rangle.
\end{aligned}$$

Next, focus on a single operator from the left in each term and move its partner successively to its right, multiplying by σ upon each exchange of operators, until all pairs are coupled as in

$$\begin{aligned}
&\langle\hat{\psi}^\dagger(\xi_1')\hat{\psi}^\dagger(\xi_2')\hat{\psi}^\dagger(\xi_3')\hat{\psi}(\xi_3)\hat{\psi}(\xi_2)\hat{\psi}(\xi_1)\rangle \\
&\quad= \sigma^{4+2}\langle\dot{\hat{\psi}}^\dagger(\xi_1')\dot{\hat{\psi}}(\xi_1)\ddot{\hat{\psi}}^\dagger(\xi_2')\ddot{\hat{\psi}}(\xi_2)\acute{\hat{\psi}}^\dagger(\xi_3')\acute{\hat{\psi}}(\xi_3)\rangle \\
&\qquad+\sigma^{4+1}\langle\dot{\hat{\psi}}^\dagger(\xi_1')\dot{\hat{\psi}}(\xi_1)\ddot{\hat{\psi}}^\dagger(\xi_2')\ddot{\hat{\psi}}(\xi_3)\acute{\hat{\psi}}^\dagger(\xi_3')\acute{\hat{\psi}}(\xi_2)\rangle \\
&\qquad+\sigma^{3+2}\langle\dot{\hat{\psi}}^\dagger(\xi_1')\dot{\hat{\psi}}(\xi_2)\ddot{\hat{\psi}}^\dagger(\xi_2')\ddot{\hat{\psi}}(\xi_1)\acute{\hat{\psi}}^\dagger(\xi_3')\acute{\hat{\psi}}(\xi_3)\rangle \\
&\qquad+\sigma^{3+1}\langle\dot{\hat{\psi}}^\dagger(\xi_1')\dot{\hat{\psi}}(\xi_2)\ddot{\hat{\psi}}^\dagger(\xi_2')\ddot{\hat{\psi}}(\xi_3)\acute{\hat{\psi}}^\dagger(\xi_3')\acute{\hat{\psi}}(\xi_1)\rangle \\
&\qquad+\sigma^{2+1}\langle\dot{\hat{\psi}}^\dagger(\xi_1')\dot{\hat{\psi}}(\xi_3)\ddot{\hat{\psi}}^\dagger(\xi_2')\ddot{\hat{\psi}}(\xi_2)\acute{\hat{\psi}}^\dagger(\xi_3')\acute{\hat{\psi}}(\xi_1)\rangle \\
&\qquad+\sigma^{2+2}\langle\dot{\hat{\psi}}^\dagger(\xi_1')\dot{\hat{\psi}}(\xi_3)\ddot{\hat{\psi}}^\dagger(\xi_2')\ddot{\hat{\psi}}(\xi_1)\acute{\hat{\psi}}^\dagger(\xi_3')\acute{\hat{\psi}}(\xi_2)\rangle.
\end{aligned}$$

Finally, we replace identical symbols on each pair with angle brackets around them, i.e.,

$$\langle\hat{\psi}^\dagger(\xi_1')\hat{\psi}^\dagger(\xi_2')\hat{\psi}^\dagger(\xi_3')\hat{\psi}(\xi_3)\hat{\psi}(\xi_2)\hat{\psi}(\xi_1)\rangle$$

$$
\begin{aligned}
&= \langle\hat{\psi}^\dagger(\xi_1')\hat{\psi}(\xi_1)\rangle\langle\hat{\psi}^\dagger(\xi_2')\hat{\psi}(\xi_2)\rangle\langle\hat{\psi}^\dagger(\xi_3')\hat{\psi}(\xi_3)\rangle \\
&+\sigma\langle\hat{\psi}^\dagger(\xi_1')\hat{\psi}(\xi_1)\rangle\langle\hat{\psi}^\dagger(\xi_2')\hat{\psi}(\xi_3)\rangle\langle\hat{\psi}^\dagger(\xi_3')\hat{\psi}(\xi_2)\rangle \\
&+\sigma\langle\hat{\psi}^\dagger(\xi_1')\hat{\psi}(\xi_2)\rangle\langle\hat{\psi}^\dagger(\xi_2')\hat{\psi}(\xi_1)\rangle\langle\hat{\psi}^\dagger(\xi_3')\hat{\psi}(\xi_3)\rangle \\
&+\langle\hat{\psi}^\dagger(\xi_1')\hat{\psi}(\xi_2)\rangle\langle\hat{\psi}^\dagger(\xi_2')\hat{\psi}(\xi_3)\rangle\langle\hat{\psi}^\dagger(\xi_3')\hat{\psi}(\xi_1)\rangle \\
&+\sigma\langle\hat{\psi}^\dagger(\xi_1')\hat{\psi}(\xi_3)\rangle\langle\hat{\psi}^\dagger(\xi_2')\hat{\psi}(\xi_2)\rangle\langle\hat{\psi}^\dagger(\xi_3')\hat{\psi}(\xi_1)\rangle \\
&+\langle\hat{\psi}^\dagger(\xi_1')\hat{\psi}(\xi_3)\rangle\langle\hat{\psi}^\dagger(\xi_2')\hat{\psi}(\xi_1)\rangle\langle\hat{\psi}^\dagger(\xi_3')\hat{\psi}(\xi_2)\rangle.
\end{aligned}
$$

5.2 The formula can be derived in the following manner:

$$
\begin{aligned}
\ell(k_Q r) &\xrightarrow{T\to 0} \frac{\pi}{4}\int_0^{\tilde{\varepsilon}_F} \frac{\sin(k_Q r\tilde{\epsilon}^{1/2})}{k_Q r}\mathrm{d}\tilde{\epsilon} && \tilde{\epsilon}^{1/2} \equiv \frac{t}{k_Q r} \\
&= \frac{\pi}{2(k_Q r)^3}\int_0^{k_F r} t\sin t\,\mathrm{d}t && k_F = k_Q\tilde{\varepsilon}_F^{1/2} = k_Q\left(\frac{6}{\pi}\right)^{1/3} \\
&= \frac{\pi}{2(\pi/6)(k_F r)^3}\left(-t\cos t\Big|_0^{k_F r} + \int_0^{k_F r}\cos t\,\mathrm{d}t\right) \\
&= \frac{3\,(-k_F r\cos k_F r + \sin k_F r)}{(k_F r)^3}.
\end{aligned}
$$

From the Taylor series expansions of $-x\cos x = -x + \frac{1}{2!}x^3 - \frac{1}{4!}x^5 + \cdots$ and $\sin x = x - \frac{1}{3!}x^3 + \frac{1}{5!}x^5 - \cdots$, we then obtain the expression near $r = 0$.

Problems of Chapter 6

6.1

(a) Let us multiply (6.18) by $\mathrm{e}^{-\mathrm{i}\mathbf{k}'\cdot(\mathbf{r}_1-\mathbf{r}_2)}$, integrate over $\mathbf{r} \equiv \mathbf{r}_1 - \mathbf{r}_2$, and use the orthonormality of (4.14) given by $\langle\varphi_{\mathbf{k}'}|\varphi_{\mathbf{k}}\rangle = \delta_{\mathbf{k}'\mathbf{k}}$. Then, replacing $\mathbf{k}' \to \mathbf{k}$, we obtain $\mathscr{V}_k = \int \mathscr{V}(r)\,\mathrm{e}^{-\mathrm{i}\mathbf{k}\cdot\mathbf{r}}\mathrm{d}^3 r$. This integral can be calculated analytically by choosing the z axis along $\mathbf{k}$ and adopting polar coordinates $\mathbf{r} = (r\sin\theta\cos\varphi, r\sin\theta\sin\varphi, r\cos\theta)$ $(0 \le r < \infty, 0 \le \theta \le \pi, 0 \le \varphi < 2\pi)$. We have

$$
\begin{aligned}
\mathscr{V}_k &= \int_0^{2\pi}\mathrm{d}\varphi\int_0^\infty \mathrm{d}r\, r^2\mathscr{V}(r)\int_0^\pi \mathrm{d}\theta\,\sin\theta\,\mathrm{e}^{-\mathrm{i}kr\cos\theta} \\
&= 2\pi\int_0^\infty \mathrm{d}r\, r^2\mathscr{V}(r)\int_{-1}^1 \mathrm{d}t\,\mathrm{e}^{-\mathrm{i}krt} = 2\pi\int_0^\infty \mathrm{d}r\, r^2\mathscr{V}(r)\frac{\mathrm{e}^{\mathrm{i}kr}-\mathrm{e}^{-\mathrm{i}kr}}{\mathrm{i}kr}
\end{aligned}
$$

$$= \frac{2\pi U_0}{ik} \int_0^\infty dr\, r \left[e^{-(1/r_0 - ik)r} - e^{-(1/r_0 + ik)r} \right]$$

$$= \frac{2\pi U_0}{ik} \left[\frac{1}{(1/r_0 - ik)^2} - \frac{1}{(1/r_0 + ik)^2} \right] = \frac{8\pi U_0 r_0^3}{(1 + r_0^2 k^2)^2}.$$

(b) Expansion (6.51) can be expressed in this instance as

$$\frac{8\pi U_0 r_0^3}{[1 + r_0^2(k^2 + k'^2 - 2kk'x)]^2} = \sum_{\ell'=0}^{\infty} (2\ell' + 1) \mathscr{V}_{\ell'}(k, k') P_{\ell'}(x),$$

where $x \equiv \cos\theta_{\mathbf{kk}'}$. Let us multiply the equation by $P_\ell(x)$, integrate over $-1 \le x \le 1$, and use (6.36) to obtain

$$\mathscr{V}_\ell(k, k') = \frac{1}{2} \int_{-1}^{1} dx \frac{8\pi U_0 r_0^3 P_\ell(x)}{[1 + r_0^2(k^2 + k'^2 - 2kk'x)]^2}.$$

This integral for $\ell = 0$ can be calculated analytically with $P_0(x) = 1$,

$$\mathscr{V}_0(k, k') = \frac{8\pi U_0 r_0^3}{\left(1 + r_0^2 k^2 + r_0^2 k'^2\right)^2 - 4 r_0^4 k^2 k'^2}.$$

(c) Substitute the results of (a) and (b) into (6.52) to obtain

$$F_0^{\mathrm{s}} = \frac{8\pi U_0 r_0^3 D(\varepsilon_{\mathrm{F}})}{2V} \left(2 - \frac{1}{1 + 4 r_0^2 k_{\mathrm{F}}^2} \right), \qquad F_0^{\mathrm{a}} = -\frac{8\pi U_0 r_0^3 D(\varepsilon_{\mathrm{F}})}{2V(1 + 4 r_0^2 k_{\mathrm{F}}^2)}.$$

Problems of Chapter 7

7.1 This Schrödinger equation forms an ordinary second-order differential equation with constant coefficients for each of $|x| < a$ and $|x| \ge a$. Hence, it can be solved most simply in terms of k and κ in (7.5),

$$\phi(x) = \begin{cases} A \cos kx + A' \sin kx & : |x| < a \\ B\, e^{-\kappa x} & : x \ge a \\ B'\, e^{\kappa x} & : x \le -a \end{cases},$$

where A, A', B, and B' are constants, and we have used the boundary conditions $|\phi(\pm\infty)| = 0$. Because the potential is even in x, eigenfunctions can be classified into even functions with $A' = 0$, $B' = B$ or odd functions with $A = 0$, $B' = -B$. The lowest eigenfunction without nodes is an even function, so we set $A' = 0$, $B' = B$ and focus on the region $x \ge 0$. We match the solutions for $0 \le x < a$ and $x \ge a$ by requiring that $\phi'(x)/\phi(x)$ be continuous at $x = a$. The condition yields

$$\eta = \xi \tan \xi,$$

where (ξ, η) are defined by (7.6). By plotting the curve in Fig. 7.2, we find that this function has an intersection with (7.8) in the first quadrant for an arbitrary $U_0 > 0$.

To obtain the solution of $U_0 \to 0$ analytically, we expand $\tan \xi = \xi + O(\xi^2)$ to approximate $\eta \approx \xi^2$ in the above equation. Substituting the result into (7.8) and noting $\xi \ll 1$, we obtain $\eta \approx (2ma^2/\hbar^2)U_0$. Moreover, (7.5) and (7.6) yield $\eta = \left(-2ma^2\varepsilon/\hbar^2\right)^{1/2}$. Equating the two expressions of η fixes the ground-state energy for $U_0 \to 0$,

$$\epsilon \approx -(2ma^2/\hbar^2)U_0^2.$$

Thus, a bound state is formed by an infinitesimal attraction also in one dimension. It follows from considerations in Sect. 7.2 that this result can be attributed to the fact that the one-dimensional density of states given by (4.54) does not vanish for $\epsilon \to 0$. Indeed, (4.54) for $d = 1$ even diverges, thereby producing a deeper bound-state energy than that of (7.26) for $d = 2$ with a constant density of states.

Problems of Chapter 8

8.1

(a) Let us substitute the expansion of the pair wave function into (8.3) and then rewrite $\hat{Q}^\dagger$ as

$$\hat{Q}^\dagger = \sum_{\mathbf{k}} \phi_{\mathbf{k}} \frac{\hat{c}^\dagger_{\mathbf{k}\uparrow}\hat{c}^\dagger_{-\mathbf{k}\downarrow} - \hat{c}^\dagger_{\mathbf{k}\downarrow}\hat{c}^\dagger_{-\mathbf{k}\uparrow}}{2} = \sum_{\mathbf{k}} \phi_{\mathbf{k}} \hat{c}^\dagger_{\mathbf{k}\uparrow}\hat{c}^\dagger_{-\mathbf{k}\downarrow}.$$

In deriving the last expression, we have used $c^\dagger_{\mathbf{k}\downarrow}c^\dagger_{-\mathbf{k}\uparrow} = -c^\dagger_{-\mathbf{k}\uparrow}c^\dagger_{\mathbf{k}\downarrow}$ and symmetry $\phi_{\mathbf{k}} = \phi_{-\mathbf{k}}$, and also made a change of variables $\mathbf{k} \to -\mathbf{k}$ to simplify the expression.

(b) Let us substitute the expression of (a) into (8.6), expand the exponent in a Taylor series, and use the identity $\left(\hat{c}^\dagger_{\mathbf{k}\alpha}\right)^2 = 0$ for fermions. We can thereby transform the condensate wave function as

$$|\Phi\rangle \equiv A \exp\left(\sum_{\mathbf{k}} \phi_{\mathbf{k}} \hat{c}^\dagger_{\mathbf{k}\uparrow}\hat{c}^\dagger_{-\mathbf{k}\downarrow}\right)|0\rangle = A \prod_{\mathbf{k}} \exp\left(\phi_{\mathbf{k}} \hat{c}^\dagger_{\mathbf{k}\uparrow}\hat{c}^\dagger_{-\mathbf{k}\downarrow}\right)|0\rangle$$

$$= A\prod_{\mathbf{k}}\left(1+\phi_{\mathbf{k}}\hat{c}^{\dagger}_{\mathbf{k}\uparrow}\hat{c}^{\dagger}_{-\mathbf{k}\downarrow}\right)|0\rangle = \prod_{\mathbf{k}}\left(u_{\mathbf{k}}+v_{\mathbf{k}}\hat{c}^{\dagger}_{\mathbf{k}\uparrow}\hat{c}^{\dagger}_{-\mathbf{k}\downarrow}\right)|0\rangle,$$

with $u_{\mathbf{k}} \equiv 1/\sqrt{1+|\phi_{\mathbf{k}}|^2}$ and $v_{\mathbf{k}} \equiv \phi_{\mathbf{k}}/\sqrt{1+|\phi_{\mathbf{k}}|^2}$.

8.2

(a) Equation (8.96) can be shown using (8.7) as follows:

$$\begin{aligned}
[\hat{Q},\hat{Q}^{\dagger}]_{+} &= \frac{1}{2}\phi^{*}(\bar{\xi}_1',\xi_2')[\hat{\psi}(\bar{\xi}_2')\hat{\psi}(\xi_1'),\hat{Q}^{\dagger}]_{+} \\
&= \frac{1}{2}\phi^{*}(\bar{\xi}_1',\xi_2')\left\{\hat{\psi}(\bar{\xi}_2')[\hat{\psi}(\xi_1'),\hat{Q}^{\dagger}]_{+}+[\hat{\psi}(\xi_2'),\hat{Q}^{\dagger}]_{+}\hat{\psi}(\bar{\xi}_1')\right\} \\
&= \frac{1}{2}\phi^{*}(\bar{\xi}_1',\xi_2')\left\{\phi(\bar{\xi}_1',\bar{\xi}_2)\hat{\psi}(\bar{\xi}_2')\hat{\psi}^{\dagger}(\bar{\xi}_2)+\phi(\bar{\xi}_2',\xi_2)\hat{\psi}^{\dagger}(\bar{\xi}_2)\hat{\psi}(\bar{\xi}_1')\right\} \\
&= \frac{1}{2}\phi^{*}(\bar{\xi}_1',\xi_2')\{\phi(\bar{\xi}_1',\xi_2')-\phi(\bar{\xi}_1',\xi_2)\hat{\psi}^{\dagger}(\bar{\xi}_2)\hat{\psi}(\bar{\xi}_2') \\
&\quad +\phi(\bar{\xi}_2',\xi_2)\hat{\psi}^{\dagger}(\bar{\xi}_2)\hat{\psi}(\bar{\xi}_1')\} \\
&= \frac{1}{2}|\phi(\bar{\xi}_1,\xi_2)|^2+\hat{\psi}^{\dagger}(\bar{\xi}_2)\phi(\bar{\xi}_2,\bar{\xi}_1)\phi^{*}(\bar{\xi}_1,\bar{\xi}_2')\hat{\psi}(\bar{\xi}_2').
\end{aligned}$$

(b) It is clear that, when the second term of (8.96) can be neglected, the relevant operator $\hat{c}^{\dagger}$ satisfies $[\hat{c},\hat{c}^{\dagger}]_{+} = 1$. Moreover, the BCS wave function can be written as $|\Psi\rangle = A\,\mathrm{e}^{\Theta\hat{c}^{\dagger}}|0\rangle$. The normalization condition $\langle\Psi|\Psi\rangle = 1$ yields $|A| = \mathrm{e}^{-|\Theta|^2/2}$.

Problems of Chapter 9

9.1 We focus on the region $E > 0$ for an even function (9.46) and express its sum over **k** using the normal density of states (6.28) as

$$D_{\mathrm{s}}(E) = \int_{-\infty}^{\infty} \mathrm{d}\varepsilon_k D(\varepsilon_k)\delta(E-E_k) \approx D(\varepsilon_{\mathrm{F}})\int_{-\infty}^{\infty} \mathrm{d}\xi_k \delta(E-E_k),$$

where we have approximated $D(\varepsilon_k) \approx D(\varepsilon_{\mathrm{F}})$ and made a change of variable, $\varepsilon_k \to \xi_k \equiv \varepsilon_k - \mu$. Subsequently, we use the relation $E_k^2 = \xi_k^2 + \Delta^2$ to make another change of variable $\xi_k \to E_k$. Because the resulting function $\xi_k = \xi_k(E_k)$ is multivalued, i.e., $\xi_k = \pm\sqrt{E_k^2-\Delta^2}$, we must choose a single branch to perform it appropriately. This is done so that the signs of ξ_k and E_k coincide as $\xi_k = \sqrt{E_k^2-\Delta^2}$ for $E_k > 0$ and $\xi_k = -\sqrt{E_k^2-\Delta^2}$ for $E_k < 0$; see the thick line in the figure below. With $E > 0$ in the integrand, we only need to perform the above integration for the positive branch $\xi_k = \sqrt{E_k^2 - \Delta^2}$,

$$D_s(E) = D(\varepsilon_F)\int_0^\infty d\xi_k \delta(E - E_k) = D(\varepsilon_F)\int_\Delta^\infty dE_k \frac{d\xi_k}{dE_k}\delta(E - E_k)$$
$$= D(\varepsilon_F)\int_\Delta^\infty dE_k \frac{E_k}{\sqrt{E_k^2 - \Delta^2}}\delta(E - E_k) = \theta(E - \Delta)D(\varepsilon_F)\frac{E}{\sqrt{E^2 - \Delta^2}}.$$

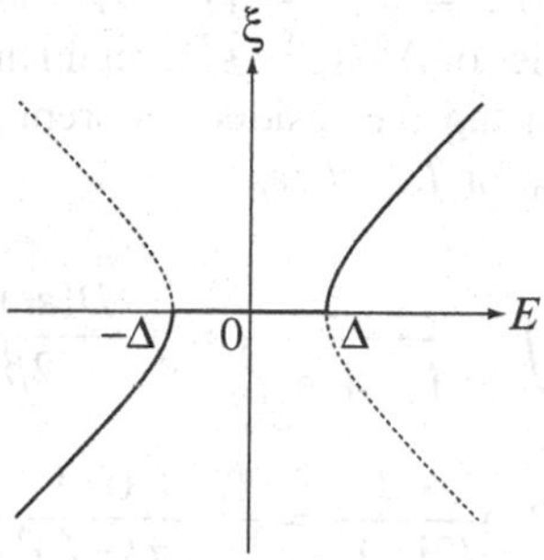

9.2 Let us substitute (9.48) into (9.47), approximate $\bar{n}(E) \approx e^{-\beta E}$ as appropriate at low temperatures, and make a change of variable $E = \Delta \cosh x$. We thereby express the heat capacity as

$$C = D(\varepsilon_F)\int_\Delta^\infty \frac{E^3}{(E^2 - \Delta^2)^{1/2}} \frac{e^{-E/k_B T}}{k_B T^2} dE$$
$$= D(\varepsilon_F)\frac{\Delta^3}{k_B T^2}\int_0^\infty e^{-(\Delta/k_B T)\cosh x} \cosh^3 x \, dx$$
$$= -D(\varepsilon_F)\frac{\Delta^3}{k_B T^2}\frac{d^3}{da^3}\int_0^\infty e^{-a\cosh x} dx\bigg|_{a=\Delta/k_B T}.$$

The last integral is exactly the modified Bessel function $K_0(a)$, which for $a \gg 1$ is approximated as [1, 2]

$$K_0(a) \equiv \int_0^\infty e^{-a\cosh x} dx \approx \left(\frac{\pi}{2a}\right)^{1/2} e^{-a}.$$

Using it, we obtain the heat capacity for $a \equiv \Delta/k_B T \gg 1$ as

$$C \approx -D(\varepsilon_F)\frac{\Delta^3}{k_B T^2}\frac{d^3}{da^3}\left(\frac{\pi}{2a}\right)^{1/2} e^{-a}\bigg|_{a=\Delta/k_B T} \approx D(\varepsilon_F)\frac{\Delta^3}{k_B T^2}\left(\frac{\pi}{2a}\right)^{1/2} e^{-a}\bigg|_{a=\Delta/k_B T},$$

which gives (9.49).

9.3 Using (9.39) and the identity in the problem, we can transform (9.51) as

$$F_{\rm sn} = \frac{D(\varepsilon_{\rm F})}{\beta}\sum_{n=0}^{\infty}\int_{-\infty}^{\infty}\left(-\ln\frac{E^2+\varepsilon_n^2}{\xi^2+\varepsilon_n^2}+\frac{\Delta^2}{E^2+\varepsilon_n^2}\right)d\xi$$
$$= \frac{D(\varepsilon_{\rm F})}{\beta}\sum_{n=0}^{\infty}\int_{-\infty}^{\infty}\left[-\ln\left(1+\frac{\Delta^2}{\xi^2+\varepsilon_n^2}\right)+\frac{\Delta^2}{\xi^2+\varepsilon_n^2+\Delta^2}\right]d\xi,$$

with $\beta \equiv 1/k_{\rm B}T$ and $\varepsilon_n \equiv (2n+1)\pi k_{\rm B}T$. Subsequently, we expand the integrand in a Taylor series of $\Delta^2/(\xi^2+\varepsilon_n^2)$, retain the leading term, and perform the integration over ξ using the residue theorem [2]. We thereby obtain the condensation energy $F_{\rm sn}$ for $T \lesssim T_{\rm c}$ as

$$F_{\rm sn} \approx -\frac{D(\varepsilon_{\rm F})\Delta^4}{2\beta}\sum_{n=0}^{\infty}\int_{-\infty}^{\infty}\frac{d\xi}{\left(\xi^2+\varepsilon_n^2\right)^2} = -\frac{D(\varepsilon_{\rm F})\Delta^4}{2\beta}\sum_{n=0}^{\infty}\lim_{\xi\to i\varepsilon_n}\frac{d}{d\xi}\frac{2\pi i}{\left(\xi+i\varepsilon_n\right)^2}$$
$$= -\frac{D(\varepsilon_{\rm F})\Delta^4}{2\beta}\sum_{n=0}^{\infty}2\pi i\frac{-2}{(2i\varepsilon_n)^3} = -\frac{D(\varepsilon_{\rm F})}{4(\pi k_{\rm B}T)^2}\frac{7}{8}\zeta(3)\Delta^4,$$

where we have performed the same transformation as (9.41). Setting $T \approx T_{\rm c}$ in the final expression and substituting (9.42), we obtain (9.53).

Problems of Chapter 10

10.1

(a) Substitute the identity

$$0 = 1 + \int_{-\infty}^{\infty} d\xi \frac{\partial}{\partial\xi}\frac{1}{e^{\beta\xi}+1}$$

into the right-hand side of (10.16), rewrite the derivative of the mean occupation number using $1/(e^x+1) = \frac{1}{2}\left(1-\tanh\frac{x}{2}\right)$ and (9.39), and perform the differentiation of the resulting expression to obtain

$$Y(T) = 1 + \frac{1}{2}\int_{-\infty}^{\infty} d\xi\left[\frac{\partial}{\partial E}\tanh\frac{\beta E}{2} - \frac{\partial}{\partial\xi}\tanh\frac{\beta\xi}{2}\right]$$
$$= 1 + \frac{2}{\beta}\sum_{n=0}^{\infty}\int_{-\infty}^{\infty} d\xi\left[\frac{\partial}{\partial E}\frac{E}{E^2+\varepsilon_n^2} - \frac{\partial}{\partial\xi}\frac{\xi}{\xi^2+\varepsilon_n^2}\right]$$

$$= 1 + \frac{2}{\beta}\sum_{n=0}^{\infty}\int_{-\infty}^{\infty} d\xi \left[\frac{-\xi^2 - |\Delta|^2 + \varepsilon_n^2}{(\xi^2 + |\Delta|^2 + \varepsilon_n^2)^2} - \frac{-\xi^2 + \varepsilon_n^2}{(\xi^2 + \varepsilon_n^2)^2}\right].$$

Performing the integration using the residue theorem, we obtain (10.17).

(b) Approximate $\varepsilon_n^2 + |\Delta|^2 \approx \varepsilon_n^2$ in the denominator of (10.17), transform the resulting expression in the same way as in (9.41), and substitute (9.42). We thereby obtain (10.47).

(c) Approximate $\bar{n}_k \approx e^{-E_k/k_B T}$ in the integrand of (10.16) and transform the resulting expression into the form

$$Y(T) = -2\int_0^{\infty} d\xi_k \frac{\partial \bar{n}_k}{\partial E_k} \approx \frac{2}{k_B T}\int_0^{\infty} d\xi_k e^{-\sqrt{\xi_k^2+\Delta^2}/k_B T} \qquad (\xi_k = \Delta \sinh x)$$

$$= \frac{2\Delta}{k_B T}\int_0^{\infty} dx\, e^{-(\Delta/k_B T)\cosh x}\cosh x = -2a\frac{d}{da}\int_0^{\infty} dx\, e^{-a\cosh x}\bigg|_{a=\Delta/k_B T}$$

$$= -2a\frac{dK_0(a)}{da}\bigg|_{a=\Delta/k_B T} \approx -2a\frac{d}{da}\sqrt{\frac{\pi}{2a}}e^{-a}\bigg|_{a=\Delta/k_B T} = \sqrt{\frac{2\pi\Delta}{k_B T}}e^{-\Delta/k_B T},$$

where $K_0(a)$ is the modified Bessel function [1, 2], and we have used its asymptotic form for $a \to \infty$.

Problems of Chapter 11

11.1 Equation (11.11) can be shown using $[e^{\beta(\Omega-\hat{\mathscr{H}})}, e^{\pm i\hat{\mathscr{H}}t/\hbar}] = 0$ and the invariance of trace under cyclic permutations as follows:

$$\langle[\hat{\mathscr{O}}_H(t), \hat{\mathscr{H}}'_{\omega H}(t')]\rangle$$

$$= \mathrm{Tr}\, e^{\beta(\Omega-\hat{\mathscr{H}})}\big[e^{i\hat{\mathscr{H}}t/\hbar}\hat{\mathscr{O}}e^{-i\hat{\mathscr{H}}(t-t')/\hbar}\hat{\mathscr{H}}'_{\omega}e^{-i\hat{\mathscr{H}}t'/\hbar}$$

$$- e^{i\hat{\mathscr{H}}t'/\hbar}\hat{\mathscr{H}}'_{\omega}e^{i\hat{\mathscr{H}}(t-t')/\hbar}\hat{\mathscr{O}}e^{-i\hat{\mathscr{H}}t/\hbar}\big]$$

$$= \mathrm{Tr}\, e^{\beta(\Omega-\hat{\mathscr{H}})}\big[e^{i\hat{\mathscr{H}}(t-t')/\hbar}\hat{\mathscr{O}}e^{-i\hat{\mathscr{H}}(t-t')/\hbar}\hat{\mathscr{H}}'_{\omega} - \hat{\mathscr{H}}'_{\omega}e^{i\hat{\mathscr{H}}(t-t')/\hbar}\hat{\mathscr{O}}e^{-i\hat{\mathscr{H}}(t-t')/\hbar}\big]$$

$$= \langle[\hat{\mathscr{O}}_H(t-t'), \hat{\mathscr{H}}'_{\omega}]\rangle.$$

11.2 It follows from $\hat{\mathscr{H}}'^{\dagger}_{\omega} = \hat{\mathscr{H}}'_{-\omega}$ and $\hat{\mathscr{H}}'^{\dagger}_{-\omega,H}(t) = [e^{i\hat{\mathscr{H}}t/\hbar}\hat{\mathscr{H}}'_{-\omega}e^{-i\hat{\mathscr{H}}t/\hbar}]^{\dagger} = e^{i\hat{\mathscr{H}}t/\hbar}\hat{\mathscr{H}}'^{\dagger}_{-\omega}e^{-i\hat{\mathscr{H}}t/\hbar} = \hat{\mathscr{H}}'_{\omega,H}(t)$ that the expectation in the integrand of (11.17) satisfies

$$\langle[\hat{\mathscr{H}}'_{-\omega,H}(t), \hat{\mathscr{H}}'_{\omega}]\rangle^* = \{\mathrm{Tr}\, e^{\beta(\Omega-\hat{\mathscr{H}})}[\hat{\mathscr{H}}'_{-\omega,H}(t)\hat{\mathscr{H}}'_{\omega} - \hat{\mathscr{H}}'_{\omega}\hat{\mathscr{H}}'_{-\omega,H}(t)]\}^*$$

$$= \mathrm{Tr}[\hat{\mathscr{H}}'^{\dagger}_{\omega}\hat{\mathscr{H}}'^{\dagger}_{-\omega,H}(t) - \hat{\mathscr{H}}'^{\dagger}_{-\omega,H}(t)\hat{\mathscr{H}}'^{\dagger}_{\omega}]e^{\beta(\Omega-\hat{\mathscr{H}})}$$

$$= \mathrm{Tr}\, \mathrm{e}^{\beta(\Omega-\hat{\mathscr{H}})}\big[\hat{\mathscr{H}}'_{-\omega}\hat{\mathscr{H}}'_{\omega,\mathrm{H}}(t) - \hat{\mathscr{H}}'_{\omega,\mathrm{H}}(t)\hat{\mathscr{H}}'_{-\omega}\big]$$
$$= -\big\langle\big[\hat{\mathscr{H}}'_{\omega,\mathrm{H}}(t), \hat{\mathscr{H}}'_{-\omega}\big]\big\rangle.$$

Using it, proof of (11.18) obtains.

Problems of Chapter 12

12.1 Writing $\bar{n}(E) = \frac{1}{2}(1 - \tanh\frac{\beta E}{2})$ in the integrand of (12.34) for $V = 0$ and using (9.39), we find

$$\frac{\bar{n}(E') - \bar{n}(E)}{E' - E} = \frac{2}{\beta}\sum_{n=0}^{\infty}\frac{EE' - \varepsilon_n^2}{(E^2 + \varepsilon_n^2)(E'^2 + \varepsilon_n^2)},$$

with $\varepsilon_n \equiv (2n+1)\pi\beta^{-1}$. The ε_n^2 term in the numerator does not contribute to the integral of (12.34) at $V = 0$ because the other function (12.30) is odd in E. Next, we make a change of variable $E \to \xi \equiv \mathrm{sgn}(E)\sqrt{E^2 - |\Delta_\mathrm{L}|^2}$ to obtain $M_{\mathrm{sL}}(E)\mathrm{d}E = M_{\mathrm{sL}}(E)(\xi/E)\mathrm{d}\xi = N_\mathrm{L}(\varepsilon_\mathrm{F})(|\Delta_\mathrm{L}|/E)\mathrm{d}\xi$. We thereby obtain an alternative expression for $I_\mathrm{c} = -I_{f\mathrm{s}}(0)$ as

$$I_\mathrm{c} = -\frac{4e}{\hbar}\langle|T_{\mathbf{kq}}|^2\rangle_\mathrm{F}\int_{-\infty}^{\infty}\mathrm{d}\xi$$
$$\times\int_{-\infty}^{\infty}\mathrm{d}\xi'\frac{2}{\beta}\sum_{n=0}^{\infty}\frac{N_\mathrm{L}(\varepsilon_\mathrm{F})N_\mathrm{R}(\varepsilon_\mathrm{F})|\Delta_\mathrm{L}||\Delta_\mathrm{R}|}{(\xi^2 + |\Delta_\mathrm{L}|^2 + \varepsilon_n^2)(\xi'^2 + |\Delta_\mathrm{R}|^2 + \varepsilon_n^2)}.$$

The integral can be calculated analytically to yield (12.40).

12.2 Sum $2\pi\beta^{-1}\sum_n$ for $T \to 0$ reduces to the integral of ε_n over $0 \le \varepsilon_n \le \infty$. It can be transformed by setting $\varepsilon_n = |\Delta_\mathrm{L}|\tan\theta$ $(0 \le \theta \le \pi/2)$ for $|\Delta_\mathrm{L}| \le |\Delta_\mathrm{R}|$ into

$$I_\mathrm{c} = \frac{|\Delta_\mathrm{L}|}{|e|R_\mathrm{N}}\int_0^{\pi/2}\frac{\mathrm{d}\theta}{\sqrt{1 - (1 - |\Delta_\mathrm{L}|^2/|\Delta_\mathrm{R}|^2)\sin^2\theta}}$$
$$= \frac{|\Delta_\mathrm{L}|}{|e|R_\mathrm{N}}K\big(\sqrt{1 - |\Delta_\mathrm{L}|^2/|\Delta_\mathrm{R}|^2}\big),$$

where K denotes the complete elliptic integral [1, 2]. Using the identity $K(x) = \frac{2}{1+y}K(\frac{1-y}{1+y})$ with $y = \sqrt{1-x^2}$, we produce a symmetric expression as in (12.42), where we have extended the formula to $|\Delta_\mathrm{L}| > |\Delta_\mathrm{R}|$ as $|\Delta_\mathrm{R}| - |\Delta_\mathrm{L}| \to \big||\Delta_\mathrm{L}| - |\Delta_\mathrm{R}|\big|$.

Problems of Chapter 13

13.1 Here, we simplify the notation of (8.79) as $\underline{\Delta}(\mathbf{k}) \to \underline{\Delta}_{\mathbf{k}}$, $\mathbf{u}_{\tilde{\alpha}}(\mathbf{k}) \to \mathbf{u}_{\mathbf{k}\tilde{\alpha}}$, etc.

(a) Let us write (8.79) for $\mathscr{\underline{H}}_{\rm HF} = \xi_k \underline{\sigma}_0$ in a form suitable for a perturbation expansion with $\underline{\Delta}_{\mathbf{k}}$ as

$$\begin{bmatrix} (E_{\mathbf{k}\tilde{\alpha}} - \xi_k)\underline{\sigma}_0 & \underline{0} \\ \underline{0} & (E_{\mathbf{k}\tilde{\alpha}} + \xi_k)\underline{\sigma}_0 \end{bmatrix} \begin{bmatrix} \mathbf{u}_{\mathbf{k}\tilde{\alpha}} \\ \mathbf{v}_{\mathbf{k}\tilde{\alpha}} \end{bmatrix} = \begin{bmatrix} \underline{0} & \underline{\Delta}_{\mathbf{k}} \\ -\underline{\Delta}^*_{-\mathbf{k}} & \underline{0} \end{bmatrix} \begin{bmatrix} \mathbf{u}_{\mathbf{k}\tilde{\alpha}} \\ \mathbf{v}_{\mathbf{k}\tilde{\alpha}} \end{bmatrix}.$$

We distinguish the pairs of eigenvalues that satisfy $E_{\mathbf{k}\tilde{\alpha}} \geq 0$ by subscripts $\tilde{\alpha} = 1, 2$. Let us solve the equation perturbatively in terms of $\underline{\Delta}$. The zeroth-order equation is given by

$$\begin{bmatrix} (E^{(0)}_{\mathbf{k}\tilde{\alpha}} - \xi_k)\underline{\sigma}_0 & \underline{0} \\ \underline{0} & (E^{(0)}_{\mathbf{k}\tilde{\alpha}} + \xi_k)\underline{\sigma}_0 \end{bmatrix} \begin{bmatrix} \mathbf{u}^{(0)}_{\mathbf{k}\tilde{\alpha}} \\ \mathbf{v}^{(0)}_{\mathbf{k}\tilde{\alpha}} \end{bmatrix} = \begin{bmatrix} \mathbf{0} \\ \mathbf{0} \end{bmatrix}.$$

Its eigenvalues $E^{(0)}_{\mathbf{k}\tilde{\alpha}} \geq 0$ and eigenvectors are obtained by considering the cases $\xi_k > 0$ and $\xi_k < 0$ separately. The results are given concisely using the step function (4.11),

$$E^{(0)}_{\mathbf{k}\tilde{\alpha}} = |\xi_k| \quad (\tilde{\alpha} = 1, 2), \qquad \begin{bmatrix} \mathbf{u}^{(0)}_{\mathbf{k}1} & \mathbf{u}^{(0)}_{\mathbf{k}2} \\ \mathbf{v}^{(0)}_{\mathbf{k}1} & \mathbf{v}^{(0)}_{\mathbf{k}2} \end{bmatrix} = \begin{bmatrix} \theta(\xi_k)\underline{\sigma}_0 \\ \theta(-\xi_k)\underline{\sigma}_0 \end{bmatrix} \equiv \begin{bmatrix} \underline{u}^{(0)}_{\mathbf{k}} \\ \underline{v}^{(0)}_{\mathbf{k}} \end{bmatrix}.$$

Hence, the positive eigenvalues are degenerate. Next, we use the above eigenvectors to obtain eigenvalue $E^{(1)}_{\mathbf{k}\tilde{\alpha}}$ to first order,

$$\begin{bmatrix} E^{(1)}_{\mathbf{k}1} & 0 \\ 0 & E^{(1)}_{\mathbf{k}2} \end{bmatrix} = \begin{bmatrix} \underline{u}^{(0)\dagger}_{\mathbf{k}} & \underline{v}^{(0)\dagger}_{\mathbf{k}} \end{bmatrix} \begin{bmatrix} \underline{0} & \underline{\Delta}_{\mathbf{k}} \\ -\underline{\Delta}^*_{-\mathbf{k}} & \underline{0} \end{bmatrix} \begin{bmatrix} \underline{u}^{(0)}_{\mathbf{k}} \\ \underline{v}^{(0)}_{\mathbf{k}} \end{bmatrix} = \underline{0}.$$

Thus, there is no first-order correction to the eigenvalues. The first-order eigenvectors are found by solving

$$\begin{bmatrix} (E^{(0)}_{\mathbf{k}\tilde{\alpha}} - \xi_k)\underline{\sigma}_0 & \underline{0} \\ \underline{0} & (E^{(0)}_{\mathbf{k}\tilde{\alpha}} + \xi_k)\underline{\sigma}_0 \end{bmatrix} \begin{bmatrix} \mathbf{u}^{(1)}_{\mathbf{k}\tilde{\alpha}} \\ \mathbf{v}^{(1)}_{\mathbf{k}\tilde{\alpha}} \end{bmatrix} = \begin{bmatrix} \underline{0} & \underline{\Delta}_{\mathbf{k}} \\ -\underline{\Delta}^*_{-\mathbf{k}} & \underline{0} \end{bmatrix} \begin{bmatrix} \mathbf{u}^{(0)}_{\mathbf{k}\tilde{\alpha}} \\ \mathbf{v}^{(0)}_{\mathbf{k}\tilde{\alpha}} \end{bmatrix}$$

as

$$\begin{bmatrix} \mathbf{u}^{(1)}_{\mathbf{k}1} & \mathbf{u}^{(1)}_{\mathbf{k}2} \\ \mathbf{v}^{(1)}_{\mathbf{k}1} & \mathbf{v}^{(1)}_{\mathbf{k}2} \end{bmatrix} = \begin{bmatrix} \theta(-\xi_k)\underline{\Delta}_{\mathbf{k}}/2|\xi_k| \\ -\theta(\xi_k)\underline{\Delta}^*_{-\mathbf{k}}/2\xi_k \end{bmatrix} \equiv \begin{bmatrix} \underline{u}^{(1)}_{\mathbf{k}} \\ \underline{v}^{(1)}_{\mathbf{k}} \end{bmatrix}.$$

Substituting the above results into (8.72), we thereby obtain $\underline{\tilde{\rho}}^{(1)}(\mathbf{k})$ to first order,

$$\underline{\tilde{\rho}}^{(1)}(\mathbf{k}) \approx \left(\underline{u}_{\mathbf{k}}^{(1)}\underline{v}_{\mathbf{k}}^{(0)\dagger} + \underline{u}_{\mathbf{k}}^{(0)}\underline{v}_{\mathbf{k}}^{(1)\dagger} - \underline{v}_{-\mathbf{k}}^{(1)*}\underline{u}_{-\mathbf{k}}^{(0)\mathrm{T}} - \underline{v}_{-\mathbf{k}}^{(0)*}\underline{u}_{-\mathbf{k}}^{(1)\mathrm{T}}\right)\frac{1}{2}\tanh\frac{\beta|\xi_k|}{2}$$
$$= \left(\underline{\Delta}_{\mathbf{k}} - \underline{\Delta}_{-\mathbf{k}}^{\mathrm{T}}\right)\frac{1}{4|\xi_k|}\tanh\frac{\beta|\xi_k|}{2} = \underline{\Delta}_{\mathbf{k}}\frac{1}{2\xi_k}\tanh\frac{\beta\xi_k}{2},$$

where we have used symmetry (8.76).

(b) Substitution of the result of (a) and (13.3) into (8.76) at $T = T_{\mathrm{c}}$ yields

$$\underline{\Delta}_{\mathbf{k}} = -4\pi\int\frac{\mathrm{d}^3k'}{(2\pi)^3}\mathscr{V}_1^{(\mathrm{eff})}(k,k')\sum_{m=-1}^{1}Y_{1m}(\hat{\mathbf{k}})Y_{1m}^*(\hat{\mathbf{k}}')\underline{\Delta}_{\mathbf{k}'}\frac{1}{2\xi_{k'}}\tanh\frac{\xi_{k'}}{2k_{\mathrm{B}}T_{\mathrm{c}}}.$$

Subsequently, we express the integral in terms of the density of states using (8.90), approximate $N(\varepsilon_{k'}) \approx N(\varepsilon_{\mathrm{F}})$, and substitute (13.2). Further, we expand the gap matrix as

$$\underline{\Delta}_{\mathbf{k}} = \sqrt{4\pi}\sum_{m=-1}^{1}\underline{\Delta}_{1m}\theta(\varepsilon_{\mathrm{c}} - |\xi_k|)Y_{1m}(\hat{\mathbf{k}})$$

and use orthogonality (8.86). We thereby obtain

$$\left[1 + N(\varepsilon_{\mathrm{F}})\mathscr{V}_1^{(\mathrm{eff})}\int_{-\varepsilon_{\mathrm{c}}}^{\varepsilon_{\mathrm{c}}}\mathrm{d}\xi_{k'}\frac{1}{2\xi_{k'}}\tanh\frac{\xi_{k'}}{2k_{\mathrm{B}}T_{\mathrm{c}}}\right]\underline{\Delta}_{1m} = \underline{0}.$$

Hence, we conclude that (13.35) holds for any internal state.

13.2 The angular integral of (13.27) is transformed by setting $t \equiv \cos\theta_{\mathbf{k}}$ into

$$\int\frac{\mathrm{d}\Omega_{\mathbf{k}}}{4\pi}\frac{\theta(|E| - |\Delta_{\mathbf{k}}|)}{(E^2 - |\Delta_{\mathbf{k}}|^2)^{1/2}} = \begin{cases}\displaystyle\int_0^1\mathrm{d}t\frac{\theta\left(|E| - \Delta_{\max}\sqrt{1-t^2}\right)}{\sqrt{E^2 - \Delta_{\max}^2(1-t^2)}} & : \text{ABM state}\\ \displaystyle\int_0^1\mathrm{d}t\frac{\theta\left(|E| - \Delta_{\max}t\right)}{\sqrt{E^2 - \Delta_{\max}^2t^2}} & : \text{polar state}\end{cases}.$$

The two integrals can be calculated using the indefinite integrals:

$$\int\frac{\mathrm{d}x}{\sqrt{x^2+c}} = \ln(x + \sqrt{x^2+c}), \qquad \int\frac{\mathrm{d}x}{\sqrt{a^2-x^2}} = \arcsin\frac{x}{a},$$

and considering the two cases $|E| < \Delta_{\max}$ and $|E| > \Delta_{\max}$ separately. To be more specific, the step function for $|E| < \Delta_{\max}$ replaces the lower (upper) limit of the integral by $\sqrt{1 - E^2/\Delta_{\max}^2}$ $\left(|E|/\Delta_{\max}\right)$ for the ABM (polar) state, whereas it is irrelevant for $|E| > \Delta_{\max}$. We thereby obtain (13.29).

Problems of Chapter 14

14.1

(a) Following the procedure described in the problem, we obtain

$$\hat{\Sigma}(\varepsilon_n, \mathbf{k}, \mathbf{r}) = \frac{1}{V^2} \sum_{\mathbf{k}'\mathbf{k}_1\mathbf{k}_2} U^{\mathrm{imp}}_{\mathbf{k}_1} \hat{\sigma}_z \hat{G}(\varepsilon_n, \mathbf{k}', \mathbf{r}) \hat{\sigma}_z U^{\mathrm{imp}}_{\mathbf{k}_2} N_a \delta_{\mathbf{k}_1+\mathbf{k}_2,\mathbf{0}} \delta_{\mathbf{k}_1,\mathbf{k}-\mathbf{k}'}.$$

Subsequently, we use relation $U^{\mathrm{imp}}_{-\mathbf{k}} = (U^{\mathrm{imp}}_{\mathbf{k}})^*$, which follows from $U^*_{\mathrm{imp}}(\mathbf{r}) = U_{\mathrm{imp}}(\mathbf{r})$, to obtain (14.102).

(b) One may prove (14.71) by following the procedure in the problem.

Problems of Chapter 15

15.1 Imagine that the supercurrent density has changed by $\delta\mathbf{j}$ during time δt in inducing a change $\delta\mathbf{B}$ in the flux density inside the superconductor. According to Faraday's induction law, this change in the flux density produces an electric field $\mathbf{E}$,

$$\nabla \times \mathbf{E} = -\frac{\delta\mathbf{B}}{\delta t},$$

which in turn will perform work on the external current,

$$-\delta' W = \int \mathrm{d}^3 r \, \mathbf{j}_{\mathrm{ext}} \cdot \mathbf{E}\, \delta t.$$

Moreover, the decrease $-\delta F$ in the free energy in reversible processes can be written as $-\delta F = -\mathbf{H} \cdot \int \mathrm{d}^3 r \delta\mathbf{B}$ according to (15.6) and (15.8), where by definition $\delta\mathbf{B} = \mathbf{0}$ outside the sample. Moving $\mathbf{H}$ into the integral over the whole space and using Faraday's induction law, $-\delta F$ is further transformed as

$$\begin{aligned}
-\delta F &= -\int \mathrm{d}^3 r\, \mathbf{H} \cdot \frac{\delta\mathbf{B}}{\delta t}\delta t = \int \mathrm{d}^3 r\, \mathbf{H} \cdot (\nabla \times \mathbf{E})\delta t \\
&= \int \mathrm{d}^3 r \, [\nabla \cdot (\mathbf{E} \times \mathbf{H}) + \mathbf{E} \cdot (\nabla \times \mathbf{H})]\, \delta t \\
&= \int \mathrm{d}\mathbf{S} \cdot (\mathbf{E} \times \mathbf{H})\delta t + \int \mathrm{d}^3 r (\nabla \times \mathbf{H}) \cdot \mathbf{E}\delta t \\
&= \int \mathrm{d}^3 r (\nabla \times \mathbf{H}) \cdot \mathbf{E}\delta t,
\end{aligned}$$

where we have used identity $(\nabla\times\mathbf{A})\cdot\mathbf{B}=\nabla\cdot(\mathbf{A}\times\mathbf{B})+\mathbf{A}\cdot(\nabla\times\mathbf{B})$ and Gauss' theorem [2]. Substituting the above two expressions into the equality $\delta F=\delta' W$ from (1.23) for reversible isothermal processes, we obtain $\nabla\times\mathbf{H}=\mathbf{j}_{\text{ext}}$.

15.2 Let us express the integrand of (15.51) as $\left[|\phi_{00}(\mathbf{r}')|^2\right]^2$. As it is a periodic function, we can expand $|\phi_{00}(\mathbf{r}')|^2$ in terms of the reciprocal lattice vector in (15.48) as

$$|\phi_{00}(\mathbf{r}')|^2=\frac{1}{V'}\sum_{\mathbf{K}'}I_{00}(\mathbf{K}')\,\mathrm{e}^{\mathrm{i}\mathbf{K}'\cdot\mathbf{r}'},$$

where $I_{00}(\mathbf{K})$ is given by (15.52) with $N_1=N_2=0$. Substituting these relations into the definition of (15.51) and performing the integration with

$$\int \mathrm{e}^{-\mathrm{i}(\mathbf{K}_1'+\mathbf{K}_2')\cdot\mathbf{r}'}\mathrm{d}^3r'=V'\delta_{\mathbf{K}_2',-\mathbf{K}_1'},$$

we obtain (15.56).

15.3 Let us expand $\mathbf{B}(\mathbf{r})$ and $\delta^2(\mathbf{r})$ as

$$\mathbf{B}(\mathbf{r})=\hat{\mathbf{z}}\int\frac{\mathrm{d}^2k}{(2\pi)^2}B_{\mathbf{k}}\mathrm{e}^{\mathrm{i}\mathbf{k}\cdot\mathbf{r}},\qquad \delta^2(\mathbf{r})=\int\frac{\mathrm{d}^2k}{(2\pi)^2}\mathrm{e}^{\mathrm{i}\mathbf{k}\cdot\mathbf{r}},$$

and substitute them into (10.46) with $n=-1$ and $\mathbf{r}_0=\mathbf{0}$. We then obtain $B_{\mathbf{k}}=\Phi_0/(\lambda_{\mathrm{L}}^2k^2+1)$. The corresponding magnetic flux density is then written in two-dimensional polar coordinates using a couple of integral representations for the Bessel functions [1, 2],

$$\begin{aligned}\mathbf{B}(\mathbf{r})&=\hat{\mathbf{z}}\Phi_0\int\frac{\mathrm{d}^2k}{(2\pi)^2}\frac{\mathrm{e}^{\mathrm{i}\mathbf{k}\cdot\mathbf{r}}}{\lambda_{\mathrm{L}}^2k^2+1}=\hat{\mathbf{z}}\frac{\Phi_0}{2\pi\lambda_{\mathrm{L}}^2}\int_0^\infty\mathrm{d}k\frac{k}{k^2+\lambda_{\mathrm{L}}^{-2}}\int_0^{2\pi}\frac{\mathrm{d}\varphi}{2\pi}\mathrm{e}^{ikr\cos\varphi}\\&=\hat{\mathbf{z}}\frac{\Phi_0}{2\pi\lambda_{\mathrm{L}}^2}\int_0^\infty\mathrm{d}k\frac{k}{k^2+\lambda_{\mathrm{L}}^{-2}}J_0(k)=\hat{\mathbf{z}}\frac{\Phi_0}{2\pi\lambda_{\mathrm{L}}^2}K_0(r/\lambda_{\mathrm{L}}).\end{aligned}$$

Problems of Chapter 16

16.1 If $\xi_{\mathbf{q}'}<0$ holds, (16.12) is replaced by

$$\frac{|C|^2}{|A_1|^2}=\frac{2E}{E-\sqrt{E^2-|\Delta_0|^2}},\qquad\frac{|C|^2}{|A_2|^2}=\frac{2E}{E+\sqrt{E^2-|\Delta_0|^2}},$$

so that $|A_1| < |A_2|$. Hence, we can identify the z element of (16.17) with coefficient $|A_2|^2$ and $k_{Fz} < 0$ as the incident current from the normal side. In addition, current (16.18) is replaced by

$$\mathbf{j}^{\mathrm{s}}_{\mathrm{qp}} = -\frac{\hbar \mathbf{k}_{\mathrm{F}}}{m}|C|^2 \frac{\sqrt{E^2_{\mathbf{q}'} - |\Delta_0|^2}}{E_{\mathbf{q}'}}.$$

With these modifications, we can express the transmission coefficient for $\xi_{\mathbf{q}'} < 0$ as

$$\mathscr{T} = \frac{|C|^2}{|A_2|^2}\frac{\sqrt{E^2 - |\Delta_0|^2}}{E}\theta(E - |\Delta_0|) = \frac{2\sqrt{E^2 - |\Delta_0|^2}}{E + \sqrt{E^2 - |\Delta_0|^2}}\theta(E - |\Delta_0|),$$

which is identical to (16.19). This is the process where a hole on the normal side, represented by the lower element of (16.5), is incident on the interface and reflected as an electron.

16.2 First, we substitute (16.70) into the differential equation in (16.66) and use $K_0'(x) = -K_1(x)$ [2]. We then obtain (16.72). Second, we integrate the differential equation of (16.66) with the initial condition (16.70) to express the flux density of $0 \le r < r_{\mathrm{c}}$ as

$$B(r) = cK_0(r_{\mathrm{c}}/\lambda_{\mathrm{L}}) + \int_r^{r_{\mathrm{c}}} \mathrm{d}r_2\, j(r_2).$$

Third, we divide the integration of (16.76) at $r = r_{\mathrm{c}}$, perform that over $0 \le r \le r_{\mathrm{c}}$ using the expression above, and perform that over $r_{\mathrm{c}} \le r \le \infty$ using the asymptotic expression $B(r) = cK_0(r/\lambda_{\mathrm{L}})$ and the identity $xK_0(x) = -[xK_1(x)]'$ [2]. Equation (16.76) thereby becomes

$$\left[\frac{r_{\mathrm{c}}^2}{2}cK_0(r_{\mathrm{c}}/\lambda_{\mathrm{L}}) + \int_0^{r_{\mathrm{c}}} \mathrm{d}r_1 r_1 \int_{r_1}^{r_{\mathrm{c}}} \mathrm{d}r_2\, j(r_2)\right] + c\lambda_{\mathrm{L}} r_{\mathrm{c}} K_1(r_{\mathrm{c}}/\lambda_{\mathrm{L}}) = 1.$$

Substitution of (16.72) into this equation yields (16.71).

16.3 Regarding the gradient term in (16.45) as a perturbation, we obtain the first-order equation

$$2\varepsilon_n f^{(1)} + \hat{\mathbf{v}}\cdot\boldsymbol{\partial} f^{(0)} = 2\Delta g^{(1)},$$

where $f^{(0)}$ denotes the homogeneous solution in (16.41), and $g^{(1)}$ for $\varepsilon_n > 0$ is obtained from (16.38) as

$$g^{(1)} = -\frac{\bar{f}^{(0)} f^{(1)} + f^{(0)}\bar{f}^{(1)}}{2g^{(0)}} = -\frac{\Delta^* f^{(1)} + \Delta \bar{f}^{(1)}}{2\varepsilon_n},$$

with $\bar{f} \equiv f^*(\varepsilon_n, -\mathbf{k}_{\rm F}, \mathbf{r})$. The above equation for $f^{(1)}$ can be solved easily by coupling it with that for $\bar{f}^{(1)}$ as

$$\begin{bmatrix} f^{(1)} \\ \bar{f}^{(1)} \end{bmatrix} = \frac{1}{4\varepsilon_n(\varepsilon_n^2 + |\Delta|^2)} \begin{bmatrix} 2\varepsilon_n^2 + |\Delta|^2 & -\Delta^2 \\ -\Delta^{*2} & 2\varepsilon_n^2 + |\Delta|^2 \end{bmatrix} \begin{bmatrix} -\hat{\mathbf{v}} \cdot \boldsymbol{\partial} f^{(0)} \\ \hat{\mathbf{v}} \cdot \boldsymbol{\partial} \bar{f}^{(0)} \end{bmatrix}.$$

Substituting $f^{(0)} = |\Delta| e^{i\varphi(\mathbf{r})}/\sqrt{\varepsilon_n^2 + |\Delta|^2}$ into the above expression for $f^{(1)}$, we obtain

$$f^{(1)} = -i \frac{\varepsilon_n \Delta}{2(\varepsilon_n^2 + |\Delta|^2)^{3/2}} \hat{\mathbf{v}} \cdot (\nabla\varphi + \mathbf{A}),$$

which yields $g^{(1)}$ above as

$$g^{(1)} = i \frac{|\Delta|^2}{2(\varepsilon_n^2 + |\Delta|^2)^{3/2}} \hat{\mathbf{v}} \cdot (\nabla\varphi + \mathbf{A}).$$

Let us insert this expression into (16.47) and perform the average over the isotropic Fermi surfaces of $d = 2, 3$. Also noting (10.17), we obtain (16.74).

References

1. M. Abramowitz, I.A. Stegun (eds.), *Handbook of Mathematical Functions: With Formulas, Graphs, and Mathematical Tables* (Dover, New York, 1965)
2. G.B. Arfken, H.J. Weber, *Mathematical Methods for Physicists* (Academic, New York, 2012)

Index

A
ABM state, 194
AC Josephson effect, 183
Andreev reflection, 251
Avogadro constant , 13

B
BCS wave function, 102
Bloch–De Dominicis theorem, 64
Bogoliubov–de Gennes equations, 110
    Andreev approximation, 248
Bogoliubov–Valatin operator, 105
Bohr magneton, 84
Boltzmann constant, 16
Boltzmann's principle, 18
Bose distribution, 45
Bose statistics, 39
Bose-Einstein condensation, 51
boson, 28
BW state, 191

C
canonical distribution, 20
canonical ensemble, 20
Caroli-de Gennes-Matricon mode, 254, 262
chemical potential, 9
Clausius inequality, 5
coherence factors, 166
coherence length, 232
coherent state, 103
condensate wave function, 41, 66
Cooper pairing, 98
Cooper's problem, 97
Cooper-pair creation operator, 102
cutoff energy, 128

D
DC Josephson effect, 183
density matrix, 61
    reduced, 61
density of states, 45
    per unit cell and spin component, 121
    per unit volume and spin component, 95
Dirac delta function, 45
dissipation, 162
dyadic, 116

E
effective mass, 81
effective pairing interaction, 130
Eilenberger equations, 220
entropy, 5, 6, 16
equation of state, 3
    ideal gas, 3
    van der Waals', 3
exchange hole, 70
expectation, 14
extensive variable, 4

F
Fermi contact interaction, 169
Fermi distribution, 45
Fermi energy, 53
Fermi sea, 53
Fermi statistics, 39
Fermi vacuum, 37, 53

© Springer Japan 2015

T. Kita, *Statistical Mechanics of Superconductivity*, Graduate Texts in Physics,
DOI 10.1007/978-4-431-55405-9

Fermi wave number, 53
Fermi-surface sum rule, 80
fermion, 28
first law of thermodynamics, 4
fluctuation, 14
fluctuation range, 140
flux quantum, 155

G

gap equation, 119
gas constant, 3
gauge invariance, 211
Gibbs entropy, 16
Gibbs-Duhem relation, 86
Ginzburg–Landau equations, 224
GL parameter, 232
Gor'kov equations, 208
gradient, 1
gradient expansion, 211
grand canonical distribution, 22
grand canonical ensemble, 22
grand partition function, 22
grand potential, 9, 22

H

Hartree–Fock equations, 77
Hartree–Fock potential, 76, 109
heat, 4
heat capacity at constant volume, 10
Heaviside step function, 45
Hebel-Slichter peak, 173
Helmholtz free energy, 7, 20
hyperfine interaction, 169

I

impact parameter, 252
intensive variable, 4
internal energy, 4, 18
isolated system, 6

K

Knight shift, 169
Korringa relation, 173
Kramer-Pesch effect, 261
Kronecker delta, 15

L

Landau level, 235
Landau parameters, 83
left-right subtraction trick, 216
Legendre transformation, 8
London equation, 154
London penetration depth, 154, 232
longitudinal magnetic relaxation time, 172

M

magnetization, 230
Matsubara frequency, 203
Matsubara Green's function, 202
Maxwell's relation, 3, 10
Maxwell-Boltzmann distribution, 45
Meissner effect, 151, 154
method of Lagrange multipliers, 17
microcanonical distribution, 18
microcanonical ensemble, 18

N

Nambu matrix, 208
Nernst's theorem, 5
normal-ordering operator, 206

O

off-diagonal long-range order, 69
order parameter, 140

P

pair distribution function, 70
pair potential, 109
particle-hole symmetry, 111
partition function, 20
Pauli exclusion principle, 36, 70
Pauli matrices, 111
permutation, 25
  cyclic, 25
  even, 26
  odd, 26
polar state, 197
potential, 1
pure state, 5, 15

Q

quasi static process, 4
quasiparticle, 88
quasiparticle field, 103

R

relaxation time, 220

retarded Green's function, 163
reversible process, 4

S
second law of thermodynamics, 5
second quantization, 34
second-order phase transition, 134, 141
Shannon entropy, 16
Slater determinant, 36
Sommerfeld expansion, 55
spin, 26
spin-statistics theorem, 28
spontaneous symmetry breaking, 140
spontaneously broken gauge symmetry, 140, 185
SQUID, 186
standard deviation, 14
state function, 3
state quantity, 1
state variables, 3
superfluid ^{3}He
  A phase, 194
  B phase, 191
superfluid density, 149

T
thermodynamic critical field, 231
thermodynamic potential, 9
third law of thermodynamics, 288
total derivative, 2
transposition, 25
type-I superconductor, 229
type-II superconductor, 229

U
upper critical field, 234

V
vacuum permeability, 151, 169
von Neumann entropy, 16

W
weak coupling, 128
Wick decomposition, 64
Wigner transform, 211
  gauge-covariant, 212
  gauge-invariant, 212
work, 4

Y
Yosida function, 147

Z
Zeeman effect, 84, 152, 169

Printed in the United States
By Bookmasters